A SCIENTIST SPEAKS OUT

A Personal Perspective on Science, Society and Change

A SCIENTIST SPEAKS OUT

A Personal Perspective on Science, Society and Change

Glenn T. Seaborg
University of California, Berkeley, USA
Ernest Orlando Lawrence
Berkeley National Laboratory, USA

World Scientific
Singapore • New Jersey • London • Hong Kong

Published by

World Scientific Publishing Co. Pte. Ltd.

P O Box 128, Farrer Road, Singapore 912805

USA office: Suite 1B, 1060 Main Street, River Edge, NJ 07661

UK office: 57 Shelton Street, Covent Garden, London WC2H 9HE

Library of Congress Cataloging-in-Publication Data

Seaborg, Glenn Theodore, 1912–

A scientist speaks out : a personal perspective on science, society and change / Glenn T. Seaborg.

p. cm.

Includes bibliographical references and index.

ISBN 9810222041

1. Science -- Social aspects. 2. Science -- Philosophy. 3. Change. I. Title.

Q175.5.S42 1996

303-48'3--dc20 96-13324

CIP

British Library Cataloguing-in-Publication Data

A catalogue record for this book is available from the British Library.

This book was produced with the assistance of Ernest Orlando Lawrence Berkeley National Laboratory, operated on behalf of the U.S. Department of Energy, by the Regents of the University of California.

This book is printed on acid-free paper.

Printed in Singapore by Uto-Print

Preface

When World Scientific Publishing Co. first approached me about compiling a volume of some of my popular speeches, I was slightly hesitant due to my still very busy schedule which I have been keeping up at my young age of 83. However, World Scientific did a very fine job on a previous volume of some of my scientific papers, *Modern Alchemy: Selected Papers of Glenn T. Seaborg*, and after giving their proposal some consideration, I decided it would be an excellent opportunity to present some different but interesting aspects of my professional life that began with a career in science. My life in science and in public service has given me the opportunity to learn about and reflect on a great many topics, from the formulation of national science policy to the promise of youth. I am glad that this volume of selected speeches and articles will allow me to share some of these thoughts, as I originally penned them, with the reader.

My introduction to national public service came when I was appointed by U.S. President Harry S. Truman to serve on the first General Advisory Committee (GAC) to the U.S. Atomic Energy Commission (AEC) for a term extending from January 1947 to August 1950. This first GAC played an important role in helping to establish a number of the basic policies of the AEC.

However, the real beginnings of diversion from my scientific career began in the fall of 1952, when Clark Kerr, who had just been appointed the first chancellor of the Berkeley campus of the University of California, called me to his office. Would I like to serve as faculty athletic representative to the Pacific Coast Intercollegiate Athletic Conference? I was taken by surprise and unable to give him an immediate answer. However, I had had a lifelong interest in athletics and had been an avid follower of the University of California's athletic teams. I soon decided to accept Kerr's offer. I served as faculty athletic representative from January 1953 until July 1958, attending all of the meetings of the Pacific Coast Conference

(PCC) and several meetings of the National Collegiate Athletic Association (NCAA). My duties included ruling on the eligibility of the athletes and presiding over the Berkeley campus athletic advisory board meetings. During the latter half of this period, I served as the press spokesman for the Pacific Coast Conference at its semiannual meetings. This was a lively responsibility since during this period the rather widespread rules violations in the conference were discovered. The violations led to the ultimate breakup of the conference. I was then cast in a leading role in putting together the successor to the PCC — the Athletic Association of Western Universities, which has in the meantime been expanded to ten members or the so-called Pacific 10 Conference (PAC 10).

In the summer of 1958, Chancellor Kerr became president of the University. Once again, Kerr called me to his office. Perhaps because of our close association in weathering, somewhat successfully, the highly publicized and difficult PCC affair, he and the regents wondered if I would be a candidate for the chancellorship. The decision was a difficult one to make. I was reluctant to step out of science to the extent that this position seemed to require; yet the larger challenge was appealing. When I was allowed to continue as an associate director of the Radiation Laboratory and to direct graduate student research in the laboratory and the College of Chemistry, I accepted. The two-and-a-half years I served as chancellor was a period of extensive development for the Berkeley campus on many fronts.

While chancellor, I was appointed by President Dwight D. Eisenhower, in 1959, to be a member of the President's Science Advisory Committee (PSAC), on which I served until January 1961, and to the National Science Board of the National Science Foundation (1960–61). While serving on the PSAC, I accepted an assignment to chair a special PSAC panel on basic research and graduate education. From the work of this panel, we issued the report "Scientific Progress, the Universities, and the Federal Government — A Statement on Basic Research and Graduate Education in the Sciences" which called for a partnership between the federal government and the universities to recognize and to support the symbiotic relationship necessary between basic research and graduate education. The report, which became known as the "Seaborg Report," gained widespread attention and laid the foundation whereby the federal government is actively engaged in support and funding of scientific inquiry at research universities.

In January 1961, President Kennedy called me to serve as the chairman of the Atomic Energy Commission, a position that I held for more than ten years while on leave of absence from the University of California, a period longer than that of any previous chairmen and spanning the terms of two presidents, John F. Kennedy and Lyndon B. Johnson, and part of the term of President Richard M. Nixon. In this capacity I also served on such bodies as the National Aeronautics and Space Council, the Federal Council for Science and Technology, the Federal Radiation Council and the President's Council on Marine Resources and Engineering Development.

As chairman of the AEC, I took on the role of spokesman and advocate for the peaceful uses of the atom. It was a role with which I was not unfamiliar. Towards the end of World War II, during which time I headed the chemistry section at the Metallurgical Laboratory at the University of Chicago which had the responsibility to extract and to purify plutonium as part of the Manhattan Engineer District efforts, I was an original signatory of the Franck Report, which was a recommendation by some Manhattan Project scientists to President Truman to demonstrate for the Japanese government the atomic bomb on an uninhabited island. Our suggestion was either ignored by or did not reach the president.

In the convening years (1955, 1958, 1964 and 1971), I was a member of the U.S. delegation to the United Nations Conference on the Peaceful Uses of Atomic Energy held in Geneva, Switzerland, serving as chairman of the American delegation in 1964 and as president of the conference in 1971. Also during my tenure, I attended the annual General Conference of the International Atomic Energy Agency in Vienna, Austria, eleven times. In my ten and one-half years as AEC chairman, I traveled to over 60 countries, speaking on diverse topics such as international cooperation in science, peaceful applications of nuclear power, predictions of the possibilities in a scientific society, and arms control. Three times a presidential plane was placed at my disposal when I circled the globe in January 1967 (Australia, India, Pakistan, Spain, and Thailand), July 1967 (Argentina, Brazil, Chile, Colombia, Peru, and Venezuela), and January 1970 (Ethiopia, Democratic Republic of Congo [now Zaire], Germany, Ghana, Kenya, Morocco, Spain, and Tunisia). I also recall fondly many trips to the former Soviet Union.

One of my proudest moments as chairman of the AEC came on 5 August 1963, when, as a member of U.S. Secretary of State Dean Rusk's delegation, I witnessed the signing of the Limited Test Ban Treaty in Catherine Hall in the Kremlin. (I have recounted the innumerable meetings, negotiations, events and activities leading to the signing of the LTBT in my book *Kennedy, Khrushchev, and the Test Ban*, written with my friend and AEC colleague Benjamin S. Loeb; we also wrote a follow-up book *Stemming the Tide: Arms Control in the Johnson Years* that describes the negotiations leading to the Nonproliferation Treaty of 1968.) To this day, I still support, lobby, write, and speak about the need for a comprehensive test ban.

Since the 1950's, I have been quite interested in the improvement of the teaching of science, the need to interest young people in careers in science, the impact of science on society, the necessity for greater scientific literacy among the general public and the need for scientists to understand social problems and the humanities and to help bridge the gap between the two cultures. Thus I have served as chairman of the Steering Committee of CHEM Study. I had the privilege of serving on the Commission on the Humanities (1962–65), whose report played an important role in the establishment of the National Foundation on the Arts and the Humanities. I also served for nearly 30 years as chairman of the Board of Trustees of Science Service, a Washington-based organization with a strong national program for the popularization of science; it conducts the annual Science Talent Search in Washington, D. C., and the annual International Science and Engineering Fair, and publishes the very effective weekly *Science News* magazine. I also served as a member of the Board of Directors of the National Educational Television and Radio Center. This organization later became the Public Broadcasting Service.

My interest in the broad implications of science led to an increasing number of public speeches covering this subject, which expanded to even broader areas. Although I made many speeches on scientific subjects at seminars and meetings of scientific societies, I did not feel myself capable of venturing beyond this domain for a long time, and then I approached it gradually. Actually my first speech in a broader, philosophical vein was not made until I had passed the age of forty; this was entitled "Dawn of the Nuclear Age" and was delivered at the Charter Day banquet of the alumni of the University of California, Riverside in March 1953. This had been preceded by a number of talks to University of California, Berkeley alumni groups, including an alumni tour in 1952 with UC President Robert

Gordon Sproul up and down the state of California, on the "Peaceful Uses of Atomic Energy"; and followed by a similar alumni tour in 1956 with Chancellor Kerr, on which I spoke on the shortage of scientists; and by numerous talks to Berkeley alumni groups on these topics and other topics such as athletics and the Berkeley campus. These beginning attempts were followed by an escalation of topics and number of speeches, including the period of my Berkeley chancellorship, culminating in my Washington period, which resulted in hundreds of speeches beginning in 1961. Such activity led to my election in 1967 as a member of the Board of Governors of the International Platform Association, an organization of professional speakers, and in 1968–1969 and in 1981–1986 as its president.

Soon after my return to the University of California at Berkeley, I was appointed University Professor of Chemistry (a title now held by some 15 professors throughout the statewide nine-campus University of California with its 7,500 faculty members). I was also appointed head of the Nuclear Chemistry Division (soon to become the Nuclear Science Division) of the Ernest Orlando Lawrence Berkeley National Laboratory (the former Lawrence Radiation Laboratory) in which position I served until 1975, when I was appointed Associate Director-at-Large of the Ernest Orlando Lawrence Berkeley National Laboratory, a position that I have held since.

During 1972 I served as president of the American Association for the Advancement of Science and during 1973 as chairman of the Board of Directors of the AAAS. While serving in these offices I was instrumental in establishing the Office of International Science. It was largely in this connection that the Interciencia Association (which encourages collaboration with our neighbors in Latin America) was established. In 1975, I served as president-elect of the American Chemical Society. I then served as president of the ACS in 1976 (the year of its centennial) and as past-president in 1977. I served as a member of the ACS Board of Directors during each of these three years. At the centennial banquet held in New York in April 1976, I played host to a gathering of heads of chemical societies throughout the world, and at that time I proposed the creation of an International Chemical Society. Although such a society never came to fruition, one consequence of my proposal was the creation of the Affiliate Scheme of the International Union of Pure and Applied Chemistry, which came into effect in February 1985.

From 24 May to 10 June 1973, my wife Helen and I visited the People's Republic of China. We were members of a team including physical and biological scientists and representatives of the social sciences and the humanities (and a few of their wives) sponsored by the U.S. Committee for Scholarly Communication with the People's Republic of China. Ours was the first group to visit the People's Republic of China on a semi-official basis for the purpose of negotiating agreements for exchange of scholars and visitors between our two countries. While in Beijing we had the pleasure of attending a meeting in the Great Hall of the People (National People's Congress Building) in Tiananmen Square in Beijing. Of special interest was the attendance at this meeting of Chinese Premier Zhou Enlai, whom we all met and who played an active role in negotiating the exchange agreement with the members of our delegation. During this visit to the People's Republic of China, we visited, besides Beijing, Shenzhen, Guangzhou, Nanjing, Wuxi, Xuzhou, Shanghai and Changsha. I kept a complete journal during this visit, which was published in a document entitled "China Journal."

I visited the People's Republic of China again from 17 May–11 June 1978, as chairman of a delegation for "Pure and Applied Chemistry." During this trip we visited Beijing, Dalian, Changchun, Shenyang, Fushun, Shanghai, Hangzhou, Xian, and Langzhou. Again I kept a complete journal which was published in a document entitled, "China Revisited." (A book, *Chemistry and Chemical Engineering in the People's Republic of China*, was also published by the American Chemical Society.)

I served as founding president of the International Organization for Chemical Sciences in Development (IOCD) beginning in 1981 and continuing until 1992 and with which I am still affiliated. The aim of this organization is to help to provide the benefits of chemistry to people in developing countries of the world. IOCD's working groups draw scientists from Third World countries into collaboration with scientists from industrial countries into research focused on areas of vital developmental concern: unchecked population growth, persisting parasitic diseases, and declining agricultural productivity. The scientists in IOCD's working groups represent the broad geographic regions of Africa, Asia, Latin America, and the Middle East.

Having been involved with the Lawrence Hall of Science on the Berkeley campus since its inception in the fall of 1958, I immediately resumed an active role on its Faculty Advisory Committee upon my return from Washington in 1971, later serving as the chairman of this committee.

I then served as director of the Lawrence Hall of Science from 1982–1984 and have served as chairman since 1984. The Lawrence Hall of Science has played and is playing a leadership role in solving the national crisis in precollege science and math education.

I served as a member of the National Commission on Excellence in Education (NCEE), which worked under the direction of Chairman David P. Gardner from the fall of 1981 until April 1983 to produce its report, "A Nation At Risk." We presented this report to President Ronald Reagan on 26 April 1983. This report has had a substantial impact on the national reform movement toward improving the status of precollege education, especially science and math education, in the United States.

My public speaking schedule included about 700 speeches on a wide variety of topics. Included were talks each year (1972–1993) at the Science Talent Search in Washington, talks at seven of the annual International Science and Engineering Fairs, 19 commencement addresses, 40 lectureships, building dedicatory addresses, talks and lectures every year to freshman chemistry classes, and numerous speeches to professional associations and public groups.

In retrospect, I can see my stepwise progress, how each phase of my career prepared me for the next. The extraordinary, I called it magic, Berkeley atmosphere of the 1930's offered unparalleled opportunity. My close association with Gilbert Newton Lewis showed me firsthand how the world's best chemist worked and thought, and gave me, a young chemist rather lacking in self-confidence, the confidence that was essential to future progress. The association with the Radiation Laboratory, and the amazing Ernest O. Lawrence, made possible the discoveries with which I was involved before the war. This experience made it possible to carry out my subsequent responsibilities at the wartime Metallurgical Laboratory of the University of Chicago. These experiences, in turn, made it possible for me to carry out my research role in the Radiation Laboratory during the productive period 1946 to 1958. All of these experiences, and especially my experience as the faculty athletic representative to the PCC, prepared me for the Berkeley chancellorship. And the administrative experience gained in the latter position, together with all of my research experience, prepared me for the chairmanship of the AEC.

I conclude this introduction and overview with a comment related to the time span of the selected speeches and articles, which cover a period of some forty years. Some of the early speeches used the term "man," an obsolescent reference to people of both sexes. Some make predictions about the future that can now be compared with the actual situation today. These are often rather accurate but some miss the mark ranging from a moderate to a large degree.

Glenn T. Seaborg
October 1995

Contents

Acknowledgement

My thanks to several people who helped bring this book to publication. World Scientific editors Dr. Grace Tan guided the book through its critical phases, and Yanan Fan followed through to production; University of California at Berkeley Archivist William M. Roberts was exceptionally resourceful in identifying photographs; Anne Fleming, Rita McLean and Ruby Tebelak of E. O. Lawrence Berkeley National Laboratory Technical and Electronic Information Department expedited processes and information. I am especially grateful for the efforts of my staff: Sherrill Whyte, Kristin Balder-Froid, David Yan, Carol Harris Earls, Perry Hall, Nancy Lockhart, and Jean Suzuki.

Glenn T. Seaborg

The Role of Basic Research

Luncheon talk given at joint meeting of Atomic Industrial Forum and Stanford Research Institute at Mark Hopkins Hotel, San Francisco, April 5, 1955.

I suppose that I should begin with a definition of what I mean by the expression "basic science". The science of semantics resembles a sort of non-physical theory of relativity which exposes to our view the frames of reference underlying our words, whose meaning we assume, often unwittingly, is understood. Certainly our attitude toward the expression "basic science" depends on our educational and scientific frame of reference. Person A who speaks of himself as a scientist with an inflection indicating his deep pride in being so regarded would, no doubt, define whatever he is doing as basic science and anything with more recognizable utility as applied science or, even worse, as engineering. Another and purer scientist B might regard what Mr. A is doing as the crassest commercial prostitution of a scientific education.

Scientists A and B belong to the halo school of definition of basic science, as shown by their frequent substitution of the expression pure science. The undercurrents of suggestion here of unadulterated, dedicated, uncontaminated, unfettered, unrestrained, unsullied search for truth provides a real lift to many an underpaid scientist and a real substitute for money — provided he has married the right kind of woman.*

*In 1955, we assumed a scientist was male. Please accept my remarks in the context of the times.

But the world consists of many types of people and many scientists do not look at the term, basic science, with glasses of the same color. Some actually use the expression as an epithet. The distinction between a Ph. D. and an M. D. is sometimes made that a Ph. D. is the kind of doctor who doesn't do anybody any good. Or we can recall that a prominent director of one of our largest industrial research laboratories is supposed to have said that basic science is that scientific activity which cannot possibly provide any financial return to the people paying the cost of doing it.

Most of us here today probably associate ourselves more closely with the halo school of definition of basic science than we do with the epithet school. I know that I personally do. Yet most of us have been closely enough associated with practical problems in wartime military applications, in industrial research, in the development of new chemo-therapeutic agents, or with other matters so that our purity has been badly sullied. For most of us it is symptomatic of our real attitude that when we try to defend basic research to visiting state legislators or federal budget review officials or boards of directors, we almost universally point to the most outstandingly practical applications we can single out and swear that these could not possibly have happened without the basic research of past years. Clearly, one of the major peacetime uses of the atomic bomb has been to blast loose purse strings to obtain money for basic research in physics. Even so, it is surprising how many equivalent megatons of persuasion this takes.

Hence we might think that in a sense basic research is not really different from applied research in its real purpose — it is merely applied research with a time delay. How happy a person is in doing basic research is frequently a matter of how long a time delay he is adjusted to or is willing to tolerate. Willard Gibbs, perhaps the greatest scientist America has produced, was able to see very little practical influence of his monumental contributions to thermodynamics during his lifetime, but it is doubtful that this ever caused him any concern or deterred him from the most vigorous prosecution of his theoretical studies. *His* time delay was set at infinity. We all know persons among our personal acquaintances who spend about a decade in basic research pursuing studies which appealed to them for their intellectual stimulation and with no distinct utilitarian purpose in mind and then reoriented their careers to exploit some promising discovery of their basic science days. Frequently such a person is quite happy in both phases of his scientific career and would have become distinctly unhappy had he attempted to stay oriented in basic science for his entire life. The time delay for such a person is about ten years. At the end of the scale there must be industrial divisions of "Research" where the new Ph. D. staff member may be allowed a couple of months to brood on and attack some fundamental problem of wide scientific interest provided he comes up with some real moneymaker by the end of this time. The time delay here falls in the range of months down to microseconds.

I hasten to state that I have no intention of trying to name a time delay that distinguishes between basic and applied science. For I was not really serious in suggesting that there was no real distinction between basic and applied science other than the provision that we be more tolerant of the basic scientist and allow him more time to produce some useful results. Actually utility is not a real basis for talking about basic science. There is no reason why basic research results cannot have immediate, important, financially rewarding application. X-rays were in use in a Vienna hospital within 3 months of their discovery, in 1895, but this does not mean that Röntgen was not doing basic research. Nor is there any reason why they need ever have clearly discernible practical results. The distinction, I think between basic and applied research lies in the motivation behind the research and the criteria that are applied to determine what work shall be undertaken and what changes shall be made in the lines of investigation as the study develops. In basic research the motivating force is not utilitarian goals, but a search for a deeper understanding of the universe and of the living and inorganic phenomena within it. Its keynote is intellectual curiosity.

Now I realize that a great amount of flowery nonsense has been written on this point. The scientist has been portrayed as a dedicated seeker after truth, bending his every effort to extract the last ounce of understanding from reluctant Nature, scornful of utilitarian applications of his work, completely altruistic, and receiving a bountiful reward of inner satisfaction quite sufficient to his needs, scornful of monetary gain, fame or outward rewards of any kind.

This description does not fit me. Probably it *does* fit you, but you will undoubtedly agree with me that it is not an accurate description of the person on your left or on your right here today. Scientists are by and large quite ordinary people. Within their specialties their natural intellectual capacities are greater than the average man's and their trained competence is certainly greater, but as human beings they are subject to the same shortcomings, the same wants, desires and drives as anyone else. Nevertheless, I think we are correct in saying that the underlying motivating force in basic research is intellectual curiosity, and that this curiosity is to be rated with the highest qualities of mankind. The investigator may be a thoroughly disreputable character in many ways, the intellectual curiosity may belong to his boss or his professor rather than to him, but somewhere behind the work he does is the desire to find out *why* or *how* or *what*.

We can use some clues to determine how basic a research program is. If the final goal is very precisely stated, the program is probably not too basic. If the investigator is not free to make radical changes in his program, and to pursue some unexpected question which has arisen in his work and which excites his curiosity as to *why* or *how*, the program is probably not basic.

With these comments by way of outlining what I have in mind when I talk of basic science, let us get on to consideration of the role of basic science.

At the risk of appearing inconsistent, let me state that the arguments we present to state legislators or federal budget reviewers or boards of directors are undoubtedly valid and that a good reason for supporting basic research is because of the practical results such support ensures. This is a ticklish argument because it has to be made clear that these results may be a long time coming, that most of the projects for which support is requested should not be expected to be fruitful but that a broad-gauge support toward goals defined in only a very broad sense are required if major discoveries are to be expected.

Insofar as an appreciation of this rather subtle point is produced in the government administrators, university officials and industrial managers, a climate is created in which basic science can thrive. Certainly, great strides have been made during the past few decades, particularly in the post-war years, and mutual understanding of scientists and administrators has been good. Much remains to be done, of course. A poor appreciation of this point by administrators of research funds can mean the stifling of basic studies by the requirement of detailed program goals, project justifications, etc. which suggest to the investigator that unless he achieves the stated goal by the end of his contract period a renewal of his contract is not likely to be forthcoming. The result can be programs representing rather unimaginative extrapolation of known results which waste the scientific powers of the applicant.

The university by its very nature and purpose is and will remain the natural center for basic research. In our major universities the administration of funds supplied from the general university budget has been generally favorable to the fostering of fundamental studies. The rising costs of research studies, particularly in the physical sciences, and the growth of the participation of private corporations, private foundations, the National Science Foundation and State and Federal Government agencies [particularly the ONR (Office of Naval Research) and the AEC (Atomic Energy Commission)] have created a new problem. The sizeable amounts of money these agencies have put into research and development have unquestionably been a very strong positive influence on the growth of basic research. Furthermore, the men who have administered these funds have had a fine appreciation of the role of basic studies. However, in their dual responsibility to the scientist on the one hand and to the taxpayer or stockholder on the other, it has not yet been possible to work out mechanisms for the disbursement of research funds which avoid the tendency of the small-project system, with all its red tape for reviewing and referring and project justification, to take away the freedom of the talented investigator to wander off into really uncharted regions of intellectual and scientific endeavor. It should be possible to say to more scientists of demonstrated competence and responsibility, "Here is some money to keep you going. Run along now and do whatever you want. Don't bother us anymore about it. If you find out anything worthwhile send it into a scientific journal. All we ask is that you work hard. In fact, don't even do that if you can get more accomplished in another way." Some grants at the present time are in effect administered in this way with very good results. In the end, putting money in basic research is betting on the person who is doing the work, not on the worth of the program which he foresaw before he began it.

There is little doubt in my mind that we would achieve more in the long run per million dollars expended in research if this system were more widely used. I feel that basic research conducted by universities would be more efficiently done and more likely to reach beyond limited objective studies if lump sums of money were provided by external agencies to the general university budget or to general departmental budgets for administration in the manner traditional to university research rather than disbursed piecemeal to small projects bearing specific titles. I am certain we will make improvements in our present financial arrangements. The whole business of wide scale support of university research by industrial and governmental agencies has grown so fast that we have been forced to use improvised contractual arrangements which in many cases are more suited to the limited objective developmental project than to the fostering of the fundamental research.

I should hasten to add that a generous supply of funds for basic research is not in itself sufficient. Money does not directly beget new ideas. We need to attract increasing numbers of people who are capable of creative thinking. Traditionally, the occupation of unfettered research has been centered in the university professor who also has the dual role of a teacher. The union of these two activities has in my estimation been a happy one because, on the one hand, advanced instruction is surely more inspiring if given by someone who has first-hand knowledge of the subject irrespective of his abilities as an orator, and on the other hand, there is nothing more stimulating to uninhibited thinking than the uninhibited questioning of successive generations of students. We are now faced with the delicate dilemma of being able to foresee the availability of funds for creative activity without the likelihood of the increased number of teaching professors to make the best use of them. It seems to me that the universities must have the support for creating more positions which are comparable to that of the professor in freedom, prestige and tenure. Whether these "research professorships" also entail formal teaching or not is not quite as important as is the contact with graduate students and the general university atmosphere. In any case, the point I wish to emphasize is that a healthy sustained increase in basic research in a university cannot be built upon people hired on the basis of one-year contracts, generous as these might be. There must be a general increase in the permanent staff, even though this might mean its size goes beyond current teaching needs.

The role of basic research in the training of scientists and engineers is a very important one. Beyond question the university graduate school is the most effective device we have for the cultivation of the intellectual powers of a potential scientific investigator. The paramount business of the university graduate school is to make learning possible. In principle there are no boundaries on the questions to be asked and the answers to be sought. An atmosphere of this nature is ideally suited for the inspiration of the superior human intellect and for challenging this intellect to its greatest exertion. We can cite many examples from our own experience of very promising men who did poorly through the grade school and high school system because the educational system was geared to the mediocre student and failed to strike a spark of interest in them. Some of these failed to see this inspiration in college because of dull and uninspired professors and teaching methods and wandered off into life to leave their potential intellectual power dormant for ever. Others of this kind happily survive to go on to graduate school and there catch fire when they learn the true meaning of scientific research. There is no human intellect, however great in genius, which cannot find a challenge worthy of its powers in contemplating the structure of the atomic nucleus, the nature of the chemical bond, the physiology of a living organism, the meaning of life, or the origin and the future of the universe. The more a university or any other agency engaged in basic research can stick to the goal of unrestricted human thought, the less likely the great intellects are to be deterred from direct contact with the great unanswered questions of science.

Whether or not we follow our ideals closely in graduate school training, we do try to orient our research toward basic problems and feel that the student has a better development of his scientific abilities and intellectual curiosity because of it. The companies and laboratories to which we send our graduates may not always share our enthusiasm for this system and for its sometimes almost complete divorcement from economic and engineering problems. Frequently, when a fresh Ph. D. goes to his first job an agonizing reappraisal takes place. Whether it is more agonizing for the Ph. D. or for his new employer is sometimes hard to say.

There is a basic difference between undergraduate and graduate because of the fact that the undergraduate is primarily concerned with learning what others have done while the graduate student is trying to obtain new knowledge. His later activities convince him how much there is yet to be learned and give him confidence that he can help in unveiling the unknown. The point is that without a first-hand, rolled-up-sleeves contact with the frontier dividing the known and the unknown, a student is likely to have a gross misconception of the lush lands beyond the frontier which restrains him from trying to get there. Our scientific predecessors have been very diligent. Our libraries are so crammed with their findings and our daily lives so surrounded by the marvelous tools, machines, gadgets and conveniences based on their work that it is hard to avoid the feeling that most everything has been learned. Our textbook system which tends to present science in an overly organized and dogmatic manner reinforces this view. We strive by books, lectures, TV shows and demonstrations to show how incorrect this is, but the belief is deeply ingrained in our subconscious and strenuous efforts are needed to overcome it. Research in pure science is a most effective antidote for this. Let me give you an example. No one doing research today on the new subatomic particles in physics can have the slightest doubt that what we know about atomic nuclei is but a tiny speck compared to our ignorance. We used to think we had reached our ultimate building blocks of atomic nuclei in the proton and the neutron, and for a few years we rested thinking that a little theoretical work would clean up the problem and send us looking for new problem to solve. But look what happened. We now have pi mesons and mu mesons and tau mesons and kappa mesons and theta mesons and V particles and many more, some of which come in three varieties, positive, negative and neutral. In the last few years the number of fundamental particles — and here I use the term very loosely — has grown to more than twenty. These days it has become almost as important to "undiscover" a fundamental particle as it is to discover one.

The activity in this area can only be described as feverish. Here is basic science at its best. Here is a problem with no foreseeable answer which is making an irresistible appeal to the best efforts of physicists all over the world. Bristol, Milan, Brussels, Ecole Polytechnique, in Paris, Padua, Manchester, Bombay, Columbia, Brookhaven, Cornell, MIT, California, Chicago. These are exciting meaningful names to the workers in this field. Most of these men are paid to do this work. But no pay could account for the eagerness, the industry, the ingenuity with which these men attack their illusory goal. Intellectual curiosity and the thrill of helping to unravel secrets which cannot help but be of the most profound importance for our understanding of the universe in which we live are what drive them on. Here also we see exemplified the teaching role of basic science, for at each one of these international laboratories you will find an eager group of young graduate students participating fully in the program. No need to lecture to them on the scientific method, to urge them to master the mathematical tools or the experimental tools of their trade, no need to tell them that the physical world has not been completely explored and explained. Discarded hypotheses lie scattered all around them. Mathematical descriptions of great complexity enter into their daily talk, the fast electronic circuits, the emulsion techniques, the cloud chambers, the bubble chambers are the apparatus without which the data could not be collected. The challenge to their skill is great and the development of their manual and mental powers will stand them in good stead when their graduate school training is complete, even though their careers may take them far from the world of subatomic physics. In this connection, I am tempted to tell you the story I heard somewhere of the physicist who was working on the design of the reactor to be placed in the submarine Nautilus. He complained that he had gotten his training in subatomic physics but had had to switch over into atomic sub physics.

I have been deliberately speaking about the role of basic research broadly, rather than solely in relation to atomic energy, which, of course, is of especial interest to you here today. However, the interests of the atomic energy field are so broad that I feel that this is justified. The Atomic Energy Commission, I believe correctly, is supporting basic research over a very wide spectrum, not limited to fields which seem to have foreseeable application to present endeavors. I do not believe that it would be profitable to use some of the limited time at my disposal for any kind of description or summary of these programs. As you know, these cover broad areas of physics, mathematics, chemistry, metallurgy, biology, medicine and related fields. Much of this research is directly related to reactors and weapons, but much of it is completely uncommitted and is basic science in the purest sense. Although the greatest proportion of the research budget is allocated to the Atomic Energy Commission's own large laboratories, a reasonable proportion of the available money is used to support basic research through hundreds of contracts with universities, colleges, industrial and private foundation laboratories. Since the primary interest of most of you here today probably has to do with the progress of the development of the atomic power industry based on nuclear fission, I suppose that this might be considered a good occasion to try to point out some specific areas where increased activity in basic research might be required. High temperature thermodynamics is certainly one such field. A number of areas in the engineering sciences, such as heat transfer, are also in this category (forgetting for the moment our attempted definitions of basic research). In my opinion, a limit to the extent to which atomic power can be used in the future will be set by the ultimate solution to the waste disposal problem, and here, it seems to me, the path for the ideal solution must lie outside of any ideas presently contemplated.

I would next like to comment more specifically on the overall amount and proportion of basic research and the budgets expended for this purpose in our country. The total research and development expenditure, by government, industry, universities and non-profit institutions, is presently at a rate of about 3 billion dollars a year. Any attempt to estimate the proportion of this devoted to basic research emphasizes our difficulty of definition again. However, even a generously broad definition indicates that no more than 5 percent of this total annual budget is used for the support of basic research. It is my opinion that this proportion and amount of money expended for basic research is much too small! I believe that we need and should work toward an expansion by something like a factor of two as soon as possible. The efficient use of money for basic research at such an increased level is, I believe, easily possible in spite of the shortage of scientists. The results would not go up by a factor of two with a doubled budget but the increase would be so large that this would be the greatest bargain that the American people ever received for their money! For example, the budget of the National Science Foundation should, in my opinion, be rapidly built up to the proportions envisaged in the original Bush report and the amount of money spent by the Atomic Energy Commission on basic research, some 40 million dollars per year, could very profitably be doubled. Correction of the present inadequate ratio of basic to applied work would soon lead to such a broad improvement in technological advance that the return on the overall budget for research and development, on any objective basis for evaluation, would be increased way out of proportion to the small relative additional cost. The problem of a properly balanced research effort is paramount today not only because of the importance of our defense research, upon which our very existence depends, but also because our whole national economy is rapidly becoming tied closely to continued technological advance.

I also want to make a comment or two about our critical shortage of scientists and engineers. Obviously, my comments on the needs for larger budgets for basic research apply with even greater force to the need for more people in this area. Although I do not have the time to discuss this serious question in any detail, I do want to remind you that practically all studies show that a serious loss of potential scientists occurs in high school and in the step from high school to college or university. We simply must do something to inspire more high school students through the improvement of the quality of science teaching, which can only be done by improving the position of such teachers so as to lure more able people into this profession. In order to entice a larger proportion of high school graduates into scientific fields it will be necessary to improve the status of the scientist to the point where the attraction will compete with that of careers in other professions and in business. Such unfavorable comparisons do not seem particularly to bother the scientists themselves, and a certain proportion of high school graduates will continue to choose scientific careers under present circumstances, but it will probably be impossible to effect the needed increase without some added incentives.

I would like to conclude with another remark on the subject of improvement of education in science, in this case general education in science. I believe that a good case can be made for the thesis that a better education and a greater understanding of science by the general public may very well be a prerequisite to working our way out of the present international dilemma brought about by the amazing weapons which science has produced for both sides of the conflict of political ideologies. This was put in different language recently by the head of one of the large industrial concerns in our country when he said, after partially defining culture as "a particular stage in civilization and the particular features of that stage," and a cultured man therefore as "one who has sufficient knowledge to fit his environment," that science should now be considered one of the humanities and "therefore, there is no reason why all men of culture cannot and should not have a reasonable grounding in basic science and an appreciation of its problems."

The Future Through Science

This article is based on an address given 27 December 1956, during the annual meeting of the AAAS in New York, at a banquet held to honor the 25th anniversary of the Gordon Research Conferences of the AAAS and to pay tribute to the memory of Neil E. Gordon.

As we consider the future of our country and of the world, we can perceive that our human resources of trained brainpower in all fields, including those of science and engineering, are foreordained to be of crucial importance to our destiny. It is my belief that we only dimly perceive the extent to which this is true, and that by no means are we taking the necessary steps to encourage the adequate development of this brainpower.

We, the people of the United States, are the most fortunate large group of people on the face of our planet in terms of our physical well-being, personal liberty, material comfort, and the opportunities open to us for social and intellectual development. We occupy only 7 percent of the world's total land area and number only 6 percent of the world's population, yet we are privileged to use 40 percent of the world's total electric power, to drive three-quarters of the world's automobiles, and to communicate through 60 percent of the world's telephones. We own 30 percent of the world's radio and television sets. In going about our daily business, we manage to use up two-thirds of the world's supply of newsprint, 60 percent of the world's annual production of aluminum, 40 percent of the world's annual supply of chromium, one-third of the world's annual supply of tin,

Reprinted with permission from *Science*,
Vol. 124, No. 3235, pp. 1275–1278, December 28, 1956.

and 50 percent of the world's annual supply of copper. Our gross national product of goods and services has soared from $56 billion in 1933 to $90 billion in 1939 to $375 billion in 1955.

Although we may think we enjoy this great prosperity simply because we richly deserve it, other more substantial reasons can be uncovered. We live in a land highly favored in climate and natural resources. Our shores have great natural harbors. We have extensive systems of rivers and lakes for internal transportation and irrigation. We possess extensive fertile range and farm lands situated in several climate zones, permitting the production of large yields of greatly varied grains, fruits, and vegetables. We have great forests, large amounts of water power, large deposits of coal, oil, copper, iron, uranium, mercury, zinc, phosphates, sulfur, and a host of other important metals.

We owe much to our democratic political system. We can be thankful to our tradition of religious toleration and the separation of church and state, which allows one man to pursue those goals and engage in those activities that could have been prohibited by the religious convictions of another. During the early decades of our history, the geographical isolation of the United States aided greatly in the evolution of our governmental and social system without serious molestation from without. Our particular system of capitalistic business enterprise, regulated in many important ways by government, provides, by and large, in an excellent way for the interests of the stockholder, the manager, the working man, and the consumer. The development of our public school system and the growth of our private and state institutions of higher learning have had an important role in all phases of our history in providing a literate and intelligent electorate, a group of skilled workers, and the core of leadership in the professions, business, religion, and public life.

Debt to Science

To all these and other fundamental reasons for our present prosperity that are widely understood must be added another factor which has achieved its greatest importance only in the past two decades and is not so widely recognized. This factor is the firm establishment of scientific research and development as an important base element of our industrial system. The importance of the scientist and engineer in the development of the weapons of war is well recognized because of their impressive record in World War II, but the contribution of the scientist in times of peace is even more important.

The farmer of 1900 farmed with methods not greatly different from those of previous centuries except for the introduction of the horse-drawn steel plow, the reaper, and a few other simple machine tools. Our farmer of 1900 would not recognize the practice of farming as it is carried on in 1956. Indeed, most city people do not realize the extent of the revolution in farming methods that has occurred in the past 15 years. The introduction of more and more specialized motor-driven tools, of scientifically developed strains of crops and scientifically bred cattle, of new irrigation techniques, of weed-controlling chemicals, of insecticides, of new fertilizers and fertilization techniques, and of chemicals for the control of growth and disease in poultry and cattle has allowed great expansion in productivity in the face of a sharply declining farm population.

In medicine, we have seen the development of the sulfa drugs, the antibiotics such as penicillin, streptomycin, aureomycin, erythromycin, and the sterol derivations such as cortisone; the introduction of new operative techniques, a better understanding of viruses, and polio vaccine; the development of new drugs for the treatment of mental disease; and other developments too numerous or too technical to mention.

Every household is familiar with the miracle fabrics Nylon, Dacron, and Orlon. Soap, which remained for centuries mankind's principal cleansing agent, now runs second best to its tersely labeled synthetic successors, Dreft, Treet, Trend, Tide, Surf, All, Add, Fab, Dash, Vel, Joy, and Cheer.

Space allows only a brief recitation of important industrial products and accomplishments: synthetic rubber, plywood, plastics, foamed plastics, television, high-fidelity sound reproduction, synthetic adhesives, paints and corrosion-resistant coatings, new methods of printing and reproduction, color photography, titanium tantalum and zirconium metal, transistors, magnetic tape, solar batteries, electronic computers, nuclear reactors, jet aircraft, atomic-powered submarines, and so forth.

Each phase of our everyday life, whether it concerns food production, heavy industry, transportation, communication, commerce, entertainment, public health, medicine, home life, or our social behavior, depends very directly on the development of these new products. It is hard to establish in cold dollars and cents what part of our gross national product depends directly or indirectly on the scientific applications that have been developed to commercial usefulness only within the past decade, but it certainly is a substantial percentage. Two of our largest concerns, Dow Chemical Corporation and Monsanto Chemical Corporation, trace 30 to 40 percent of their 1956 sales to products that have been developed in the last 10 years. The new agricultural chemicals alone — the fungicides, herbicides and insecticides — constitute a $400 million-a-year business. Eighty thousand employees of the General Electric Company work on products that did not exist before the end of World War II. A spokesman for the National Science Foundation estimates that United States industry as a whole gains back $20 to $50 for each $1 that has been spent on research during the past 25 years.

Now, while it is true that a rather large proportion of our population is involved in the manufacture, distribution, and sale of these products, it is also true that the original conception and development of the ideas upon which these products are based were carried out by a comparatively miniscule fraction of the total population. In this age of specialization, we all find it possible to enjoy the benefits that stem from the ideas of the few without understanding what those ideas are. We find it sufficient for our purposes that some among us are able to synthesize penicillin, design an atomic reactor, solve a system of simultaneous equations on an electronic computer, trace the course of photosynthesis, or take a photomicrograph of a living virus. To enjoy the advantages of our complex industrial society, we recognize the necessity for this interdependence and give up the greater self-reliance we possessed in an earlier age. We give thanks that the specialists somehow seem to appear to solve the problems we cannot solve ourselves.

In our situation, when so many of us owe so much to the few, we would suppose that one of our chief national concerns would be for the conservation of our most important natural resource — trained brainpower. However, there are many signs that this is not the case. Let us consider some of the tasks of the future for which we shall require all the assistance we can get from trained brainpower — from the scientist, the engineer, the lawyer, the industrial executive, the economist, the social scientist, the statesman, and many others.

First of all, merely to maintain our present economy we shall have to carry out scientific research and industrial development on an accelerated pace. Basic science will have to be given strong support, and the intellectual geniuses among us must be discovered, trained, encouraged, and rewarded. A large body of trained workers and technicians must be raised to fill many new types of industrial positions.

Second, we must face squarely the implications of an expanding population. The present population of the United States is 168 million, increasing at the rate of 1.5 percent annually. This means a 35-percent increase in 20 years and an increase of almost 100 percent by the year 2000 — only 43 years hence. Our power consumption will go up at a much greater rate. The total installed capacity for power generation in this country is roughly 100 million kilowatts The Federal Power Commission estimates that we shall need three times as much by 1980. We are rich in coal, oil, and water power, but not nearly so rich that we could meet this staggering increase with these sources of power alone. Hence we may expect to see a great development in nuclear power generating capacity. In a few decades, atomic power will be for us an absolute necessity. In several other major industrial countries, such as England and Japan, which are even now running out of reserves of fossil fuels, atomic power is a necessity within the coming 20 years. There is also hope that solar power or the use of controlled thermonuclear reactions will add eventually to our supply of energy from other sources.

Our population growth of the next 50 years poses other problems of no mean magnitude. Our present farm surpluses will disappear, and we shall require all the skill of our soil scientists, agronomists, and agricultural experts to raise productivity to the required levels. New methods of processing food to secure greater utilization of our plant and animal crops and to reduce waste and spoilage will have to be devised. We shall be faced with increasing difficulties in obtaining sufficient forest and mineral products. The adoption of adequate conservation policies and the development of substitute materials will assume great importance. Our normal sources of fresh water will have to be supplemented on a large scale by fresh water obtained from the sea. Congestion in our cities will intensify our present problems and create new ones in transportation, utilities, water supply, waste disposal, public health and recreation, and mental health and social behavior. We may be forced to reduce drastically the waste of resources entailed in the unlimited production of consumer goods that is symbolized best by the American automobile. The automobile is a peculiarly fertile species that reproduces freely and appears to have no natural enemies sufficiently powerful to hold its growth in check. Furthermore, its reproduction has the peculiar feature that the offspring is always 6 inches longer 3 inches wider, 10 percent more powerful, and 20 percent shinier than its parent. It will be interesting indeed to note the breeding habits of the automobile 50 years hence.

School Crisis

These are but a few of the internal problems that we must face and that will have to be solved largely by the young men and women whom we are training in our schools today for intellectual leadership. If my analysis means anything at all, we need to ensure that a high percentage of our most gifted young people be trained to a very high degree in intellectual and professional fields. The graduates of the near future must exceed those of the past generation not only in numbers but in quality. Unfortunately, the alarming crisis in our school system makes it certain that we shall fall far short of this goal. Without strong emergency measures, the results may be extremely serious.

We have a school crisis because we have an almost explosive increase in the numbers of students at the same time that the social and economic position of the teacher relative to the rest of society has fallen so precipitously that only a small fraction of the needed qualified teachers can be found. The birth rate was abnormally low during the depression of the 1930's, but it has risen steadily since. In 1956 the number of children born was more than 4 million, or twice that of 1935. The tidal wave of new students has reached the primary schools and is reaching the high schools, resulting in the overcrowding and split sessions so common in most communities. Six million children were in secondary schools in 1946. 8 million are there today, and in 1966 there will be 12 million with the peak still not reached.

Our colleges are just now feeling the upswing after the downturn following the expiration of the GI Bill. Not only is the college-age population growing, but the percentage of that population seeking a college education is also increasing. Our present college enrollment of about 3 million will increase to at least 4.5 million by 1966 and could well climb to 5.5 million. Certain areas such as my home state of California will be particularly affected. In California the population will increase 30 percent in the next 10 years. The school population at all levels will increase by as much or more. If we were prepared to maintain educational standards for such an expanded school population, these statistics would be very heartening, for universal education has always been one of our cherished dreams. Considering the great deficiencies in our preparation to meet these phenomenal increases, our dream will have some nightmarish aspects.

The physical space simply does not exist to accommodate these students. We need something like 700,000 new classrooms for the next few years, not to mention laboratories, libraries, and recreational and other facilities. Finding the teachers to staff these classrooms is an even more difficult problem. We shall need 600,000 new teachers by the end of this decade. At the college level we shall need 25,000 new faculty members per year. To supply these totals over the next few years would require that one half of the recipients of college degrees accept jobs as teachers. At the present time, only 20 percent do so, and the percentage is likely to decrease rather than increase, particularly among graduates with training in the sciences. Hence the quality of teaching, particularly at the high-school and grade-school levels, is certain to decrease. Even if teachers were well rewarded, we could not hire enough highly qualified teachers because the potential supply is so much less than the total demands of industry, government, and education.

But teaching, far from being a well-paying profession, is one of the most poorly rewarded occupations. The starting salaries are substantially less than those available in industry for college graduates and after many years of service will not increase substantially, whereas a college graduate in industry will typically double his salary in 5 years. Electricians, plumbers, auto workers, and railroad conductors and engineers earn more than teachers. Considering the depreciation of the dollar, the real income of the teachers has actually fallen since 1940. Every year, thousands of capable, effective teachers who love the work they are doing leave the profession because they simply cannot support their families in any decent standard of living on the pay that they receive. It is pure fantasy to believe we can bring about any real change in our school systems without a drastic change for the better in the economic incentives for teachers.

These factors have led to a severe decline in the quality of our secondary-school education. Half the American public high schools offer no foreign language at all, and one-quarter teach no physics or chemistry or geometry. Students get less and less contact with basic fields of science. In too many schools, the inspiration of a gifted, inspiring science teacher has disappeared completely. Students of unusual intellectual endowment are not encouraged, or are actually repelled, and if later they do develop an interest in science they find that their high-school preparation is inadequate to permit them to elect a science or professional major field of study in college.

These school problems are of such proportions that they cannot and will not be solved for many, many years, but determined steps can be taken to check a complete breakdown in the quality of the education of our youth while we rebuild for the future. Limited space does not permit me to review the many recent actions of local communities, of private foundations and corporations, of scientific societies and other social and political groups which indicate an awareness of this problem and the initiation of steps to alleviate the evils. I should like to add a few personal comments on possible solutions.

First, a great many more of us are going to have to realize how important it is to our continued national well-being and progress to maintain and expand our public school system. Our forebears had no sooner set foot on the soil of New England or entered the forests of Pennsylvania or Kentucky than they entered upon the construction of schools and procured a schoolmaster so that their sons should achieve the elementary literary skills so necessary and prized even in a pioneer community. Our own sacrifices to this ideal are small by comparison, and surely there is room for additional support on our part in the form of taxes and effort. Surely a nation that can afford $5 billion annually for tobacco, $1.7 billion for newspapers and magazines, $10 billion for hard liquor, $2 billion for television sets, $1.3 billion for movie tickets, and $1.2 billion for foreign travel will not be brought to its knees by some billions of additional expenditures for education.

We must immediately put out sizable funds for classroom construction and for drastic increases in the pay scales of our teachers. I do not see how we can avoid the conclusion that a large part of the money for this purpose must come from the Federal Government. I see no fundamental reason why methods for the dispersal of such funds cannot be worked out without endangering local control of our schools. We need to establish the principle that it is in the national interest that every student above a certain level of ability, regardless of his origin and the status of his family's finances, is privileged to carry his education as far as his personal motivation and his scholastic ability will carry him. This principle requires that a system of scholarships paying a major fraction of tuition, book, and living expenses be established. These scholarships should give the student wide limits in his choice of school and choice of field of study.

We need to explore new teaching techniques. We are not quite sure of the most effective role of television, audiovisual aids, and other methods in and outside of our modern classrooms, but certainly their role is important and should be explored thoroughly at once so that our limited numbers of teachers can be used more effectively. We need to relieve our present teachers of the clerical and menial tasks that take up one to two-thirds of their time. For the gifted student, we should explore techniques of self-teaching and individual study with much less detailed guidance from the teacher. We must reexamine our licensing procedures, which now contain much nonsense. We need to do a much better job of counseling the young and instilling a better attitude or motivation toward intellectual pursuits in general. In this regard it is not enough to emphasize the importance of advanced intellectual training; we must also present a clearer picture of what a scientist actually does and try to convey some appreciation of the fun, the thrill, and the inner satisfaction which a scientist usually experiences in his career. Professional men within each community can assist greatly in this problem of counseling, and indeed I am fully aware that many member organizations of the AAAS have active programs in this area.

Many reforms are also necessary in the university science departments to maintain the high position of university research as a fountainhead of new ideas. I believe we need to establish more research professorships so that outstanding research scientists who may not be needed or suited for formal teaching are able to assist in the training of graduate students. We need to set up new central campus facilities for electronics development and service, radioactivity measurements, computing equipment, and so forth, so that the research man can spend his time on new experimental ideas rather than on the menial tasks of equipment building better done by technicians. My space does not permit me to expand here on the many ways in which we can foster the work of the creative thinker on the campus.

World Situation

I have indicated that the serious educational problems we presently face are of considerable concern because of our internal situation arising from an expanding population. Our concern is greater when we consider also the problems of foreign policy. The problems raised by rising population in our country are minor compared with those in many nonindustrial areas of the world. We are in an era of rising nationalism, an era of what Chester Bowles has called rising expectations, among the backward and colonial peoples of the world. The transition of these peoples to stable democratic self-government and to the status of modern industrial states which most of them aspire to become would be a herculean task, even if it were approached in a completely rational manner by all parties concerned. We know only too well that confused, emotional, and irrational steps leading frequently to bitter and ugly armed strife have too frequently marred the attempted transition to a better way of life.

These problems would certainly have been with us even without an ideological conflict with the Soviet Union. This conflict, of course, adds greatly to the problem of the emergence of the former colonial and backward nations. It also adds greatly to our own insecurity. For the first time in our history we are faced with the certainty of extensive damage to our own cities and civilian population in the event of a major war. In quantitative terms, this change is almost the most extreme that could be imagined. During and at the end of World War II, no nation had the power to inflict serious physical destruction on the continental United States. Today or in the near future, the Soviet Union has, or will have, the bombers and the atomic and thermonuclear bombs to lay waste the centers of all of our major cities, within a very short period of time, and to annihilate a sizable fraction of our population. We have defenses against such an attack, but our defense is only partial. Within a few years — perhaps 5, perhaps 20 — both the United States and the Soviet Union will have perfected the intercontinental ballistic missile with a thermonuclear-bomb warhead. Against this ultimate weapon an effective defense is extremely unlikely.

Our main program at the present time is the grim and uncomforting reliance on the mutual deterrent power of annihilation, the threat of counter blows so completely damaging that no rational government would initiate a large-scale war. In lieu of any real agreement with Russia and with other nations on arms limitation and inspection, this is our chief national policy. Plausible arguments can be given for believing that peace can be maintained on this basis of more and bigger bombs and surer means for their delivery. I regard this policy as somewhat disturbing, and as several more nations are added to the present three in possession of nuclear weapons, I shall regard the policy with even greater uneasiness. Countries such as France, Sweden, Belgium, Germany, and Japan clearly have it within their scientific and industrial resources to develop atomic bombs within a few years.

I am not at all sure that nuclear disarmament with adequate safeguards against treachery can be achieved, but it is vital that we maintain a continuing evaluation of every possible approach. I see no fundamental reason why we of the West and our ideological opponents behind the Iron Curtain cannot ultimately adjust our differences sufficiently to live side by side in mutual trust. The appalling waste of the nuclear arms race and its threat of potential utter disaster are two of the central problems that face us in the next few decades.

The Future

This problem and the others I have all too briefly sketched will not be solved by scientists alone or by an educated elite alone. But no solution will be successful which ignores the scientific and intellectual basis of modern civilization. We need highly skilled specialists to attack the agricultural, industrial, social, and governmental problems not only of our own country but of every part of the globe. We need many new specialists in fields to which we have paid scant attention on a national scale in the past. We need to understand the language and customs, the history, the hopes and fears of other nations so that our technical skill can be effectively applied to the problems of development in other lands. The solution of these problems is vital to our own security and to the emergence of a more satisfactory world order.

We need great leadership in all fields by men who are cognizant of the values of our civilization and the factors which influence it. The liberal education that prepares men for such leadership must include science as an integral part, for science is too central a part of our modern culture to be ignored.

We may look forward to many wonders in the coming 50 years. Large earth satellites will continuously circle the earth and monitor the weather, making weather prediction a much more certain matter. We may even effect large-scale control of the weather itself in some areas. Manned satellites may be possible, and space travel to the moon may be achieved. Large ships will be propelled almost exclusively by nuclear power plants. Air travel to any part of the world will be a matter of a few hours. We shall have TV phone communication. Our knowledge of photosynthesis and the processes of the living cell and bodily functions will be enormously more complete, giving us greater control of disease. We shall exercise greater control over human fertility. Problems of mental health, delinquency, and crime should yield to better understanding of biology and of mental processes and social behavior.

We cannot take it for granted that these things will come to pass, that our country will continue to prosper, that the continuing international crises can be overcome without disaster. We must recognize the hazards and prepare ourselves to meet them. The last 25 years have emphasized to any discerning person the importance of trained brainpower to our future — to our scientific future in which our economic prosperity and military security are by no means assured. It would be perilous indeed to neglect the discovery, training and encouragement of the intellectual leaders of this generation and of the generations of the future.

Learning in the World of Change

Address given on the occasion of Dr. Seaborg's inauguration as Chancellor of the University of California, Berkeley, on March 20, 1959.

In a time of history when man's destiny more than ever is linked to learning, the assumption of the leadership of a great company of scholars is an event worthy of serious thought. It is a moment of humility for the selectee, I assure you, in the presence, corporeal or spiritual, of past leaders who have so ably built an institution that has few academic equals. Having for some months presided, or at least sat in the chair, as Chancellor, I have only the greatest awe and admiration for my immediate predecessors, President Kerr and President Sproul.

It is my honor to attempt to evaluate the place of the university and of learning in relationship to the needs of society. I shall claim, additionally, the privilege of making some personal observations.

Since my appointment as Chancellor, friends have asked me this question: "Why did you leave science at a time when we need all our science hands, to take an administrative job for which our University has many qualified people?"

Believe me, as I have become acquainted with the complexities, the difficulties and the great responsibilities of the office, the question has also crossed my mind. But, as you know, broad problems of society are not a new preoccupation with me and with other scientists. For some years, now I have talked on the subject of science and education; too much, it appears, to have escaped the acceptance of the challenge the Chancellorship offered.

One of my beliefs is that there should be a blending of experience from the various fields of learning in the leadership of modern society. Surely the stewardship of a great university today is a critical forward position.

When, therefore, President Kerr and The Regents approached me concerning the Chancellorship, this thought occurred to me: considering the speed with which our society is, and for some time has been, proceeding toward ultimate solutions to its problems, it seems, at the worst, unlikely that the addition of one scientist to the vanguard will materially retard the pace.

If I may, in the words of Francis Bacon, "intermingle … jest with earnest," I should like to point out that I have, to my personal delight, not given up the laboratory entirely. For the university is a community affair, a proper concern of dedicated men and women from diverse parts of society. Within the University itself, I am fortunate in having the forbearance and guidance of able faculty associates in many disciplines, which not only permit me to engage to some extent in the pursuit of scientific knowledge,

Audience photo of the inauguration of Glenn T. Seaborg as the second Chancellor of the University of California, Berkeley, at the Hearst Greek Theater on the UC campus, 20 March 1959. (Reproduced by courtesy of Ed Kirwan Graphic Arts, Berkeley, CA.)

but also incorporate broad representation into the conduct of academic affairs. The helpful spirit of President Kerr, The Regents, the students, the alumni, and friends of the University are sources of great strength and inspiration.

The state and nation today have special cause to support the universities and to look to them for educational leadership through an era, generated by the universities themselves, that has no precedent in history. For thousands of years, man learned but slowly and empirically about his surroundings, his natural curiosity muted by superstition and authoritarian sanction. During long periods of history, each generation acquired from its predecessor a pattern of living and thinking, lived out its life, and transmitted the old ways, virtually unchanged, to its successor. Even the Industrial Revolution, explosive in terms of previous history, was, by our standards leisurely.

In contrast, we live in a world of change, a time of constant, rapid revolution. We are, and shall remain, passionately dedicated to environmental metamorphosis, as fast, in as many ways, and as drastically as possible. During the past two decades we have experienced centuries of environmental evolution, as measured by ancient criteria. We speak, too glibly but not entirely without significance, of Ages, which tumble one upon another, and from which only future historians will make sense: the Electrical, the Electronic, the Atomic, the Thermonuclear Ages; the Air, the Jet, the Space Ages; the Age of Automation; the Technological, the Medical Revolutions. And the future promises ever faster evolution. Before the century is finished, today's civilization may seem almost primitive.

These upheavals are accompanied by political, socio-economic and moral contortions. Old forms and old ways yield to the new, and the new are replaced almost before our eyes. Value systems are summoned to new judgments. In the years ahead, as science and technology wrench society, millions of men and women will need to adapt to new and more complicated ways of earning their livelihood. Change is not a stranger to men, but the speed of today's change is unequaled by similar periods of the past.

In this world of induced ferment, how can we equip ourselves for the trials of adaptation ahead? How can we preserve, enrich and extend the values we prize? What is the place of learning?

First, let us look briefly backward, to the road we have traveled to our world of change, to the values we most highly esteem, and to some of the problems we face. During the Renaissance, men, questioning authority and the concepts of the universe itself, unleashed forces that destroyed ancient superstitions, and opened up new worlds of thought and conduct. No longer was it necessary for man to abjure aspirations to comprehend his environment. No longer needed he despair of influencing his destiny. Through exploration and education, man could understand. By understanding, man could improve himself infinitely, and accept individual responsibility for the conduct of society.

In America, and in the Western world generally, these ideas led us to insist that the glory of mankind lies in the highest development of the individual. We treasure personal dignity and the search for truth. We exalt reason. Through a liberal democracy, we maintain, a people enlightened can govern themselves better than they can be governed by one man or a few. While we seldom delude ourselves into believing we have perfected our ideals, we remain firm in the conviction that a liberal democracy is more promising for the betterment of mankind than any other system of which we have knowledge. And since the beginning of the Republic, we have held it to be axiomatic that the essential qualification of the individual in a democracy is education.

In the beginnings of the Republic, of course, the franchise was by no means universal, and the early proprietors of our government were for the most part well educated. But as the franchise broadened, it became accepted that common sense, together with education consisting of the three R's, was necessary for intelligent participation in a democratic society. In a slower and simpler world such an education may have served. But it is not adequate today. Neither is the education of a quarter of a century ago sufficient equipment for today's and tomorrow's world of change.

Knowledge has mushroomed out of all bounds, and we are only at the beginning of knowledge-gathering. What we know is only enough to spur us on to greater effort. How can the individual citizen act intelligently on issues that are conditioned by forces that are outside his understanding?

If we had only to cope with change within our own national boundaries, we would have more than enough problems. We could readily find enough diversion in the consequences of the advances in technology and medicine, which have brought material wealth, longer life and a population explosion. But our Age of Change embraces the whole world, and the world is now a small place. The problems of others are our problems. The non-Western peoples of the world, technologically asleep several ages behind, have awakened with a start. They demand to share equally and without delay in material wealth with their Western brothers. That they will succeed, perhaps not as quickly as they or we would hope, we all acknowledge. How they succeed may well determine the nature of world civilization of the future.

For the world is now engaged in a contest, perhaps the final contest of civilization, in which the major alternatives are the ideologies of Western liberal democracy and communism. In this contest, there is, of course, still the menace of fearful war. Yet despite occasional frightening episodes to the contrary, the awareness appears to be growing that war today is a more irrational method than ever of settling international disputes. The longer we can delay violence the keener will be the awareness, and the greater the chance of reason prevailing.

Given peace, we must still look to a decisive conflict. This is a contest of ideas and of intellectual achievement and performance, in which the "blood, sweat and tears" will be shed in the classrooms, the libraries and the laboratories. The skirmishes of this war, which mercifully does not involve nuclear weapons, have already begun. There was no signal, no Redcoats clanking along the road to Lexington, no bombers over Pearl Harbor, no sudden horror to arouse the people to intellectual arms. This final war came upon us silently, and, I fear, while we were complacently distracted with our material pleasures and somewhat indifferent to the need for constant nourishment of our liberties. This war will end gradually, without signal, and perhaps without the defeated knowing it for a long time. For in the hearts and practices of the uncommitted peoples will be the seeds of humanism or the seeds of authoritarianism, implanted by the ideological and intellectual performances of one or the other.

What must we do to adapt ourselves to our world of change? I should like to select for mention five interrelated factors that are cogent to our success: time, complexity, values, education and research.

Time, the lubricant of smooth operation of democracy a century and a half ago, is our least abundant commodity. Democratic judgments must be more quickly and more surely exercised than before.

Such judgments must be exercised on more complex problems than liberal democracies have ever faced. The progressive fragmentation and growth of knowledge makes it increasingly difficult for the individual to know enough to act intelligently.

Through inspired leadership of committed and articulate men and women from all sections of society, I believe our citizens must be convinced of the reality of the nonmilitary trials that lie ahead. Once persuaded, I believe we will reexamine some of our values. Can we afford the present emphasis on leisure and material goods? In the harsh days of the 1930's, we dreamed of a nirvana of wealth, security and leisure. As a people we have gone far toward achieving a material paradise. But are we meanwhile doing what we must to preserve the ideals upon which our nirvana is based? As our modern patrons, the machines, free us more from physical labor, must we not do more mental work to cope with the challenges of the world of change? Do we not need motivations in this direction as intense as those imparted by dangers from shooting wars?

If we need to decide more complex matters more quickly, and incidentally operate an increasingly complicated industrial and technological apparatus, we must look to the more adequate preparation of citizens. Our nation and our educational system, led by the universities, must gird themselves for greater effort in several ways: the upward movement of the educational level of the whole population, which has been going on for some time, must be accelerated; we must search out and stimulate the fullest development of gifted young people in all creative fields; we must impart the conviction that education is a lifelong job; and we must seek a balance in educational content that will equip the intellect for the world of change.

We are assailed on all sides by evidence of the failure of individuals to develop their talents fully.

At the present time, of the American high school graduates who rank in the top 25 percent scholastically, apparently only about one-half go to college. Similarly, large numbers of college graduates capable of advanced study take no further academic training. Nor can our labor force be neglected; properly trained, the majority are capable of moving, and must move, up to higher, more rewarding and necessary skills. According to our creed, the individual must be free to cultivate to the utmost his own talents. If we are to survive in the contest of the intellect, we must persuade young people that fullest development of their talents will be rewarding to them and is essential to the survival of humanist ideals.

To the same ends, we must extend our efforts to rescue lost talent, among women and among minority groups. We must do more research in the development of new educational techniques, not only to get more information into people but to sharpen the individual's critical sense.

In seeking better intellectual equipment for our world of change, we must seek a balance of knowledge that will make us more effective citizens of our time. At one time, a scholar was by definition a student of religion; at another, a student of Greek, Latin and the classics. Today knowledge is valued throughout the continuum of human experience. There is beauty in discovery, mathematics in music, a kinship of science and poetry in the description of nature, and exquisite form in a molecule. Attempts to place different disciplines in different camps are revealed as artificial in the face of the unity of knowledge. All literate men are sustained by the philosopher, the historian, the political analyst, the economist, the scientist, the poet, the artist and the musician. Above all, if it is true that the proper study of man is man, then we must always emphasize efforts to understand people in their great wonder and variety.

As we measure our educational curricula against our heritage of knowledge, we must be constantly alert to imbalances. At the present time, it seems clear that one modification should be made in the field of science, whose onrush to such a critical position in our affairs has caught our educational system unaware.

Science provides the momentum of our world of change. Upon decisions about science and technology hang the fates of peoples and ideologies, our own included. If it is true that every man in a democracy need not be an originator of policy but must be a good judge of it, then our citizens must be well enough acquainted with the general principles of science to be good judges of policy based upon it, or at least to be able to reject spurious policies founded on pseudoscience or nonscience.

I would wager that a Gallup poll would disclose a few Americans who feel they do not know everything about politics. A majority of us, surely, would take pride in our judgment of economics and business, although we might confess to being a little short on a few of the more complex technicalities. However, an overwhelming majority, I wager, would disclaim unabashedly, perhaps even with some pride, any knowledge of science.

It would be folly to suggest that every person become a scientist or be expected to comprehend the intricacies of arcane theories, any more than we would suggest that everyone, while we would like to have him appreciate poetry, should become a poet.

But until our people become as much at home with science as with other phases of our national life, I fear that, in the continuing world of change, we shall have increasing difficulty living up to the ideal of an informed people as the proprietors of liberal democracy.

When I say we must seek a citizenry with scientific literacy by adding science to the curriculum, that science should be truly a part of a liberal education, you may reply, "What shall we cut out?" Not any of the humanities, I will agree. And I can only say that science must be added. Our age of intellect requires more of us than ever before. Do we not have some margin of leisure to use up without suggesting the draconian extremes of authoritarian systems? What I would suggest is not, I believe, an overemphasis on science in general, but a modest correction of what seems to me to be an underemphasis that can be fatal in our world of change.

In addition to the teaching obligation, the university has the special charge to create new knowledge. A condition of our successful adaptation to the world of change is the cultivation, in university centers, of incisive minds to extend the quarrying of new meanings from nature, the perception of new relationships in human experience, and the capture of new insights into beauty. We have become accustomed to what may seem at first glance large outlays from the federal government for scientific research. Yet when we measure the total funds for such investigations against the potential and the need, we find excellent estimates that support of basic science is less than one-half what could be invested to our great profit. If basic scientific research is impoverished, consider how poorly supported is the creation of knowledge in the arts, humanities and the social sciences, whose association with the national welfare, unlike science, has been poorly recognized. There is here an imbalance in the stimulation of knowledge-seeking, which, I believe, we must work to correct. For example, just as our University now has many sources of funds to provide scientists with periods of freedom to create, so should we be able to afford the same opportunity to scholars in the arts, humanities and social sciences.

Given the conditions of our world of change, in which learning will determine the rise or fall of our culture, we as a nation can shrink from adequate support of education only at our peril. The bill for teaching and for knowledge-seeking sometimes may seem high. Yet we are only at the beginning of the needs of a nation bursting at the seams with talented new generations whose success will depend upon their depth of knowledge and their opportunity for intellectual exploration. We have a pioneering tradition of finding ways to support the development of our frontiers. We gave away free lands in the West. We have subsidized the railroads, the airlines, the shipping industry. We cannot fail, in our great wealth, and our equally great need, to find the means to develop our brainpower, the critical resource of our age. For if it were nothing more, and it is a great deal more, education is, in the words of Burke, "the cheap defense of nations."

Perhaps the spirit of our pioneering past is the spirit we must seek for our present and future world of change. For as we search the world for horizons, wherever we turn we see them through education. Learning and discovery are the New Worlds and the Old West, the lands of opportunity in the 20th century. In intellectual activity there is always virgin soil to be tilled by the mind.

With the will to meet the future and the motivation and support for maximum fulfillment of individual talents, we can nourish the creativity and powers of judgment which perpetuate our liberal democratic ideals in the world of change.

Science and Liberal Education in the Space Age

Centennial lecture
at Augustana College, October 7, 1959.

A hundred years ago, when this college was born, our young democratic form of government faced a terrible test. We saw the main issue as one of human freedom. We came through our crisis. Our humane society, with its emphasis on the value of the individual, survived. But only a civil war could confirm our democratic principles.

We are, I believe, approaching another test of our liberal democracy in the years ahead. Although our problem is different in kind, and the time scale is longer, it is not less important than the one we faced a century ago. The issue, in the minds of many, is once again individual human freedom. The question is whether science, with all its bounty and mercy, will mock us by placing chains upon the individual. Serious students of our society cannot dismiss lightly such nightmares as those suggested by George Orwell.

Certainly we live in an age in which science seems to be the most dynamic force in our world. Men have propelled objects to the moon and into orbit around the sun. We have stepped dramatically into the space age. The course we have charted for the future challenges the imagination. It is a future lined with fateful decisions involving scientific knowledge and principles. Yet, paradoxically, while the influence of science in our

Reprinted with permission from *Augustana College, 1860–1960. Centennial Lectures. Occasional Paper* No. 5, ed. Ernest Espelie, (Augustana College, Rock Island, IL, 1959).

lives has been growing, a general knowledge of the fundamentals of science among our liberally educated leadership and citizenry has been standing still at best, and perhaps shrinking. We are relatively less informed about critical issues involving science and upon which judgment should be exercised by a broad base of our citizenry. The forces of science and technology are controllable in a democratic society by an informed electorate. But unless we attain a wider foundation of knowledge of the general nature and principles of science, we may in the future lose democratic control of our destinies. Through ignorance, and only through ignorance, we could take the dark road to *1984*. By redressing the past relative underemphasis on science in education, which is particularly a challenge to liberal education, we can strengthen democratic principles in this fast-moving age of space.

It is fitting to discuss this problem here today as the Augustana College Division of Science makes its contribution to the College's centennial celebration, with the appropriate theme, "Constancy and Change in a Century of Christian Higher Education." Augustana began as a small institution, and it has remained small by such standards as those of my own Berkeley campus of the University of California, where some 20,000 students enrolled this fall. But if Augustana has grown relatively little in size, it has grown in stature, until today it has a secure academic ranking among institutions of its kind.

Augustana brings together much of the best that personifies our nation. It was founded by emigrants who sank their roots deep into the new soil, and, like our other pioneers, were willing to make great sacrifices that their sons and daughters might have the learning they needed for effective participation in a democratic society. Moreover, Augustana, founded by graduates of the Universities of Uppsala and Lund, is bound by strong and good ties to the old culture. Parenthetically, in speaking of these matters, I am proud of my particular bond with Augustana's founders; for my people, too, were emigrants from Sweden; and they, without hesitation, sacrificed that I might obtain that prized possession — an education.

Augustana, too, is a product of the Christian thinking that came out of the Renaissance and Reformation. Christian concepts, interpreted especially by Locke, glorified the individual and established the concept of the rights of man. And, in the hands of Jefferson and others, the grand ideas of American democracy took shape: that man is educable and infinitely perfectable; that educated men can govern themselves better than can one man or a few. Moreover, liberal Christian views, encouraging the questioning of authority and the search for the truth, gave wings to science, to the exploration of nature, and to rapid change in man's way of thinking and living.

There are, then, many reasons for discussing here, at the proud century mark of Augustana, the interrelated ideas of liberal education, democracy, Christian principles, science, and constancy and change.

We are no less convinced today than was Thomas Jefferson that education, and we must qualify it as liberal, is the essential qualification for a citizen of a democratic society.

Let us consider the meaning of a liberal education. Most of us, I think, would agree that a liberal education should endow the individual with the broad intellectual and spiritual heritage of mankind. A person so educated should have a large knowledge of man's history, of his failings, and his successes. He should have a comprehensive view of philosophy, literature and the arts. He should be as well acquainted with science, including the scientific method and the spirit, the cultural and aesthetic values of science, as he should be with the great classics. He should have a love of and a trust in the truth. With all these riches at his disposal, our ideally liberally educated man will be a sensitive, tolerant, humane individual who can build a creative relationship with his community. He can provide enlightened, constructive leadership in a democratic society.

Augustana performs this function well; it sends men and women into all walks of life, including science; and these graduates more than justify the high purposes to which this college was, and is, dedicated.

Incidentally, the small liberal arts college has been shown to be the training ground of a surprising number of outstanding scientists, among whom, from your own institution, was the distinguished physiologist, Anton Carlson. I trust the small liberal arts colleges will continue to hatch such creative individuals, while at the same time performing their broader function.

One of the problems we face is the difficulty of imparting an ideal liberal education for the modern world in the time society now allocates. There are constant demands from many sources for new additions to the liberal arts curriculum, at Berkeley, as well as, I can imagine, Rock Island. Liberal arts college face many a challenge. It is my feeling that in our world of science and technology none of these challenges is more urgent than the provision of scientific literacy for the liberal arts graduate. Such literacy, I believe, is essential for the survival of the broader, humane values to which the liberal arts college and democracy itself are dedicated.

Let me say, parenthetically, that I am not advocating that every person become a scientist, even if this were possible. What I do advocate is that in an age dominated by science and technology, surely the liberal arts graduate should be as literate in science as he is in other fields of knowledge. This is particularly important to the college graduate, since we must rely upon him for the bulk of our leadership.

The original concept of a liberal education gave science an equal place with the humanities and the arts. But all too often today the liberal arts are considered everything *but* science. In too many universities and colleges, the game among liberal arts students is to see how little science it is possible to take. The effect is that the student is shortchanged. He goes into the world inadequately prepared to discharge his responsibilities relative to the great issues he must face as a citizen of a space age democracy.

Why should there be, in our exciting age of science, a decline of science in the liberal arts? I do not pretend to have the complete answer, but several things seem clear. First, the student is not usually given a good introduction to science and arithmetic in the lower schools. He has little cause to be interested in mathematics and science, and he is more likely to regard them on the basis of past experience, with dread or distaste. Secondly, the colleges and universities, intent upon their important job of providing students with a grounding in science as a prelude to specialization, too often overlook the different needs of the liberal arts student who is not majoring in science. In many of the courses designed for specialization, the general student loses his sense of perspective. He gains no understanding of fundamentals and of meanings which he can integrate into his larger experience. I fear that teachers of science too often in the past have been inattentive to cultural courses that give the general student an overall view of science, of scientific method and fundamentals. Such courses need not and must not be shallow but must be designed with as much or more care than those for future specialists. The correction of these shortcomings, which is now being undertaken in many institutions by faculty and administration, should be accelerated.

The general decline of scientific knowledge in our educational system during a period of rising influence of science has brought about what must be interpreted as a narrowing of true participation in the democratic process. Decisions must be made less and less by the electorate and less and less even by the electorate's representatives. More and more reliance must be placed upon a relatively small number of people who have scientific knowledge. And however capable or good their intentions, the disproportionate exercise of responsibility by a few is not healthy in a democracy. There is no substitute for a broad base of informed and intelligent responsibility as a bulwark against oligarchy and other forms of government that are alien to our traditions.

Let us consider the unhappy harvest of our lack of sufficient knowledge about science. We have made or are making many costly miscalculations as a result. One miscalculation is a widespread attitude, which holds that science is going too fast, that the clock should be stopped or even turned back. This attitude, often displayed by educated people, is based squarely on a lack of appreciation of the nature of science. No deep knowledge of intricate scientific theories is needed to know that stopping the clock is only wishful thinking. Once men dared to question the untested dogmas that ruled them, there was and is no turning from the unrelenting pursuit of truth. Each answer to a riddle of nature raises new questions which men with curiosity cannot and should not resist. Indeed, rather than hope for a lessening of the pursuit, we should look forward eagerly to an acceleration of knowledge-gathering.

In 1916, Alfred North Whitehead summarized the matter as follows:

> "In the conditions of modern life the rule is absolute: The race which does not value trained intelligence is doomed. Not all your heroism, not all your social charm, not all your wit, not all your victories on land and sea, can move back the finger of fate. Today we must maintain ourselves. Tomorrow science will have moved forward yet one more step, and there will be no appeal from the judgment which will then be pronounced on the uneducated."

A corollary of the regressive "stop the clock" idea is the notion that science is morally sterile and inherently evil. What a strange notion, that the pursuit of truth could be evil! The mistake lies in an interpretation. Knowledge, of course, whether it be scientific or otherwise, is power. And scientific knowledge, like all power, has been applied for both the good of mankind and to his injury. The history of man is one long, bloody record of such injury. Men have been unprejudiced about the weapons they have used against one another. Nor is there any certainty that, even among those who advocate a regression to an earlier state of society, the inevitable accompanying conditions would be acceptable: the miserable living conditions, the drudgery, the disease, the hunger, the short lifespan. No, there is no turning back for mankind. Our fate lies ahead. And the cure for the delusion that we might arrest the perversion of power by arresting scientific research lies in knowledge of the nature of science and of the scientific pursuit of truth. The concept of continued exploration should be an accepted part of the frame of reference of every citizen. The extent to which regressive ideas inhibit the support of fundamental research can only be guessed.

An offshoot of the notion that science is evil is the concept that if we continue scientific development we are inevitably doomed to an Orwellian, *1984*-type society. It is true that the perversion of scientific knowledge could so enslave a people ignorant of science and its potential. And it is partially to prevent such a result and to preserve democracy that I plead for the enlargement of the base of scientific knowledge. The civilized man, an individualist, liberally educated, with knowledge of scientific principles, will be an intelligent and dedicated foe of steps in the direction of *1984*.

There are still other miscalculations that can be laid to a lack of literacy about science. On the international scene, we are witnessing the rise of peoples of more primitive, or at least non-technical, cultures. They desire the greater dignity and the higher standard of living which Western nations have attained. Among them nationalism is on the rise. The West can be instrumental in influencing the kind of societies they will build. But I wonder if one of our big problems in influencing such peoples is not a lack of elementary knowledge of scientific principles. How widespread, for example, is the appreciation of the fact that intelligence among the races is fairly evenly distributed? The fact that a man lives in a hut in a jungle is not a measure of his potential intelligence, and persons acquainted with elementary scientific principles know this. Moreover, technical ability is not a trait peculiar to our own country. Rather, technical skills can be acquired by any people, and rather quickly — more quickly than many of us believe. And this is a fact that has been and continues to be illustrated. The colonial administrator who protests that an African is a thousand years behind the development of more cultured people, and that an equivalent period must pass before the African achieves the same advanced stage of development, cannot be well informed on scientific principles. A wider appreciation, then, of the capabilities of underdeveloped peoples, which derives from scientific principles that are not difficult to understand, might do much to inspire the Western culture to give more intelligent guidance in the peaceful development of rational governments in new countries.

We make other miscalculations. For example, although it appears that scientific research is being supported handsomely, the support is spotty and far below need in the most important area, namely, basic research. This is the kind of research that people who do not understand science call impractical and useless. It cannot be applied immediately, nor is there often much immediate prospect of its application. All this research does is to enlarge our understanding of some aspect of nature! Yet it is the most practical of all research. For basic research, and only basic research, expands the ultimate capabilities of mankind. We are now finding applications of basic research, results obtained half a century ago and which seemed of no practical value at the time. A deficiency in the support of this kind of research, the kind that tells us "why the grass is green," if you will, is another miscalculation based upon our national scientific illiteracy.

What should a liberal education provide in the way of scientific learning? Certainly, every educated person should have knowledge of the scientific method: that way of thinking by the postulation and testing of theory and the drawing of conclusions based upon the evidence of experiment. Such thinking is not limited to usefulness in science; it is present in competent people in every field; and its systematic use heightens the effectiveness of the artist, the writer, the political scientist, and others.

Every educated person should know something about the great ideas of science: for example, the laws of motion, the structure of matter, the theories of evolution and relativity, the principles of genetic inheritance, the chemical bases of life, and the origin of the universe. He should know something of the history of scientific ideas and how they have influenced his civilization and his thinking about himself and his surroundings. For scientific knowledge, far from being sterile and cellophane-wrapped, lies close to the heart of philosophy and culture.

Indeed, the cultural aspects of science, intimately intertwined with human thought, in the long run may be the most important of all. Since the earliest times, men have wondered about their origins and about the origins of the earth and the universe. Every primitive culture has its own set of myths and legends, usually anthropomorphic, which explain how the world and man were born. Such questions have always troubled man, a stranger in overwhelming surroundings. In our own culture, it has not been long since the dominant belief was that the earth was the center of the universe, that the sun revolved around the earth, and that the sky was a great bowl over the earth. Through the power of science, these fantasies have been destroyed forever. We have gone on to learn that the earth is a rare kind of cinder in an isolated part of one of uncounted billions of galaxies, each containing billions of stars like the sun. More recently, we learned that the universe is twice as big as we had thought, and we are now pondering whether the universe is a great machine that was born in a twinkling and is now running downhill, or whether creation is a continuous process, with the universe renewing itself. Can there be more basic or more exciting questions than these? And does not the discovery of new knowledge, the erasure of ignorance and superstition, lend dignity and security to man's existence in a bewildering and awesome universe?

A person who can embrace such concepts and appreciate the spirit of the scientific quest to know more will possess still two other qualifications of scientific literacy: an appreciation of the dynamic unfinished nature of science; and insight into the joy of discovery, the creative portion of science that is akin to the work of the artist, the writer, the poet.

Some liberally educated people who lack exposure to science unjustly assert that science lacks aesthetic value and that it is iconoclastic. But I wonder if the revelation of the physical principles that produce a sunset can reduce the poet's appreciation of the sunset's grandeur? Will not such understanding, rather, lend depth and a new order of appreciation of nature?

Certainly science is iconoclastic, but it does not destroy truth. Science confirms that man is insignificant physically. But at the same time, science itself is a glorification of man's power of thought. In the whole vast universe, so far as we know now, only man knows anything. Only man can shape ideas and describe and explore and form images of the strange depths of space. To scientists, this is a precious, a unique thing, and man and his preservation become of utmost value.

Science, in short, is at the core of the heritage the liberally educated person must possess; it has humane and aesthetic values of the highest order; it has power beyond any tool ever discovered by man; and without a knowledge of its general principles, we are ill-equipped to function properly in a democratic society in a space age.

I should like to speak about science and the ethics of our Christian heritage. Science owes much to the Reformation, which was led by Martin Luther. As science has searched for the truth, wherever it has led, its findings have naturally come into conflict with and overturned many old teachings. And quite naturally the world has witnessed the sound and fury of conflict between science and religion. Yet I wonder if most of the teachings that have been overturned have not been those that could be classified as dogma. Science has not harmed the basic core of Christian ethic. I know of no scientist or scientific principle which injures the ancient injunction to do unto others as you would have them do unto you; or, quoting more exactly from the Book of Matthew: "... all things whatsoever ye would that man should do to you, do ye even so to them ..." I believe that as a group there are no stronger believers in the brotherhood of man and in the equality of man than scientists. I know of no scientist who places little value on his fellow man. I do not believe that the search for truth can be harmful to ethical principles men have discovered, through experience, to be a humane code of conduct. If truth is good, and I believe it is, I cannot see how the search for and discovery of it can be harmful. Therefore, I do not believe there is essential conflict between science and a religious ethic which does not fear truth. There are, therefore, constant and universal values for a humane ethic in our age of science-induced change.

I have spoken at length about what appear to be deficiencies in scientific learning in a liberal education and how those shortcomings affect the effectiveness of our democratic society. But the liberal college graduate is not alone in need of a greater understanding of science. The larger number of our citizens who have not had a higher education have had even less opportunity to obtain a cultural appreciation of science. Their voices, too, are important in our democracy. They have no less need of an understanding of the forces that dominate our destinies. Nor will a changed curriculum in some future help here. We must educate and provide understanding of science now. I cannot summarize the need better than was done in the report, "Education for the Age of Science," issued by President Eisenhower's Science Advisory Committee, of which I have the honor to be a member.

The report said,

> "A democratic citizenry today must understand science in order to have a wide and intelligent democratic participation in many national decisions. Such decisions are being made now. They cannot be postponed for 20 years while we are improving our present educational system so that its products will constitute a significant fraction of the mature voting population."
>
> "There is, therefore, no escape from the urgency of providing high-grade and plentiful adult education in science, now, planned for those who are unprepared even in the fundamentals ..."

In summary, science and technology, which are both bountiful and merciful, also bring serious problems to a people dedicated to democratic principles. The principles of science are controlling many of the decisive issues of today and tomorrow. If the essence of democracy is the exercise of influence by an informed people, as I believe it is, then an understanding of the elementary principles of science must be broadly based in society. For some time, however, while science has grown in importance in our destinies, we have lagged in the provision by our educational system of an adequate appreciation of these elementary principles to students at all levels. Unless we reverse this trend, we shall witness a continually narrowing participation in the critical decisions that affect our future. This could lead to the downfall of our democratic form of government. Also, knowledge of science and a broadly-based participation in decisions on scientifically-oriented problems are essential to control and to avert the perversion of the great power inherent in science. The task should be pondered especially by those of you in our liberal arts colleges, large and small, which provide us with the bulk of our leadership and which are so closely identified with humane, democratic principles.

A Scientific Society — The Beginnings

The John Wesley Powell Lecture, presented at the 128th Annual Meeting of the American Association for the Advancement of Science, Denver-Hilton Hotel, Denver, Colorado, December 27, 1961.

John Wesley Powell was a man who stands large in the history of American science. He believed in the frontier, and he lived on it vigorously and adventurously, whether exploring the Colorado or insisting upon good science policy in Washington. Powell was a man of great vision. He saw clearly how science and engineering could develop the vast potential of the West to help make ours a great nation. He understood the nature of science and technology, and his Geological Survey demonstrated the usefulness of properly administered government science. It is with pride, therefore, that I speak in Powell's name, here in the West he knew so well.

As I prepared for this lecture and considered some of the developments in science since Powell's time, my thoughts drifted to personal reminiscence. I recalled that in this season 21 years ago, my colleagues and I were doing the experiment which resulted in the discovery of plutonium. Needless to say, my world has not been the same since. Nor has my experience been unusual. The same forces operating in my case have markedly altered the lives of many millions of people and, indeed, society itself. Allowing for the lack of perspective that accompanies our closeness to the events, it

Reprinted with permission from *Science*,
Vol. 135, p. 505 ff., (1961).

still seems pardonable to judge the past two decades to be one of the most portentous periods in human history. And this has been made so by science and technology.

I believe these things to be true not alone because of the novel dilemma revolving about nuclear weapons and the very survival of modern civilization; but also because of the general scientific-technological progress most dramatically exemplified by the peaceful atom and space exploration.

What is perhaps more important in the long run, granted our ability to avert total nuclear war, is the fact that in these two decades science and technology have become a dominant force in our social order. Much has been written about the scientific society, usually in the future tense. I believe we are warranted in changing the tense to the present. Although it is in its infancy, the scientific society has arrived; it has crossed the threshold in its relationship to society as a whole.

Science and technology are now part of the fabric of government, industry and business, and of our social institutions. The destinies of individuals and peoples are irrevocably associated, from day to day, with the growth and use of scientific knowledge.

As was to be expected, the birth of the new infant has not been an easy one. Nor will its development be untroubled. It seems clear that science and technology are the most powerful forces for material advancement unleashed by man. The changes these forces bring — and will continue to bring — run wide and deep through society. Men as a whole are not friendly to such changes and forces. But to scientists, these developments may seem clearer than to most men.

The conception of our infant scientific society can best be assigned to the Renaissance. At that time, men challenged authority and the dogma that had ruled for centuries, and questioned the nature of the universe and man's place in it. The spirit of questioning in the Western world occurred on many fronts — in religion and philosophy and political theory, in art and literature, and in science. One important result was the expression in the Declaration of Independence and the Bill of Rights of the Constitution of the idea of individual personal, political, and intellectual freedom as controlling in an organized society. The same forces that liberated men politically, and in other ways, also produced the scientific method. With the growth of freedom of inquiry and the development of techniques for discovery, there began an acceleration of our ideas about nature. And the knowledge gained became highly significant when translated by technologists into tools.

Through our privileged perspective, we can see that — given the conditions of the last five centuries — everything that has happened has been virtually inevitable. For the achievement by men of the right to search for truth was the critical breakthrough. When this right was established on a continuing basis, it was only a matter of time until bacteria were discovered, electricity identified, and nuclear fission revealed. In a word, modern scientific knowledge and its application are a consequence of the vigorous exercise of the freedoms that arose in Western Europe and America.

I should like to introduce my stocktaking of the twenty years now ending by recalling some personal experiences to illustrate, in an anecdotal way, something of the changes within science and its new relationship to society.

In the fall of 1940, I was a young chemist at the University of California. We had been trained to believe that a deep gulf ran between pure and applied science. I was "pure," of course, searching for knowledge for its own sake. We were also poor — a property which followed purity like the night the day. But being pure, we could accept poverty in good grace and even with some pride. Our poverty, of course, pervaded our research operations. Research funds were almost unknown. We built as much of our own equipment as we could, or coaxed our more talented friends into helping with it. Laboratory space was hard to come by. I can recall, as a graduate student, adopting the squatter's rights technique to obtain some space in an abandoned and condemned old wooden structure. But these were the accepted conditions of research science in those days, and we were hardly aware that our difficulties were difficulties.

The Lawrence Radiation Laboratory was a new kind of thing on the scientific horizon. It gave us a foretaste of things to come in some fields. The equipment was huge — by 1939, there were two cyclotrons that were giants among scientific equipment. Scientists from a variety of fields found it profitable frequently to pool their talents in working with the cyclotrons and their products. In this way, many of us encountered the emerging concept of group research. The laboratory budget, mostly from private sources, was considered enormous for the time, although it might arouse some amusement today.

Of course, we were not unaware of what was happening in the world — of the war that had started, and of the power-mad dictator who was a threat to our ideals and who aspired to engulf humanity in his medieval social order. But, like many research scientists, I did not then relate my work very much to practical things, and certainly not to war.

Until 1940, my research had been concerned with the identification — with J. J. Livingood, primarily — of new radioisotopes. In the spring of that year, Edwin M. McMillan and Philip H. Abelson, opened the transuranium field with their brilliant discovery of element 93, neptunium. It is an interesting commentary on the thinking and the priorities of the time that McMillan, who had started work aimed toward the discovery of the next higher element, was called away to do defense research on radar at the Massachusetts Institute of Technology.

With the assent of McMillan, three of us, my associate the late Joseph W. Kennedy, Arthur C. Wahl, at that time a graduate student, and I, undertook to continue the research. It seems doubtful that many theses have been written that contain significance to rival that of Wahl. A few days before Christmas, in 1940 — just 21 years ago — the cyclotron bombardment was made which, in the succeeding few weeks, resulted in the chemical identification of plutonium. Plutonium may be said to have "come of age" as this meeting takes place.

Even at Christmastime in 1940, our work was not done in an atmosphere heavy with historical import, but rather in the carefree manner of young adventurers breaking new ground. It is true that fission and its implications were then known, and that some steps were being taken to learn how to exploit this discovery, using uranium-235. It was theoretically postulated, too, that an isotope of element 94 might be fissionable. Yet there was not, 21 years ago, a clear idea of how the then-identified element, if discovered, could be practically made in quantity and how it could be put to use as a military weapon.

Subsequently, with Emilio Segrè, we created and identified the fissionable isotope, plutonium-239, in March 1941. And a way was soon visualized to make this element in quantity and to use it as a weapon. In a short time, the knowledge gained in the search for truth became a formidable bulwark of national defense.

We crossed the divide between science and technology, and our work became useful in many ways, including its significant contributions to our arsenal of defense. We went from poverty to relative riches. Instead of working alone or with a colleague or two, we banded together in the team research pattern now so well established.

At times, during the war, we dreamed of a kind of scientific V-Day, after which we would return to the old ways, most especially the pursuit of knowledge for the sake of knowledge alone and divorced from application. Some of you probably were with me in the great hegira to fundamental research which actually did occur at the end of the war.

A large number of us found that the conditions of science had changed, in varying ways and to varying degrees. Perhaps the central point is that two decades ago science was called up, as it had been in the Civil War and World War I, to fight a five-alarm blaze. But this time, in a sense, science did not return to the firehouse.

The use of the nuclear bomb crystallized, as never before and on a world stage, the enormous power of science and technology. But this power was not to be confined to war alone, but was to be used for man's benefit in the expansion of industrial productivity and the advancement of our economic system generally. Later in the two decades of which I speak, Sputnik further dramatized the lesson.

Moreover, the realization grew among us and among industrial and political leaders that the time fuse between discovery and application had become short and was growing shorter. The gulf between basic and applied science had narrowed, and in some instances had become imperceptible. This realization was expressed in many ways: for example, while the government after World War II continued the development of nuclear weapons, it dared not risk failure at the same time to support the fundamental research in particle physics. In addition, under the conditions of modern competition between great nations, the prestige and power of a society came to be measured in part by its accomplishments in the growth of all knowledge.

In the past two decades, then, science has come to stay, as a regular, essential and pervasive activity in modern society. The signs that ours has become a scientific society are all around us. Suffice it to say here that government, business and industry are dependent for survival and expansion not alone on technology, but upon an accelerating growth of knowledge deriving from research that once was sometimes described as "pure". Moreover, it appears that nearly everyone is aware of this fact.

Let me give just one example of these developments, relating to the governmental agency of which I have the honor to be Chairman. In 1940, there was no such thing as atomic energy. Today, atomic energy is one of our biggest enterprises. The capital investment of the Atomic Energy Commission is seven and one-half billion dollars before depreciation. Its annual budget is two and one-half billion dollars. It is true that approximately 75 percent of this is devoted to defense activities. Yet, some six hundred million dollars per year are also dedicated to peaceful arts — to the development of productive industries for the present and the future, such as power reactors and research on controlled fusion; to the advance of medicine and its application; to the growth of knowledge in many areas of fundamental research; to the export of materials and techniques as a part our international relations program. In addition, there is the private atomic energy industry, involving non-governmental expenditures of fifty million dollars annually on development, and with a capital investment of four hundred to five hundred million dollars. And we can hardly visualize the ultimate potential of this great private industry. And yet, all of this emanated from one discovery in basic research.

The new relationship between society and science is reflected in the spectacular growth of the numbers of people who are doing research and development or who play supporting roles in these efforts. It is to be seen in the federal budget for research and development — some nine billion dollars annually, today, compared with about four hundred million dollars in 1940. Even more important are the new attitudes — of society in general toward science, and of scientists toward society.

The former is symbolized by the policies of government and industry. Recognition by the government of the need to support basic research across a broad spectrum was slow and spotty after World War II. The tendency has been — and continues to be to a large extent — to support fashionable or dramatic areas and those that might have some early, foreseeable technological value. Considerable progress was made, however, in the early post-war days as a result of the enlightened policies of the Office of Naval Research and the later policies of the Atomic Energy Commission. The National Science Foundation has significantly expanded the concept of governmental support for broad advances in fundamental knowledge, and I believe this trend will continue and will increasingly embrace the policies of special agencies that support research. Today, about

12 percent of the federal funds for research and development are used to support basic research fields. In other words, we can detect a fairly general recognition of the fact that the growth of fundamental knowledge, even though it may not have specific foreseeable application, contributes to the general welfare. Perhaps we can even hope for an appreciation for the more subtle cultural values of basic research.

The enormous impact of the past two decades on the scientific community reflects significant integration of science into society.

I do not detect any qualitative change in the spirit of scientific inquiry, fortunately. But it would appear that there is an important alteration in the attitude of scientists about the relationship of their work to the larger social environment. Many of us can recall a fairly general feeling of pride among scientists in the isolation of their work from the practical affairs of men. Indeed, it was not difficult to find resentment at any implication that a piece of research should have more than the remotest connection with application. Now, with the reduction of the time gap between basic and applied research, and with growing general appreciation of the value of knowledge, scientists seem more willing to relate themselves and their work to social objectives.

The material conditions have been modified, too. More and more, scientists find that they are supported adequately, if not opulently, and for sustained programs. Funds are available for "elegant" equipment that saves time and gives greater power to investigators. Money can be obtained for assistants to do detailed work, giving researchers more time for creative effort. The improvements are not uniform, of course. Space to work is still in short supply, especially in our graduate schools; yet, new governmental policies promise some alleviation. And the personal rewards are still relatively less for those who train our scientists and generate much of our knowledge, than for many others in our society who play much less significant roles.

The consolidation of science into society is striking in the field of governmental policy and international relations. The government has become increasingly dependent upon scientists for advice. This is true not only in the sphere of the administration of government science, but in a much more comprehensive way. Any evaluation of the future of the economy must embrace scientific and technological knowledge. Decisions in military matters are intimately involved with science and technology.

And any commitment of portions of our national resources for science and technology themselves must be decided with the help of men of wide knowledge in these fields.

The entry of scientists into important national advisory capacities is an inevitable concomitant of the events of the last 20 years. I believe it is a healthy and essential development, and I have advocated it for many years. It does not seem to me that the influence of scientists in this respect is greater than it should be; indeed, in the national interest, I believe it must increase.

The question of the place of science in government touches upon some of the critical questions about the future evolution of a scientific society in a democratic context. Our aim must be to use science to strengthen democracy, not weaken it; to expand the potential fulfillment of the individual, not decrease it. We must avoid any erosion of the broad base of informed participation by the electorate. In the past two decades, our democracy has ingested science, but has not yet digested it — a measure of the infancy of our scientific society. This is not surprising, since our previous experience had not prepared us for anything like the explosion of those 20 years. We must expect the next 20 years to be even more dynamic. Therefore, it is urgent that we accelerate the process of assimilation.

A central problem in assimilation, it seems to me, is the extent to which men, including the otherwise well-educated, fail to identify freedom of scientific inquiry with our political and other freedoms. In the somewhat less complicated world of the 18th century, a great thinker like Thomas Jefferson could be all at once a political theorist and practitioner, a philosopher and a scientist. His mind could embrace and integrate a very large part of human knowledge. He had, therefore, a clear appreciation of the broadly humanistic values which are the common heritage of all men who pursue the truth.

But, as knowledge grew and fragmented, the specialties went their separate ways. Science has seemed to walk more apart than other fields, perhaps because the details of scientific truth touch infrequently a community of intellectual experience. Science became a stranger even to many intellectuals.

This estrangement has resulted in the paradox with which we are familiar: as science became more important to society, it apparently became less important in the curricula of liberal education. This fact was noticed in the last century by Thomas Huxley, who pleaded with contemporaries holding a narrow view of humanism to include a more generous helping of science in liberal education. A cultured, or liberally educated person, Huxley maintained, is one capable of making a criticism of life — of evaluating the environment and making enlightened judgments.

Thirty years ago George Sarton wrote in the same vein in his volume, "The History of Science and the New Humanism." He stated the issue, which remains central for our nascent democratic-scientific society, as follow: "The main issue does not simply concern humanism but the whole of education from the cradle to the grave. And the real question is: will education include science, or will it exclude it? The intellectual elite is at present divided into two hostile groups — which we might call for short the literary and scientific — who do not speak the same language nor think in the same way. If nothing is done, the gap separating them must necessarily increase, together with the steady and irresistible progress of science. Shall we deliberately widen the gap as the old humanists would have it, or shall we take special pains to reduce it as much as we can?"

In our own time, C. P. Snow has eloquently drawn attention to the same problem, in his discourses on the "two cultures."

To summarize the matter, I should like to ask a question paraphrasing Huxley: Who in our times can make an adequate criticism of life without knowledge of the ideals, the methods, and the dynamics of science?

The remedies have been widely discussed: a large content of science in the lower schools and in the universities and colleges; a wide range of efforts to give the public some appreciation of science; a greater effort by scientists to explain their work in popular terms.

All of these measures are needed. It is necessary to bring about a larger understanding of scientific principles. But in striving toward this goal, it may be even more important to promote a greater consciousness of the common heritage of all who pursue the truth. The philosopher, the social scientist, the artist, the writer, the natural scientist — all are intellectual brothers under the skin. Whether their technique involves the distillation of human experience or the ordering of measurable phenomena into statements of principles, their motivations, the quality of their experiences, and their satisfactions are rooted in a broadly defined humanism.

I am sure intellectuals generally know this to be true. Yet it would appear that it is often far back in the consciousness. I wonder if this fact is not responsible for much of the inability of Snow's two cultures to communicate? I wonder if there is not a common language, deriving from a community of basic ideals and purposes, whatever the details of different bodies of knowledge, that is the foundation for communication? I wonder if the barriers are not superficial, even as language is a superficial obstacle between men who share common bonds?

The achievement of a conscious, working realization of the common heritage of truth-seekers — among scientists as well as other intellectuals — can be significant in the successful evolution of our new kind of society. It should make it clearer that the free and uninhibited pursuit of truth in science is a natural part of the right of free inquiry that is inherent in democracy. It should do much to abolish fruitless discussions over whether we should continue doing science and whether scientists should not withhold scientific truths that may be used destructively. It should give wider acceptance of the inevitable growth of knowledge and of its continual change. It should force us to a greater awareness of the need to prepare for and to cope with the hazards that are a paradoxical by-product of the expansion of knowledge.

It has seemed natural to lay some emphasis on science in this discussion of the society that has developed in the last 20 years. I do not wish to give the impression, however, that I believe this new kind of society is the property of science. We cannot, of course, proceed intelligently without integrating into our thinking and our acting the full range of human wisdom. If you have noticed carefully, I have asked primarily for men generally, and intellectuals in particular, to return science to the fold of humanism. It is unthinkable that a democratic-scientific society could evolve constructively without a wide endowment among its people of art, music, history, literature and social dynamics.

We can hardly, discuss the future of the scientific society without relating it to the world struggle and the terrible dilemma confronting man as the result of the development of nuclear weapons.

I am reminded of the reaction of many scientists, including some of us who worked on nuclear weapons, to this dilemma — when it became a reality in 1945. Scientists sometimes have been called too optimistic and naive by social scientists. As a group, they are not lacking in idealism. Perhaps it was natural that many of us, recognizing from close at hand the significance of nuclear weapons, set out to advise the world that nuclear war was out of the question. To us, the data were unequivocal, the conclusions indisputable, and the course of action clear. We felt the world would quickly see this — and, seeing it, do something about it.

The half-life of disillusionment varied from individual to individual. Few have changed their minds about nuclear war. But, many have become more sophisticated, if less idealistic. Much of what has been described as naivete has rubbed off. But we should remember that idealism, happily, has not been limited to scientists. In the period following World War I, experienced statesmen, imperceptibly influenced by scientists, solemnly signed unrealistic treaties outlawing war. Perhaps sophisticated statesmen, aided by sophisticated scientists in an age of science, may be able to combine realism and idealism.

My own instruction in these matters includes the experience, earlier this year, of being appointed by President Kennedy to head the US. Delegation to the Fifth Annual Conference of the International Atomic Energy Agency in Vienna. It is an agency established to spread the peaceful uses of atomic energy throughout the world. Its problems, I found, are hardly less difficult than those of the United Nations.

I was impressed with the enormous difficulty of finding common solutions to problems — when the effort had to be made with individuals who seem to speak a different language, not only linguistically, but ideologically, and some of whom appear to possess a deterministic faith that is alien to our humanism.

While I found no basis for arrant optimism, neither did I find reason to stop trying. In the absence of any foreseeable breakthrough in diplomacy, it would appear that the best condition of the world we can hope for is a continuing crisis. In the competition of ideas which will accompany the crisis, the victory may be won by the successful evolution, here, of a society combining science and freedom.

Scientists and engineers can continue, as they have in the past, to make a major contribution in this contest, not only by achievements in the laboratory but also by their participation in exchange programs and international meetings, and other contacts with Iron Curtain nations through the medium of basic research when and if the occasions arise. All of these activities are essential to help keep the channels of communication and understanding open.

I believe each of us, scientist and non-scientist alike, must be aware of the importance of his own effort to the preservation of a libertarian society in the continuing crisis. Each of us needs a sense of responsibility and urgency — for the total of our efforts will be decisive, however remote from combat our work may seem. We must not do too little. We cannot delay. We must have both determination and good intentions; and what is most important — *we must act*. As I have advocated in the past, we must expand and raise the level of education all along the line. We must, especially, search out and cultivate the gifted and creative — for it is these that usually make the great breakthroughs in knowledge and understanding. We must mine every vein of our human resources and exploit our talents in the fullest measure.

The democratic-scientific society has taken root in the past two decades, combining the values of freedom and individual worth with the promise of growing material well-being. Can we preserve it — not only for ourselves, but as a choice for other peoples?

I believe we can and will, partly because of the moral strength of freedom and partly because of the material power of our new society. We cannot be blind to the fact that freedom needs strength and determination as well as a good heart. Generosity has its place in relations between men, but it is, unfortunately, a quality not uniformly respected by all nations in relations between themselves. This is why, for example, we must be prepared to negotiate from a position of unquestioned strength as well as undoubted good faith; and negotiate we must: to turn our back on this most hopeful and sensible solution of the differences between East and West would be as foolish as it could perhaps be fatal. But we must recognize that until all nations can proceed from the same definition of right and truth, international agreements which involve our vital interests must incorporate provision for adequate controls against violations as well as provision recognizing the other's rights. We must be firm when our own security is at stake, as well as fair when another's is. I cannot help but recall, in this vein, that eloquent passage from President Kennedy's Inaugural Address: "... civility is not a sign of weakness, and sincerity is always subject to proof. Let us never negotiate out of fear. But let us never fear to negotiate."

Beyond these principles, my confidence in freedom is based upon a personal faith, originating in my interpretation of human experience, to which one must appeal when scientific data are absent or inconclusive. Many times in history the future has not looked bright. However, the things most feared have not always comes to pass. Man's native faith and hope in his own destiny have motivated solutions to awesome problems. History does, we know, repeat itself — both in crises and in their resolution — and so, we must trust it will again.

Science and the General Welfare in a Democracy

The Harrelson Lecture, delivered at North Carolina State College, Raleigh, North Carolina, March 11, 1964.

I am greatly honored to be chosen as one of the lecturers in the Harrelson series. It is a great pleasure also to be here on the campus of North Carolina State College, a school that has been for several years so intimately and productively involved in the national atomic energy program. North Carolina State has pioneered in the field of nuclear education and training. Here at Raleigh, a proposal for construction of the first reactor to be built on any campus for purposes of nuclear training and research was initiated early in 1949 and operation of the reactor was achieved in 1953. At North Carolina State, in 1950, undergraduate and graduate courses in nuclear engineering were offered for the first time in any college. In view of these notable beginnings, one might have anticipated that the first doctoral degree in nuclear engineering would be awarded by North Carolina State, as it was to the Brazilian scientist Hervasio Guimaraes de Carvalho in 1954.

I should like to note especially the contribution made by this school in helping to establish the program of international training that later evolved as the International Institute for Nuclear Science and Engineering at the Argonne National Laboratory. Many of you may be aware that this program, which went through its embryonic stages at North Carolina State, has been of great significance in furthering the Atoms-for-Peace program in that the cadres of a number of national atomic energy organizations received their training at the Institute. Dr. de Carvalho, I might say, is now a professor of physical chemistry at the Brazilian Center for Physical Research in Rio de Janeiro.

We also admire the contribution to American social science made by men like Howard Odum and Rupert Vance of your sister institution, the University of North Carolina, and we know that their work has been important since the Thirties in helping to mold our national social and economic viewpoints. I might add, by the way, that North Carolina does not limit herself to offering ideas. Many of her very capable citizens have provided their services to forwarding our national programs in a direct way, among whom is my valued friend, Secretary of Commerce, Luther H. Hodges.

Another significant contribution of North Carolina, during this era of scientific and technological achievement when the arts and the humanities are in danger of being neglected, has been the work of Paul Green, Kermit Hunter, and their associates in developing and bringing into sustained production the genre of the outdoor drama. The work of this group has been significant both for the arts and the humanities since these dramas have vividly portrayed critical eras and events in American history. These and other achievements, I believe, suggest the great potentialities for growth of the system of higher education in North Carolina.

This evening I want to consider with you some of the relations between science and the general welfare in a democracy such as ours. I have spoken on other occasions of the Third, or Scientific, Revolution and how it is affecting in the most pervasive and intensive way the fabric of modern society. It is the Third Revolution of Science that concerns us today, as in earlier days the Democratic Revolution called into service the architects of our national political institutions, and as the Industrial Revolution engaged the best talents and energies of those who fashioned the great industrial system of America during the last century. With the advent of the Third Revolution, science has become so important to our national welfare, not only in terms of national defense but also for our economic growth and social well-being, that the relationship between science and government is for us in our day a matter of profound significance.

There was the foreshadowing of such a relationship as early as the Civil War, and a yet stronger indication during the first World War. No one, however, could fail to see at the conclusion of World War II that the contributions of science had been fundamental to national survival in our greatest crisis. Later, the shifting of alignments and the prolongation of latent hostilities in the shape of the Cold War meant that the federal government had perforce to become more and more heavily involved in the support of science. These factors account for much of the government's continued role in setting national objectives in science and influencing progress toward these objectives.

But as we responded to the persisting demands of the Cold War and the needs of national security, we awakened also to a more significant realization. We found that we could not abandon the productive fusion of forces that emerged during the war years. We discovered that the powerful focusing of basic and applied science, engineering development, and industrial production which accomplished so much for us in wartime had equally profound implications for our peacetime future. The advancement of peaceful national goals, for example, economic expansion, the improvement of health, the development of adequate energy resources and assistance to other nations — all these were recognized to rest upon a continuation of the machinery of concerted scientific–technological efforts.

Before considering more directly the problems that confront us with regard to the structure of our national science policy as it affects our general welfare, I want to explore in somewhat greater depth the circumstances of the Scientific Revolution, the manifold possibilities we are offered, and the sobering responsibilities with which we must grapple as modern people in a changing world. Though the forces that have brought about this manifold transformation of human capabilities have been developing for several centuries, it was the discovery of nuclear fission that unmistakably marked our transition into a new world. The sinister aspect of this event was that man had now become capable of drastically modifying the total ecology of his native planet in less time than it will take me to deliver this speech. We have in our hands negative potentialities for destruction that approach infinity; the constructive potentialities for amelioration of the human condition are no less. Perhaps we should look at both in their broad outlines.

As a general condition applicable more or less across the board, we should note first the factor of *amplification*. In every phase of modern technology, the potentiality for phenomenal degrees of amplification forces us to consider the whole planet and even its surrounding regions in space as a closely coupled system. On the terrestrial level, a devastating flood or earthquake in the remotest region of the globe can almost instantly, through modern means of communication and transportation, marshall the enormous resources of our own and other technologically advanced nations to assist the beleaguered inhabitants of that region. The effects of a drought in some part of the world resonate throughout the economies of the major grain-producing nations with an almost imperceptible time-lag. In the reaches of outer space, a scientific experiment by one nation becomes the concern of the international community of scientists even before it can be performed. Today, in the strictest sense of their truth, we can quote John Donne's words that "no man is an island unto himself." Certainly, with the increasing unity of world science, no nation of great power can shape the future of its own scientific and technological development without profoundly affecting also the future of all mankind.

The policy such a nation adopts with regard to the development of its energy resources, for example, will of necessity be watched with interest by developing nations in need of the help that new forms of energy can provide for their emerging economies. We have some distance to go with the development of nuclear power, and later perhaps with controlled thermonuclear power, before every inhabitant of the world can flick a switch and have electrical energy derived from the atom at his command. Nevertheless we are certain such a day will come if humanity is sufficiently endowed with patience, good will and wisdom.

The widespread use of radioisotopes has already done much through medical applications to alleviate human suffering and prolong life, and, in addition, these new materials now available in substantial quantities provide tools of great value for the improvement of industrial processes and agricultural productivity. What we have accomplished in the span of only a few years should convince us that nuclear energy can also revolutionize maritime transportation, quench the thirst of arid nations, lengthen the shelf life of perishable foods, provide the electrical power in space satellites needed to make worldwide radio and television a reality for this generation, and in the not-too-distant future propel us on missions of exploration to other planets.

The consequences of the misuse of nuclear energy, on the other hand, can place us all in jeopardy. The threat of rockets armed with nuclear warheads imposes new burdens upon man's capacity for rational conduct. In the present conflict it is not men who are in jeopardy — it is man.

Powerful and dangerous as nuclear energy can be when pointed toward destructive ends, it is not strange that the applications of this force in peaceful pursuits also require of man that he proceed always on the basis of precise knowledge and with the greatest respect for the possible consequences of his undertakings. With the exploitation of nuclear energy for power, we must concern ourselves increasingly with the safe disposal of large quantities of radioactive waste materials. With the ever-widening employment of radioisotopes for industrial, medical and research purposes, we need widespread sophistication among the scientists, physicians, and technicians who handle these remarkably potent substances. As we prepare to use the enormous potentialities of nuclear explosions for such projects as forming canals or cutting passes through mountains, we cannot safely neglect any expense or care needed to assure that we will accurately and reliably achieve the objectives we have planned. At the same time we should have the courage to make the best use of these new instruments in their right relation to other technological developments when the need arises.

Not only do we have these tremendous new energy resources well on their way to successful exploitation, we are also rapidly mastering the revolutionary technology of high-speed computers and automation. Amplification is very much in evidence in this field: unprecedented amplification of the rate of production and equally unprecedented amplification of our thought processes, leading to the solution of problems of remarkable scope and complexity. The potential effects of this technology are of such significance that we must restudy, and perhaps may have to reshape, what we had thought were unchanging economic concepts and principles.

Equally profound changes have taken place in the life sciences, especially in their relationships to physics and biochemistry. The possibilities of transforming microorganisms, plants and animals in such a manner as to improve our ecological and economic situations are so nearly infinite as to dwarf description. There are millions of species, and the routes of artificial evolution through which any one of them may be educed into new forms are diverse, ever-branching and virtually endless. We may expect that inevitably our recently acquired knowledge of the genetic code will be applied also to the improvement of the human species which Julian Huxley said recently so badly needs to be done. He says that our species is deteriorating "thanks to the keeping alive of genetic defectives … and thanks to new mutation-causing agents." But once we have the ability to determine the genetic characteristics of a human infant, who will step forward with sufficient wisdom to choose those traits most beneficial to the man of the future?

These possibilities, together with the urgent problems of an exploding world population, force us to consider the fundamental question — what are people for? We will need to answer this question at the same time that we are struggling with the multitude of other problems brought about by the population explosion. Science must play the key role in the solution of these problems. We may expect, for example, that our present farm surpluses will soon disappear and we shall require all the skill of our soil scientists, agronomists and agricultural experts to raise productivity to the required levels. These difficulties are sure to be further increased by the withdrawal of appreciable acreages of our best farm lands from cultivation to be used for urban developments, freeways, military installations, factories, etc. New methods of food processing and means to reduce waste and spoilage will have to be devised so that we may more completely utilize our plant and animal crops. Our diminishing reserves of forests and minerals will force the adoption of adequate conservation policies, and the development of substitute materials will assume a new urgency. We shall find ourselves increasingly dependent on the desalting of sea water to fill urban reservoirs. The congestion of our cities will intensify present problems and create new ones in transportation, utilities, water supply, waste disposal, public health and recreation, mental health and the entire range of social behavior.

Fortunately, we have the potent instrumentality of the new techniques in automation to help us with the solution of many of these problems. Not only does this new computer-based technology provide the means of tremendously increasing production but it also will enable us to operate the systems of great complexity which we will need for the simultaneous control and harmonizing of the many factors affecting our expanding urban civilization. The extremely great capability of the new computer technology brings with it, however, the threat of major social and economic distortions. One free-wheeling mathematician believes that the rise of automation will propel us into an entirely novel kind of world where 2% of our population, working in factory and on farm, will produce all the goods and food that the other 98% can possibly consume. Faced with this overabundance of leisure, an economist predicts that we may have to keep the unemployed portion of our population under more or less constant sedation unless we can figure out something better for them to do.

Sedation as an answer may already be old hat because of the startling development of new chemotherapeutic drugs which, upon further development, hold promise for the alleviation of mental suffering, and, perhaps also, for the beneficial readjustment of personality and our more transient moods. We would hope that, valuable as these drugs may prove themselves to be, their use on any extensive scale will be no more than a passing phase in our history, giving way in due time to a finer adjustment in the relations between the individual and society. It is possible that, by the combined attack of improved human genetics and the development of more sophisticated social sciences and psychology, we may someday be able to empty our medicine cabinets. However this may be, we can imagine the grave political consequences, even apart from war, should these psychotherapeutic drugs be used as psychochemical weapons for coercion and control. Since self-control is essential for non-violent resistance, peaceful Gandhian methods could be rendered ineffectual by mood-altering drugs; in such a nightmare existence, brainwashing might become a specialty of chemists.

We need not accept the probability of such extreme predictions concerning our future to realize that, by the most conservative estimates, profound social and economic changes are in progress through the impact of science. There can be no doubt that the kinds of jobs people will be doing in the future will be very different from those being performed by the majority of us today. Indeed, there has already been a great increase in the number of jobs based on providing services and recreation — a trend that is certain to be magnified with further increases in national productivity and the accompanying increase of leisure time.

The most remarkable testimony to the unsinkable buoyancy and optimism of mankind is that, over the centuries with possibilities and problems of this magnitude facing him, man has nevertheless begun to feel restless within the confines of his local planet and has sent spacecraft with their human cargo beyond the atmosphere. We have now mustered and coordinated the massive resources of our whole technology in an effort to reach our moon and ultimately the neighboring planets of our solar system. The horizons of human vision have shifted from a horizontal to a vertical orientation. More and more often as we think of the future, our gaze turns toward the immense regions of outer space.

Against this background of the impact of science on our future, perhaps we are ready now to consider the question — how well are we equipped as a nation to enter upon this new Age of Science, the Third Revolution? There is at this point, I think, no question but that we must face the problems of our time and must assume the responsibilities that our new knowledge and technology are forcing upon us ever more urgently. Before we become too scholarly about the question, however, I must say that some have managed to look at the lighter side of our situation. A few of the more irreverent individuals, scientists I suspect, have compared our management of science policy in recent years to a ship with a thousand helms all connected to one rudder with rubber bands. Another of these whimsical images would have federal science and its leaders as a colony of ants riding a rolling, tossing log down a flooded stream. As each new twist of the log brings a different bunch of ants out of the water and on top of the log, the ants then able to see daylight proclaim to the others that they have the situation well in hand and know exactly which way to steer the vehicle.

As to where we presently stand with respect to a national policy toward science, there is no better way of gaining a perspective than to review briefly the genesis of science in our government. The growth of our national policy affecting science and our programs for furthering this policy have been the result of many forces and events over the span of our history. Looking at the course its development has followed, some have described the evolution of our national policy as the product of a struggle between factions aiming at unified control — those tending toward a monolithic system — and opposing forces striving to preserve pluralism.

One of the earliest visions of a national science, the national university as envisaged by Thomas Jefferson, Benjamin Rush and Joel Barlow, would have been both centralized and comprehensive — including many functions now performed not by universities but by research organizations within the government. But, like Adams's dream of a national academy to restrict and regulate the vagaries of the American language, this plan came to naught. What has happened since bears some resemblance to a slow-motion ants-on-the-log situation.

Beginning with the establishment of the Coast Survey in 1843, the history of federal science except for periods of crisis has been mainly the rise of a single agency predominant in one field of science, its longer or shorter period of dominance, and the relinquishment of its dominant position in the end to another agency born into the glamorous aura of another emerging field. A. Hunter Dupree says of the Geological Survey which succeeded the Coast Survey that it "gained its hegemony by creatively joining a live and developing science to the expansions of a nation's horizons in the great basin and Colorado plateau," that it was "still there when most people had ceased to get a thrill out of the problems of the trans-Mississippi West, and it was still there when the need for uranium ore brought prospecting on its classic ground back into fashion." In the line of further succession we would include the Forest Service of the Progressive Era and later the Department of Agriculture.

With the continuing crisis of recent years, the picture in terms of dominance has become a bit confused. The Department of Defense has unquestionably been the dominant agency from the standpoint of expenditures and perhaps even in the multiplicity of its interests. But it was the Atomic Energy Commission that captured and held the national imagination until the post-Sputnik era when the National Aeronautics and Space Administration quickly became the focus of all eyes. I would say that both AEC and NASA continue to engage us in our dreams of a better future. During these same years the National Institutes of Health (NIH) has entered the galaxy of dominant agencies, both in terms of funding and the scope of its programs. Lesser in budgetary stature but of growing significance, especially in its influence on the future of basic research, is the National Science Foundation.

Nothing in this line of development, however, speaks of the events which have from time to time pushed us harder in the direction of a unitary national policy for science, for none of these agencies aspired to leadership over the whole of American science. The nearest approach in our history to completely centralized federal science came during World War II with the establishment of the Office of Scientific Research and Development. It was both a predominant agency and a central scientific organization. It had a straight line organization within the government culminating in its director, Vannevar Bush. The director had an immediate access to the President and a relationship to Congress that resulted in hardly any limitation on money or on the timing of actions. In this great emergency, the director and his associates worked closely with the military but could proceed, if they chose to do so, on their own initiative. They brought into the service of the government an array of laboratories undreamed of earlier by using contractual relationships to exploit existing resources in the unversities. When one part of the job proved to be beyond the scale of other endeavors — I am speaking of the wartime nuclear program — they showed great administrative adaptability in assigning it as the Manhattan Project to the Army Engineers. The most remarkable aspect of the whole arrangement, however, was the assumption that when the emergency was over the OSRD would be dissolved. Implementation of this was like no other decision in all American political history.

The decision was made in good faith, and pluralism again was dominant in the American scientific community. However, the complete dominance of pluralism and the complete absence of any unified direction by the federal government was to last only a short time. Already, as I have shown, some few had recognized that our country was caught up in the beginnings of a Third Revolution in science at the same time that we found ourselves becoming engaged in a scientific and technological struggle with the East. We began to recognize the importance of expanded scientific and engineering efforts on our economic growth and social well-being. As these realizations dawned, they motivated large-scale federal support for research and development. Equally important, they brought about a somewhat more unified approach to the formulation and execution of a national policy for science and technology. The effect of Sputnik in October 1957 ensured that there would be recognition and some guidance of this major aspect at the highest levels — the President and the Congress — to meet the challenge to both our national security and our international prestige. In a less conscious but perhaps no less effective way, we thus began to meet also the more fundamental challenge of the Third Revolution.

As in other times of crisis, the nation responded effectively and with impressive accomplishments. The history of the past year or two, I believe, reflects the result of this response in some lessening of international tensions. We have been given time to pause, the time to think at least briefly, and to reflect on what our future course should be. And as we have almost invariably done in like situations of the past, we have stopped to examine more closely our national pocketbook.

We discovered, of course, that the national expenditures for science and technology had mounted on a super-planetary trajectory with no sign of deceleration. We found that over the period of years from 1955 to 1964 the expense of federally sponsored research and development had grown from about $3 billion to $15 billion annually — or expressed in terms of percentage of gross national product, an increase from a little over one-and-one-half percent to nearly three percent. To put this more concretely, our present rate of federal expenditure amounts to nearly $100 for every man, woman, and child in the United States. For a family of four, this is equivalent to a new color television or one or two major appliances for the home each year. If our research and development expenditures were to continue to grow at the present rate, we could expect them to nearly equal the gross national product at the turn of the century in the year 2000.

Considered thus, the absurdity of such a projection becomes apparent. It is equally absurd to extrapolate the present rate of growth of federal research and development expenditures and to determine thereby that such expenditures would be equal to the total of *all* government spending by the late 1970's. Obviously, there must be a limit. Even though our economy continues to grow at a substantial rate, we can reasonably expect that the proportion of our gross national product invested in science must taper off at some point in its expansion. We have already reached a position of having to make some hard choices and we are beginning to make them.

Thus while our system of modified pluralism with all of its imperfections has worked remarkably well during a number of crucial years in our nation's history, we have come to a point at which we must grapple with basic issues, questions both as to the management and the support of federal science. The degree of our preparedness as a people to meet the new challenge of the Third Revolution can be measured by our ability to cope with these issues of the management of science in a democracy. For this reason, I want to consider with you the pattern of the problems we are now contending with. They are not problems of the kind that one man, however capable, or one committee, however distinguished its membership, can hope to solve in the span of a few days or months. I present these questions rather so that we may consider the shape of our future as it may be implied in our present national situation in science. They are questions that are sure to be with us in one guise or another for many years to come.

I have called the national program of science as it is presently operating a form of modified pluralism. Dr. Donald F. Hornig, recently appointed

Science Adviser to the President, expressed the matter very well the other day in his appearance before the Joint Committee on Atomic Energy. He made a general observation concerning the use of the words "science policy" in connection with national planning and programming. Dr. Hornig said, "It is obvious that no simple broad policy declaration can meaningfully guide the planning of science programs. There are in fact many national science policies designed to satisfy a number of national objectives."

By virtue of his position as the President's Science Adviser, no one is better able to appreciate the nature and problems of our system of modified pluralism than Dr. Hornig. The view from where he sits encompasses some half dozen federal agencies — the National Science Foundation, the Department of Defense, the National Aeronautics and the Space Administration, the National Institutes of Health, the Office of Education, and the Atomic Energy Commission — all supporting programs of broad scope with mounting budgets and multitudes of projects competing for growing space in the financial sunlight of federal support. In his efforts to help coordinate this diverse array of federal programs, the Science Adviser must wear several hats. Primarily, of course, he advises the President on scientific questions of national importance, including their budgetary aspects. But in order to pull together the variety of data and considerations that support such advice, the President's Science Adviser is also Director of the Office of Science and Technology which provides him with staff assistance and a headquarters. He also serves as Chairman of the President's Science Advisory Committee, which represents the scientific community, especially the universities, and as Chairman of the federal Council of Science and Technology with its representatives from all the scientific agencies of the government. This pattern, which evolved out of the appointment of Dr. James R. Killian in November 1957 as Special Assistant to the President for science, has obviously been an influence working toward the better coordination of our national goals in science.

There continue to be stubborn difficulties, however, even in this arrangement which represents considerable progress. I believe we may with continued thought and consideration be able to achieve a more nearly optimum balance between unity and pluralism. For example, much of the present coordination of the science program is accomplished through ad hoc panels, and though there may be good reason for continuing to employ this approach in many phases of our activities, there is also danger in the lack of continuity and the possibility of failure to achieve an adequate integration of programs from a national standpoint. We should give serious thought to ways of improving our management of science by the Executive Branch of Government.

If the present form of pluralism within the Executive Branch impresses us as needing further improvement, the Legislative Branch of our Government, the Congress, is at least up to this point even less adequately equipped to regard science from the viewpoint of an overall national policy. This is not to say there has been no encouraging progress during recent years. The Joint Committee on Atomic Energy has for some time worked closely with the Atomic Energy Commission, both in reviewing program accomplishments and in developing needed legislation. The House Science and Astronautics Committee and Senate Committee on Aeronautical and Space Sciences have performed a similar function in connection with NASA programs and have lately begun to review broader questions. Perhaps the most encouraging of recent developments, however, has been the work undertaken by the Elliott Select Committee in its efforts to develop a comprehensive view of the whole range of national science programs.

It is plain that Congress is becoming sensitive to the need for a more inclusive and better informed view of science, but, though a number of proposals have been made to accomplish this by providing legislators with full-time science advisers, so far these efforts have met with little enthusiasm. One bill recently introduced calls for the establishment of a panel of three science advisers for the House and a similar staff for the Senate. Another would create a Congressional Office of Science and Technology patterned after the President's scientific office. As things now stand, whatever coordination exists is mostly of an informal kind among the various committees of the House and Senate, and the advice received by Congress is either a general variety from witnesses called before committees or else it stems from advice generated within the Executive Branch.

Thus we may conclude with regard to our present arrangements for the management of national science programs that, despite welcome progress, we may yet have a considerable distance to go before the desired balance between pluralism and a more nearly integrated viewpoint can be achieved. Though our current practices have the pluralistic advantage of great flexibility, we continue as a nation to be plagued by a persistent myopia as to our long-term national goals in science.

As we turn now to examine a number of very thorny questions concerned with the financial support of science by government, the perspective shifts in some degree toward a distribution of problems occupying a foreshortened time scale. In talking of financial questions I believe we would do well to think in terms of the strains to occur during the next few years. I make this observation because it seems to me that the next several years will witness economic and technological readjustments tending to ease the present financial strains imposed by science, readjustments brought about largely by the gains accrued through further scientific and technological developments. We nevertheless have this period of stress to be weathered, and during this period our major decisions will be concerned with the wise parceling out of available funds among competing programs. We must decide what programs to emphasize and — since emphasis is unquestionably translated into terms of financial support — with what degree of financial support we shall emphasize them.

We need especially in our present situation to develop effective and rational criteria for assigning priorities. Someone, for example, will have to make decisions between the competing claims of big science and little science. Would we do well to give greater heed to the support of the independent university investigator whose other job besides research is to train the next generation of scientists and engineers, or is it better to keep a few great centers growing at the rate of 15% a year? We will need to apportion funds among the different fields; weigh the relative merits of spending for research, for facilities, and for the improvement of teaching; and decide how much is justified for high-energy physics, for oceanography, for radio astronomy, and for other areas of work in which the costs for basic facilities are high. If we cannot afford to advance each major field at the optimum pace, should we stress the biological sciences and let the physical sciences lag or vice versa?

The time has passed when every competent scientist can have all the money he wants for any reasonably worthwhile project. The time has also passed when we can expect that very large and costly projects will be financially underwritten by the Executive and Congress without having the support of a consensus of the scientists working in the fields for which the projects are intended. The means by which such consensus is established may in the end have some influence on the functions and responsibilities of our scientific advisory structure in the federal government. Perhaps we have not yet arrived at the best means of obtaining an objective consensus from any segment of the scientific community, if such a consensus can in fact be reached. Is there a truly effective way to do this which will at the same time avoid certain predictable dangers? Are we to use some kind of measure weighted according to the prestige of the people voting? These are not trivial questions for a society whose future is so greatly predicated upon the future of its advancement in science.

It is important also, at this transitional point in the development of our technological society, that we preserve clearly in our minds the distinction between basic scientific research and applied research and development. I say at this transitional point for the reason that it is a point marked by budgetary problems and we need to keep in mind that the great bulk of our $15 billion or more in federal expenditures for science is spent on applied research and development, especially development. I think it is fair to say that up to the present time the great majority of our citizens, including many of those most influential in government, have been able to justify expenditures for basic research only by keeping somewhere in the back of their thinking that the money spent would be returned many-fold in the form of useful applications. In almost every case this expectation is realized. The results of basic research are the raw material for the practical applications which now follow so swiftly. But as a people we should be beginning to learn what scientists have felt from the beginning, that new knowledge is vital even if its only purpose is to extend the boundaries of reality, to give us new insights into the circumstances that surround our lives. Basic research is the life-blood of our time. We cannot neglect it, absolutely or relatively, without danger of retrogression.

In this same vein we must recognize that for our time and the foreseeable future the encouragement of excellence merits our sustained recognition and support. The core of my own philosophy on this point coincides with the statement *Scientific Progress, the Universities and the Federal Government* issued by the President's Science Advisory Committee in November 1960. In this report the Panel states: "In science, the excellent is not just better than the ordinary; it is almost all that matters. It is therefore fundamental that this country should energetically sustain and strongly reinforce first-rate work where it now exists." The report states further that "It is of equal importance to increase support for rising centers of excellence." How are we to accomplish this increasing support for rising centers of excellence in the face of our increasingly difficult budget situation? The costs of excellence can be very high, both for staff and for the major items of equipment demanded by Big Science. How do we prevent the competition for establishing such new centers in the various regions of our country from degenerating into squabbles reminiscent of our saltier pioneer political eras? How can we prevent a science pork barrel?

We must not let our national support of science and technology degenerate to the point where no state — no Congressional district — is complete without a Post Office, a reclamation project and a new science laboratory. This does not imply a lack of merit in the considered geographical distribution of our scientific expenditures. Any such program, however, should be both soundly conceived and wisely administered if we are to build new centers of excellence in new geographical areas without tearing down or undermining other centers that have already achieved and sustained excellence. We must manage somehow to provide for the support of new centers and allocate the development of major facilities among them in such a way that we do not turn our best scientists into migratory workers. We cannot afford to have these men gravitating from one scientific mecca to another, dependent upon the fluctuating whims of federal support.

If there are problems connected with being too poor, there are some who are now saying — though I have never personally experienced this kind of discomfort — that there are dangers accompanying extreme affluence. I refer by this seeming paradox to the impact of our large federal expenditures for science upon the universities and industries which receive them. Government expenditures make up about two-thirds of the total amount of money spent in this country on research and development. With this we may compare industry's portion, amounting to about 30%, and the 3% from universities and other non-profit institutions, the latter going mostly for basic research. It should be abundantly clear that federal funds constitute a large enough part of the total to have a strongly felt impact on both the business and academic sectors.

Experience shows that for the most part the impact of these funds on the universities has been beneficial. They have brought about a significant growth in the science. They have increased the number of students entering scientific fields. They have been allotted in such a way as to foster a close involvement of basic research with graduate education. This latter point is of especial importance because the direction and force of the Third Revolution will depend upon new men and women, new scientists and engineers, new minds and ideas, and these can come only from the rising generations of students.

There have also, unfortunately, been drawbacks associated with federal support of science in the universities. Because of the influx of government funds, some universities may tend to become one-sided in their emphasis on science. The humanities may be neglected. In other schools, professors may find themselves burdened with administrative duties which preempt the time they would otherwise be able to devote to teaching. An extreme complaint occasionally heard is that a professorship may be awarded not because of an individual's scientific and educational merits, but rather because of his ability to obtain federal support for the university's research programs.

While government money does not necessarily subvert university freedoms, alter its structure of professional relationships, or distort the patterns of emphasis placed on different fields of study, hazards are certainly present. The government must bear these things in mind. But an important task of the universities is to assure that their own standards of excellence and freedom are maintained in a period of growing connection with the federal government. The universities have an obligation — which they are generally not at all hesitant to exercise — to keep government actions and methods reasonable and proper.

In the private sector there are problems of a similar nature. Much of the $15 billion of research and development funds spent by the federal government is placed in the hands of industrial contractors performing the many-faceted research and development tasks this nation pursues. This results from the fact that a major part of the total federal funds are for developmental efforts — hardware and engineering.

In industry, the one-sidedness of federal expenditure has had a noticeable effect. The civilian technologies supported largely or solely by the private sector have suffered in comparison with those which depend on the growing and thriving fields of science of interest to the government, such as aerospace. Young science and engineering graduates are naturally enough attracted to the newer, and more glamorous, fields of endeavor. We can imagine them asking the question, "Why improve the family refrigerator when you can play a role in man's quest for the stars?"

Nevertheless, the scientific and engineering problems sponsored by the government primarily for the military and space programs also may have benefits for the general industrial technology. "Spin off" and "scientific fallout" are the terms often used to remind us that ideas, processes and inventions developed in one program can greatly influence new developments in other fields.

As in the case of the universities, the more indirect relationships between government and industry are also beset with problems and questions. The unique patent position held by the government as a result of its massive support of science is an area of recurring problems. Where the government supports the work, the government understandably also often retains patent rights to the inventions resulting from this work. With the continuation of large-scale federal involvement, one can see the possibility that the government would eventually come to hold a singular and somewhat monopolistic position. There are complexities enough here to challenge our best legal talent in industry and government for a long time to come.

What is apparent in looking at government–university and government–industry relationships is that over a period of many years, a period that began with the widespread operation of the Office of Scientific Research and Development during World War II, a generally fruitful and beneficial pattern has evolved. This flexible and somewhat ingenious interpenetration of government, industry and the academic world is a distinctively American invention which has cut several Gordian knots in the past and will continue to make worthwhile contributions to our way of life.

As we reflect in this brief and very general discussion on the problems our nation confronts at this stage of its progress into the Third Revolution, I believe we cannot fail to agree on one major conclusion: if we are to make our way into the future with any degree of assurance that we are on the right path, we will have to give nothing less than our best efforts and wisdom to planning and carrying into effect our long-term national goals in science. We cannot in good conscience treat this aspect of our national well-being with the crisis-to-crisis approach that has so frequently been followed in the past.

Obviously, then we all have a lot to do. Like it or not, we are living in the 20th century. Our country and the world are undergoing a period of change in which the influence of science reaches throughout the fabric of society, shaping nearly every aspect of our lives and our institutions. In our generation a man can no more ignore science than medieval man could ignore the Christian church or the feudal system. To be fully alive today is to be a conscious participant in forwarding the creative evolution brought about by man's increasing knowledge of his environment.

For the scientist, the onus of his new social responsibility cannot be lightly regarded, and it is time that all of us as citizens helped occasionally to remind him of this fact. We should do all that we can to encourage people of scientific capability to enter government. Nor should we be deterred by the once-popular image of the scientist as a scrambled egghead or an amateur human being. For all his human failings, the scientist also is capable of having some of the talent found in people with other backgrounds and training. Not all scientists, certainly, are practical; not all of them are persons with unusually good judgment. I have even known a few who were not good administrators. But many have this potentiality, and I suggest that, if we have not already done so, we discard the stereotype of the absent-minded professor along with other relics of the past. The government will continue to need scientists of stature and competence to help administer the great technical programs on which we are embarked. Our official policy and public attitudes should be directed toward achieving this end.

In the academic world more than ever before, scientists and educators bear the responsibility for assuring that the new generations of scientists now coming up through our schools, colleges and universities receive education of the highest degree of excellence. In the pursuit of basic research, so significant for our future, the federal government and the university are partners, each necessary and neither sufficient in itself. I would repeat that basic research and graduate education should reinforce each other in a variety of ways. They belong together at every possible level. It is up to all of us involved in the affairs of science to see that this reinforcement is made a vital thing.

The universities, and in particular the scientist in the university, must go even further — they must go beyond the limited task of offering professional education to scientists and look also to the problem of educating a citizenry for an age of science. In this task, the problem is not one of seeing that every college graduate is a scientist of some kind. The task is rather to see that every person who comes out of our institutions of higher learning to become a responsible member of our national community has some background in science and understands as thoroughly and as broadly as possible the significance of science for the modern world.

More than ever before, we must be prepared to work with all the means and media at our disposal to increase the general level of scientific literacy in this country. Much progress has been made at the grammar school and high school levels, but much more needs to be done to improve the elementary and secondary school curricula and especially to enrich the scientific content at the collegiate level of liberal education.

We are all in this together, and I, for one, believe that we shall see a growing community of thought extending throughout all groups of our increasingly mobile social structure. I would not underestimate the present growing degree of sophistication of the man in the street about scientific matters. In addition to the excitement over our new breed of astronauts, there has been a very natural and effective dissemination of science and technology, due in part to the close association many members of our growing population of technicians have with these matters on a daily basis. We are recognizing now that there may be a need to provide an additional two years of education beyond high school for many of our young people who can profit from such training. And as we go on to develop an intermediate level of higher education, I think we shall see a beneficial interaction between this new group of students and those engaged in baccalaureate programs in the sciences and the arts and humanities.

The program of education for the future must also of necessity include the practice and support of continuing education for our adult population. We have already begun to give emphasis to the significant trends in this direction through educational radio and television and through increased offerings in university and college extension courses. It is by means of these developments in our national educational process that we may hope for genuine progress in the growth of the scientific understanding so necessary to guide our future as a nation.

I want to say, in closing, that I feel very strongly about our need also to strengthen the humanities and arts at every level in the nation's educational programs. This need is not simply a question of giving the humanities a degree of support commensurate with that given to science and technology. I have tried to emphasize throughout this discussion that the shape of our future does not depend only upon our strong pursuit of excellence and understanding in science. We are brought face to face with many alternatives by our new capabilities for controlling our environment in all its aspects — physical, social and spiritual. But the use we make of these alternatives, the means we use to direct them toward achievement of the most exalted human objectives, will depend on the values we are able to create for ourselves, as we live and grow into a newer world. If we are to achieve humanity in terms of its greatest fulfillment, individually and as a nation, we need to share deeply in the varied experiences which can be reached only through the arts and humane letters. For we know that so long as we are able to see our images reflected clearly and with great validity in these lasting mirrors of the human adventure, by just so long we need not fear to ask with the Scottish poet: "O wad some Power the giftie gie us/To see oursels as ithers see us."

The New Boundaries of Reality

The Quillian Lecture, delivered at the Ministers' Week Convocation celebrating the 50th Anniversary of the Candler School of Theology, Emory University, Atlanta, Georgia, January 22, 1964.

It is a great pleasure, and I feel it to be a special privilege, to participate as a lecturer during Ministers' Week, which celebrates the fiftieth anniversary of the Candler School of Theology. A half a century once exceeded the life expectancy of the average man. Today the Biblical "three score and ten" is often achieved. Yet even by today's standards a certain dignity attaches to the respectable period of fifty years, so that its culmination marks an appropriate point for deliberation and reflection. It seems fitting to ask at this time, as you have done: What is the Challenge of Contemporary Civilization to the Church?

Today I am here as a scientist, and as a scientist I would consider how science relates to the theme chosen for consideration. The text which suggests itself — and I hope you will be charitable if I develop it according to the modes of thought of a scientist — is found in John 8:32: "And ye shall know the truth, and the truth shall make you free." These are important words and will continue to be important so long as man retains his sense of values. The truth does make us free in many ways — whether we speak of the larger spiritual truths or the truth as regards the physical universe.

I want to consider especially the access to certain kinds of freedom that accrue to humanity with the increase of scientific knowledge. One of these may seem rather commonplace. It is simply the release of human energy that comes from the elimination of drudgery. A second freedom is one that comes from the very great amplification of human sensibilities and intelligence through the tools and instruments developed on the basis of scientific knowledge. It is in these areas of ever-increasing freedom that I believe science is proving itself to be a powerful servant to the church in the field of religious scholarship. I believe we are entering upon an era of contemporary civilization in which the theological scholar will gain the great extension of freedom that comes with this increased knowledge and the highly sophisticated tools of research that science has brought to bear on scholarly labors.

Perhaps there is no need to emphasize the tremendous amount of human effort that has been expended upon scholarship. I think of the myriads of lonely men working through Europe's dark centuries to preserve and encourage the feeble glow of learning and of how much we owe to their abnegation. I think also of the many thousands or millions of hours of human labor that must have gone into the creation of the medieval manuscript libraries.

The invention of movable type must certainly be taken as the first great technological event which lightened the labor of scholarship and spread the fruits of this labor to always greater and greater numbers of people. One examines the beautifully illuminated manuscripts predating the Gutenberg Bible with the pleasure and reverence such work naturally inspires. But a moment's reflection on the prodigious labor required to produce only one such copy tells us why modern biblical study is said to have begun in 1516 with the publication by Erasmus of *Novum Instrumentum*. There were only so many hours in a day, then as now. The invention of printing gave a scholar the time needed for study and contemplation.

It may be significant also that the invention of printing, which inaugurated a new age of theological study, is recognizable as the first example of our own manufacturing techniques: standardized mass production by means of assembling interchangeable parts designed and manufactured for precision fit. Detroit was anticipated in Mainz about 1450. Gutenberg was the first modern industrial genius.

Even so, the work of scholarship continued to be an arduous task requiring the organized efforts of many men over long periods of time. We read of the plans for producing the Authorized Version of the Bible, begun in 1604 and published in 1611, that "learned men, to the number of four and fifty" were divided into six companies and set down to their task, their numbers including the formidable scholars of that time. One of these,

Dr. Andrewes, later Bishop of Winchester, was familiar with Hebrew, Chaldee, Syriac, Greek, Latin and at least ten other languages.

The first major amendment of the Authorized Version was an even more ambitious undertaking and was one of the early examples of organized international cooperation in scholarly research. Work on the Revised Version was formally initiated by the Convocation of Canterbury in May 1870 when the Convocation decided to "nominate a body of its own members to undertake the work of revision who shall be at liberty to invite the cooperation of any eminent for scholarship, to whatever nation or religious body they may belong." The magnitude of the task is reflected in the fact that the revision of the New Testament was completed only after 407 meetings by the scholars appointed to this work and that these meetings were distributed over more than ten years.

By the time work on the Revised Standard Version of the Bible had been completed in the early 1950s (the work of a committee of 32 scholars over a period of 15 years), a scientific revolution had made new tools available to biblical scholars — in particular, electronic data processing equipment employing large-scale digital computers. These machines were not ready for use in the work of translation but they were invaluable in providing for the indexing and cross-referencing necessary to the preparation of the concordance.

The last previous complete concordance to the protestant Bible was that of James Strong, published in 1894. Strong took 30 years to index the King James Version. The Reverend John William Ellison did this job for the Revised Standard Version in a little more than two years, using only his regular time off from his duties as rector of the thousand-member Church of the Epiphany in Winchester, Massachusetts, to direct work at Remington Rand's New York computing center. The idea came to Mr. Ellison during his studies at Episcopal Theological School. "I was appalled," he said, "to find people with two or three doctor's degrees sitting around cataloguing things on their fingers, so to speak. I was sure there must be some way to use mechanical means for the same purpose."

Modern high speed computers are being employed very widely for work of this kind. Such a large project can be carried out, for example, in the study of the complete works of St. Thomas Aquinas which include approximately 13 million words. A complete concordance or index to the words in all his books would, of course, be of great use to those who would like to make a detailed study of the way he thought and wrote or of how his thinking may have been influenced by others of his time and before. It has been estimated that such a complete concordance would take 50 scholars some 40 years to complete by hand in the old traditional way. By the use of the newer machine processing of such data, such a concordance could be completed with about one-fortieth of the time and effort, permitting researchers to use more of their time for their specific scholarly capabilities, such as interpreting and organizing information so obtained.

Such a program is currently underway. A library in Italy recently has put all the words in the *Summa Theologica* of St. Thomas Aquinas on some 1,600,000 punched cards for machine processing. A similar program is under way to study the words of the famous Dead Sea Scrolls. It is interesting to note, in this instance, that a special card punch had to be devised to punch the cards from right to left, since this is the form in which the Scrolls were written. The point that stands out especially in these examples of application of computers to theological scholarship is the immense reduction of routine human labor that is achieved. Not only are the projects that would ordinarily be undertaken accomplished in much less time by the use of these computers, but other projects may be undertaken which would have been dismissed previously as requiring an inordinate amount of scholarly effort.

Even more exciting possibilities are being opened up with the rapid developments in the field of machine translation of language. It is not too much to hope that in the near future at least the laborious part of translation itself can be turned over to the computer, with the result that the scholars directing this work can be assured of a high degree of uniformity and reliability in the basic translation and will be able to give more of their own time to matters of style and interpretation. Perhaps even more important for the work of the church is the possibilty that translations of the Bible into various languages can then be achieved in a phenomenally shorter period than would now be required.

I would emphasize that the use of modern electronic equipment of this kind does not supplant the scholar, and its use does not mean that the scholar relinquishes his responsibility. It permits rather a freeing of his energies so that more of his time can be spent on the highest offices of scholarship, those that demand creative thinking and the highest form of judgment.

The second freedom brought about by science in its pursuit of truth derives from the amplification of human sensibilities and intelligence through the methods, instruments and tools made available for the tasks of scholarship. Most germane to the work of theological scholarship is the partnership that has developed between the physical sciences and the studies of biblical history and archaeology. I was particularly struck by the fruitfulness of this partnership after reading about the work of archaeologist Rabbi Nelson Glueck in a recent issue of *Time*.

Rabbi Glueck's method, as a "surface" archaeologist, combines continuing reference to the Bible for clues as to where finds may be made with an expert's ability to interpret the evidence of potsherds and other specimens of pottery found near the locations he is studying. Bible in hand, Dr. Glueck has ranged the Holy Land off and on for 36 years. "Out in the desert," he says, "there is sometimes so much mist in the morning that you cannot travel. You have to wait for the sun to burn it off. To me, archaeology is like burning the mist off the Bible."

An impressive array of scientific instruments and methods is now being used in archaeological work by both the "surface" and the "pick-and-shovel" archaeologist. Aerial cameras detect the faint outlines of walls long since demolished; airborne magnetometers ferret out forgotten battlements; measurements of minute bits of carbon establish dates beyond the reach of recorded history. Mummies are submitted to autopsy for a knowledge of ancient diseases or evidence of human foul play. Fossilized grains of pollen give clues as to the climate in which they grew.

You will have a special interest, I think, in the carbon-14 dating techniques since one of their early applications was to assist in establishing the date of the Dead Sea Scrolls. Professor Willard F. Libby of the University of California at Los Angeles, a former member of the Atomic Energy Commission, was awarded the Nobel Prize in 1960 for his contributions to the exact study of the past through the introduction and use of carbon-14 dating methods. In 1946 he concluded that nearly all neutrons produced by cosmic rays entering the earth's atmosphere end their lives by converting some of the atmospheric nitrogen into carbon-14, a radioactive isotope of carbon. He reasoned also that, because of photosynthesis, all the living animal and vegetable world (the biosphere) should be weakly radioactive owing to the uptake of this minute proportion of carbon-14 in the atmosphere. Carbon-14 dating was made possible by taking advantage of the knowledge that the radiation from this radioisotope decays at a predictable rate and measuring the amount of radioactivity remaining in organic samples after they cease to live in equilibrium with the biosphere.

In the case of the Dead Sea Scrolls, some of the linen from the decomposed cloth in which the Scrolls were bound was sent to the University of Chicago for dating. The result fixed the date of the Scrolls at 33 A. D. plus or minus 200 years, which agreed with the best estimates obtained through paleographic and other archaeological methods. Here I would emphasize again the partnership of the physical sciences and other studies. Radiocarbon dating has extended man's senses to include the detection and interpretation of radiation; his judgment in the use of this new range of perception, however, makes use of inputs from many other fields of knowledge.

I have mentioned the magnetometer as a tool available for the scientist-historian to investigate what lies beneath the surface of the earth. We now know that the tiny nucleus of the hydrogen atom is both a kind of spinning gyroscope and a tiny magnet. It is possible to use these properties of the nucleus to build a highly sensitive device to measure local magnetism to one part in 100,000. Ancient tombs deep under the ground, or fireplaces, or buried walls, each subtly disturbs the earth's natural magnetic field; and the experienced physicist-archaeologist can use this information to help locate these buried articles of irreplaceable value. An international group of scientists from the United States and Italy has been using this and other techniques to investigate the ancient valley of the Crati river. In just a few days' time they outlined half a mile of one of the ancient walls of the city of Sybaris. The sensitivity of the method may be judged by the fact that most of the wall was buried beneath about ten feet of accumulated silt of the centuries.

Again we have an addition to man's sensory perceptions, this time to include the detection of extremely weak variations in the earth's field of magnetism. I might add, the proton magnetometer can also be used for dating. Potentially, it is capable of a much higher precision than radiocarbon dating and consequently its most useful contribution will be in establishing dates for more recent times — the last 2000 years.

Still another application of science to work of significance to the theological scholar is the method of activation analysis. Neutrons from an atomic reactor can be used to make minute traces of some materials sufficiently radioactive so that we can measure their presence and amount in quantities that are much too small to see with the most powerful microscope. Scientists at the Brookhaven National Laboratory recently have been working with this method to study the chemical composition of pottery from the Mediterranean and from South and Central America. The great advantage of the method is that it is a non-destructive method of analysis; that is, the original artifact still is preserved in its original form and beauty.

Elizabeth Holtzman (left) graduate student in anthropology at U. C. Berkeley, Helen V. Michel and Frank Asaro (right) from the Lawrence Radiation Laboratory, listen while Isadore Perlman, Professor of Chemistry at U. C. Berkeley and Director of the Lawrence Radiation Laboratory Nuclear Chemistry Division, explains how neutron activation analysis can determine the origin of ancient pottery. (April 1968, Lawrence Radiation Laboratory B70) Courtesy of Doug McWilliams, Ernest Orlando Lawrence Berkeley National Laboratory.

I regret to say that some of the activation analyses performed thus far indicate that men in other ages succumbed to the temptations of ill-gotten gain almost as easily as certain ones of their modern descendants. Pieces of pottery made in an ancient city of Arezzo, Italy, were very popular in the Roman Empire because of the high quality of their workmanship. The potters even marked their ware in the same way that manufacturers trademark their goods today. Yet, by means of neutron activation analysis it has been possible to show that much of this ancient pottery must have been made elsewhere and given a false trademark, since the chemical composition of the clay is sufficiently distinct to clearly establish it was *not* made in Arezzo.

There was also the analysis of Macedonian coins, of about the period 400 B. C., which disclosed a possible and curious double standard in fiscal affairs. Two very similar varieties of silver coins were struck in this period; one variety contains less than 1/4% of copper — the others contain from 5–24%, and because of their similarity in other respects there is a suspicion that the higher alloyed varieties may have been used to deceive the unwary.

Sadly enough I must admit that not all of those who put forward their claims in the good name of science can be exempt from close scrutiny. X-ray fluorescence spectrometry is a method of analysis which also is of particular interest to the archaeologist since it is sensitive and non-destructive and can tell what elements are present in the sample examined. It has been successfully used to analyze and identify the glazes of old ceramics, and has had one other unusual application. The skull and jawbone of the Piltdown man had long been an object of suspicion and irritation to many scientists, since it just didn't fit most of the rest of what was known about ancient man. One of the final contributing bits of evidence that established it as a complete fraud was proof by X-ray fluorescence spectrometry that the brown coloration of the skull was not a result of natural staining by iron in the water, as originally claimed, but was the result of artificial treatment of the bones with chromic acid. So Piltdown man has quietly been removed from both the museum and the family tree.

These are but a few of the more interesting and useful ways in which science has helped man to sharpen and extend the powers of his original sensory endowment besides adding strange new modes of perception to challenge his imagination. But what of that great and peculiarly human gift — the mind itself? Can we expect that the powers of our intelligence also will be pushed beyond their present range of capabilities?

I believe we are already entering an era in which science will bring about startling developments tending to widen the boundaries of the human mind. I have already discussed the extent to which computers are being used to reduce the mental drudgery of recording, sorting and classifying, counting and cross-referencing and other menial thought processes. With proper instructions computers can carry out somewhat more ambitious mental adventures.

You may have read in recent weeks of the computer study of the Pauline epistles planned and carried out by the Presbyterian minister, the Rev. Andrew Q. Morton, of Culross in Scotland, who also is a mathematician. He tested the original Greek of the epistles by sentence length, by word pattern and by the occurrence of common words of Greek prose and concluded that St. Paul could not have been the author of more than five of the 14 epistles attributed to him in the New Testament.

Whatever may be decided about the results of Rev. Morton's study when these have been compared with other views from the Tübingen studies down to the present, we may agree that if studies of this kind are to be undertaken, the computer offers a valuable adjunct to the human mind. We have no qualms about giving instructions by means of various switches, levers, push buttons, etc., to machines that accomplish physical labor for us. Perhaps the same attitude will come to be accepted in our use of machines to accomplish thought processes.

Will machines ever be able to learn in the ultimate sense? Will they write their own programs, develop improved recognition and translating systems, find new patterns and new analogies and new models? How far can they go in relieving man of all but the highest orders of thought involving judgment and creative insight? Much further, perhaps, than most of us would be willing to concede at present.

Impressive as these developments have been, I am even more hopeful concerning the things that our increasing knowledge may do to improve human capabilities in a direct way — in particular, through research on the genetic code. I am proud that nuclear science has been a fundamental tool of the biochemists in their very interesting explorations of how these instructions encoded into the living cell direct it to become a sparrow, a giant sequoia, or a member of that fallible and wonderful species, man. Many avenues open into the future from this work. As memory has been called "the mother of the Muses," there is even a glimmering of hope that we may be able to lengthen the productive years of our poets and prophets.

I quote from a recent article on the subject: "The aging process itself seems to be tied up in some way with the nucleic acids. Scientists now theorize that, with the passing years, the DNA and RNA in human cells are subjected to all sorts of disturbing influences, from chemical action to random radiation. Thus in steadily increasing numbers of cells, the code message is garbled, making the organism correspondingly less efficient." As we learn more about how aging processes occur, we may have greater hope that they can be slowed down, if not halted or reversed.

Hardly less challenging in its prospect for alleviating human misery and releasing man's energy for creative endeavor is the progress of chemotherapy as an effective treatment for mental diseases. Correction of aberrations and even reconstruction of personality may be possible through the administration of new drugs and medicines.

I think we have reached a point in our consideration of science as the servant of scholarship that will allow us now to view these matters in a still broader perspective. If we recapitulate these changes brought about by science in pursuit of truth — the use of computers (and electronic systems in all their variety), radioisotopic dating, location and dating with the proton magnetometer, the examination of materials by activation analysis, the new work in biochemistry — the theme discovered running throughout these developments is the amplification and diversification of human capabilities.

Man is a limited species in many ways and much of his thinking has been concentrated on overcoming his limitations. He has invented devices to extend his senses, artificial sense organs. With the aid of telescopes and suitable photographic film he can observe events taking place billions of light years away, perhaps near the very edge of the expanding universe. In the microcosmic world he sees the detailed structure of such simple self-reproducing things as viruses, some of which are so small that billions of them could fit comfortably in a sphere no bigger across than the period at the end of a sentence. Electronic amplifiers can convert the sound of a beetle scuttling through the grass into a noise as loud as an army marching in step.

We create new senses as well as extend the power of natural senses. We have no sense organs designed to detect magnetism, but devices sensitive to magnetic fields help us locate ancient pottery kilns and the walls of buried cities. Our eyes are not designed to detect radio waves, but radio telescopes pick up static from the depths of space and reveal the existence of "radio stars," remote broadcasting centers many of which cannot be observed by our most powerful optical telescopes.

To keep our minds from being swamped by the problems of handling the vast amount of data accumulated by these artificial sense organs, we have developed and still are perfecting sophisticated thinking machines able to solve the most complex kinds of problems, translate our thoughts into various natural and artificial languages, and perhaps as time goes on, able to recognize significant patterns and interpret their significance.

The outside world of the lowest form of life is inexpressibly bleak. Creatures up to the relatively simple level of worms hear no sounds and see no colors; many of them live in eternal darkness. Even a human who is blind and deaf has a tremendously more vivid sense of life than these creatures, whose sense of touch and whose chemical sense are incomparably less keen than ours.

But man, by the use of his natural endowments, including preeminently his intelligence, has increased almost infinitely the gap between himself and these creatures on the lowest rungs of the ladder of life. It is amazing in retrospect to consider how the vision of poets and prophets has anticipated the expanding scope of human experience, how many visions have been realized.

The poetry of the spirit is built of the stuff of reality. As the growth of scientific knowledge extends, expands, and diversifies the human environment — revealing what we once thought were faint stars to be island universes, opening our eyes to the marvelous structure by means of which the chromosome contains both the history and future of the flower, letting us listen to rhythms of natural processes we had only guessed at before — as all these things happen, we would expect the poetry of the spirit to respond in kind.

Thoughtful men of our generation have remarked that the metaphors of Genesis offer the texts for which science in subsequent ages has been busy providing the detailed exegeses. Has the development of the evolutionary hypothesis, for example, been other than a prolonged proof and explication of the figure of man's creation from the dust?

The scientist and the maker of metaphors will continue to have need of each other. Science needs to have its imagination stirred by the proliferation of new images and analogies. It often juggles two or three at a time — as the physicists have been doing for some years now in the wave and particle theories of light. And from the labors of science come the enlargements of reality so valuable to the visionary at work in all of us.

Looked at in this light, the separate tasks of the scientist and seer over the period of man's history on earth have gone forward in their progression as a kind of symbiosis of the imagination. The greatly extended universe of our time is, in the sense that reality grows with the explorations of our mind and spirit, the joint product of their labors. Is it too much to hope that there will be even greater visions yet to arise from our present view of reality? The prognosis would appear to be favorable. Certainly the new powers available to man through science should help to free him for new endeavors in the realms of contemplative thought.

The Scientist as a Human Being

Remarks by Dr. Seaborg upon his acceptance of the Charles Lathrop Parsons Award of the American Chemical Society, Washington, D.C., December 5, 1964.

I can think of few events during recent years that have given me more real pleasure and satisfaction than the American Chemical Society's notification that I was to receive the Charles Lathrop Parsons Award. For one thing it was like being told that I had done a fairly presentable job of what I had been urging other scientists to do — that is, to devote more of their thought and energies to matters of public concern. It is always good to feel that you have been able in some degree to practice what you preach. There is pleasure too in knowing you have been thought of along with such people as the original recipient of this award, the grand old man of the American Chemical Society, Charlie Parsons — not to mention James Conant, Roger Adams and George Kistiakowsky. Tonight I want to speak of scientists as people — as human beings.

Old legends die hard, but none have been more persistent than the belief that the scientist is something more, or perhaps something less, than a human being. The image of the scientist as a misguided or visionary eccentric is almost as old as Western culture. In fact, Aristophanes wrote one of his most successful comedies on this theme and spoke of the scientists of his day, certain philosophers, as dwellers in Cloud-Cuckooland. But the image of the scientist as a more or less wicked wizard is perhaps even more ancient. You will recall that the demigod Prometheus stole fire from heaven and was rewarded by his superiors for this bit of insolence by being sentenced

Reprinted with permission from *Chemical & Engineering News*, Vol. 42, No. 51, pp. 60–62, December 21, 1964.

to an eternal career as a bird-feeding station. Whether or not he was wicked depended upon whether you took sides with the gods or Prometheus' human beneficiaries.

To say the least, people have always felt some uneasiness upon being confronted with new knowledge and have been inclined to distrust the purveyors of such knowledge. The quasi-legendary figure of Dr. Faustus represents one reaction — that of awe-struck fear; the dwellers in Cloud-Cuckooland and their successors in the writings of Jonathan Swift are the products of another, more positive reaction — ridicule.

The two images vacillate in time with the successes and failures of science, with some component of each usually present in our culture. I believe most of you who have had the experience of presenting testimony to a Congressional committee will agree that the picture of a scientist blissfully floating along on cloud nine has not entirely disappeared, even with the great prestige that science now enjoys. And perhaps it is a sign of good mental health on the part of the average citizen that he is still able to have some fun at our expense.

When science was young, when the distinguished amateurs of the early British scientific societies were looking into all sorts of questions, the satire directed at experimenters often became biting and remarkably detailed. Interestingly enough, some of what Swift intended for satire has turned out to be, from the viewpoint of our times, more nearly prophetic than satirical.

Still there is plenty of bite in the scenes portraying Gulliver's visit to the Grand Academy of Lagado. Here is Gulliver's account of his initial encounter with science: "The first man I saw was of meagre aspect, with sooty hands and face, his hair and beard long, ragged, and singed in several places. His clothes, shirt, and skin were all of the same color. He had been eight years upon a project for extracting sunbeams out of cucumbers, which were to be put in phials hermetically sealed, and let out to warm the air in raw inclement summers. He told me, he did not doubt, that in eight years or more, he should be able to supply the governor's gardens with sunshine, at a reasonable rate; but he complained that his stock was low, and entreated me to give him something as an encouragement to ingenuity, especially since this has been a very dear season for cucumbers."

Times have changed a great deal. I am confident that our people working on their project of producing miniature suns by means of the controlled fusion process have better prospects than the cucumber expert of the Grand Academy; science requires substantial sums for its support in these days, and we are not alarmed when our controlled thermonuclear scientists indicate at budget time that it is going to be a "very dear season for cucumbers." The whole progress of physical science separates these men from the objects of Swift's satire. We are able to point to measurable progress in our controlled thermonuclear program, giving us grounds for reasonable optimism.

"In another apartment," Gulliver says, "I was highly pleased with a projector who had found a device of ploughing the ground with hogs, to save the charge of ploughs, cattle, and labor ..." The idea, of course, was that the hogs in rooting around for delectable objects underground would pulverize the soil. "It is true," he adds, that "upon experiment, they found the charge and trouble very great, and they had little or no crop." Here Swift's satire proved to be a very accurate prediction. Large flocks of geese are now being used to keep cotton plantations and peppermint fields free of weeds with a considerable savings over traditional methods. In the case of the peppermint fields, however, the geese develop a taste for the young peppermint plants and have to be retired after a couple of years.

The tasks on which Swift had the chemists engaged would hardly bear repetition in the polite society of his day or our own. I will therefore leave these scenes to the sufficiently curious reader except to mention that one prospector was "at work to calcine ice into gunpowder" and that he had also completed a treatise concerning the "malleability of fire" — not very promising endeavors.

I cannot leave the Grand Academy of Lagado without recalling at least one other project which — with apologies to our colleagues in the field of machine translation, teaching machines and information processing — has a very modern ring to it. The head of this project explained to Gulliver that "Everyone knew how laborious the usual method is of attaining to arts and sciences; whereas by his contrivance, the most ignorant person, at a reasonable charge, and with a little bodily labor, might write books in philosophy, poetry, politics, laws, mathematics, and theology, without the least assistance from genius or study." Gulliver says: "He then led me to the frame, about the sides whereof all his pupils stood in ranks. It was twenty feet square, placed in the middle of the room. The superficies was composed of several bits of wood, about the bigness of a die, but some larger than others. They were all linked together by slender wires. These bits of wood were covered, on every square, with paper pasted on them; and on these papers were written all the words of their language, in their several moods, tenses, and declensions; but without any order."

To make a long story short, some of the pupils by turning cranks on this machine changed the pattern of the words while others transcribed the bits and pieces of sentences that appeared. Gulliver relates that he was shown several volumes in large folio, already collected, of broken sentences, which were eventually to give the world a complete body of all arts and sciences. It may be that the researchers of the Grand Academy are still ahead of our progress in this field. However, a poet of my acquaintance tells me that one of his beatnik colleagues uses a simplified version of the Academy's composition machine. A bottle filled with scraps of paper on which words are written is shaken up by this cool cat and his poems are then composed using words from the scraps of paper as these are drawn one by one from the bottle.

Swift concludes his account of Lagado and its projects by anticipating Korzybski's school of general semantics. He tells of scientists who have eschewed the use of language altogether and converse with each other by pointing to particular objects out of a collection they carry about with them for purposes of discussion. He remarks that when the universe of discourse is large the burden can get rather heavy and cumbersome. In Gulliver's opinion, however, this invention might well have been adopted "if the women, in conjunction with the vulgar and illiterate, had not threatened to raise a rebellion unless they might be allowed the liberty to speak with their tongues after the manner of their forefathers; such constant irreconcilable enemies to science are the common people!"

While, as I have admitted, the image of the scientist as an otherworldly bungler is still with us, in recent years and especially since the earthshaking advent of nuclear energy the Faustian figure has again been renewed with some potency. It is very difficult for the majority of mankind to regard those who have had a hand in the release of such an awesome power of destruction as not having some infernal motivation. The process of demonstrating to the world that the beneficial aspects of nuclear energy can in the long run outweigh its potential evil will require the devoted and patient efforts of many men over many years. We have only begun to rebuild our confidence.

Nor is nuclear energy the only cause for apprehension concerning the products of applied science. Unfortunate results from the use of some modern drugs and the deleterious effects of pesticides when improperly used have caused widespread concern. The knowledge that chemical and biological agents have been developed to refined stages of destructiveness has had a disquieting effect on many people. Even those technological applications, such as computers and automation, which without question will prove to be immensely liberating for man everywhere, have raised specters of painful economic and social dislocation. The scientist as the moving force behind a vast new array of technology must expect that he will to some extent become the focus of these anxieties, even while his status and prestige increase due to the benefits his work has achieved.

The title of a recent survey of science fiction, *New Maps of Hell*, gives the flavor of many of the stories in this burgeoning field of literature. Many of these stories deal imaginatively with the kinds of reactions, either disastrous or beneficial, that society may experience in adapting to extremes of technological advancement in the near and more remote future. But the point can now be made most forcibly that never again will the scientist appear to be a distant figure isolated from the common concerns of everyday human beings. Never again will science appear to be an esoteric philosophy pursued merely for the disinterested intellectual pleasure of a few individuals. Science is so intimately connected with all levels of our life today, from the most utilitarian technology to the highest reaches of our intellectual endeavors, that there can be no question of escaping its pervasive influence. However, it is interesting to note in this connection that Einstein's work set off a far-reaching train of reactions among philosphers even before it demonstrated its concrete potentialities in helping to provide us with a new form of energy.

Thus, while we may occasionally be able to point to seemingly ridiculous aberrations, the layman is likely to feel uncertain that even the wildest may not turn up something important. Science has demonstrated its potency over and over again, and as a result we have to put ourselves in a special frame of mind to find the experiments conducted at the Grand Academy of Lagado anything like as humorous as they seemed to Swift's contemporaries.

These then are the traditional stereotypes, and though they are waning we must admit that, since scientists include very nearly the whole range of traits found in the human species, there are elements of truth in both of our stock character versions of the scientist. The scientist as the visionary bungler stubbing his toe over the most obvious facts of life certainly has his counterpart in everyday experience. Indeed the very fact of asking a lot of questions is sure to produce a lot of wrong answers. But it is the virtue of science over the long run to put these wrong answers to the test of reality and relegate them to the junk heap of human experience.

Ordinarily it is only among the unenlightened that the more extreme errors continue their precarious day-to-day existence. Naturally there are exceptions to this rule. Even well into this century I have been told there was a professional man — the profession to remain anonymous — who explained malaria as an illness that arises from eating cornbread. Among scientists, of course, there will continue to be those worthy of the highest honors accorded them and the exceptional few who help to perpetuate the image of the theoretical bungler. I have heard one story of an investigator, one of our contemporaries, who was studying the growth rate effects of planting Irish potatoes during different phases of the moon. Perhaps by now he has succeeded in establishing a correlation.

And the scientist is human in other respects, as many a government and university administrator has discovered when trying to economize at the expense of a scientist's favorite project. His attachment to the things that are peculiarly his own is as strong as the painter's identification with his canvas or the novelist's feelings for the characters coming to life under his pen. He is human and selfish, at least to that extent.

Those familiar with the history of science are aware also that the scientist can be very human on questions concerning the priority of discoveries. There may not have been quite so many words expended on the famous Newton–Leibniz controversy as have been devoted to the authorship of Shakespeare's plays, but the human feelings involved were unmistakably intense. Today perhaps, with our emphasis on group research in large organizations, we are more inclined to recognize the likelihood of simultaneous advances and be more lenient with claims and counterclaims. Nevertheless there is still a healthy spirit of competition.

The image of the scientist as a visionary eccentric has faded away and the wicked wizard picture has been revived only now and then by the spectacular potency of modern science. For the most part, the scientist tends to look more and more like the rest of the population. His numbers are increasing as science prospers and its benefits spread to the far corners of the earth.

Most of us would be hurt and disappointed, I think, if we could easily be distinguished from the lawyer, the business executive, a professor of the humanities, or a successful football coach. The image of the scientist has in fact begun to assume ordinary human proportions. To the extent that this image looms at all larger than life, it is due to the increasing involvement of the scientific community in the affairs of state and industry. Along with this involvement and a correspondingly rapid growth in the number of people choosing science as a career, there has been a major escalation of the scientist in the public's esteem. The University of Chicago's National Opinion Research Center issued recently a draft report entitled "Occupational Prestige in the United States: 1925–1962." This report lists the first nine occupations ranked according to prestige as follows: supreme court justice, physician, nuclear physicist, scientist, government scientist, state governor, cabinet member in the federal government, college professor, and U.S. representative in Congress.

A similar study was made in 1947 and the major difference between these two studies was a rise in the prestige of scientists. The most remarkable change was in the status accorded to nuclear physicists, apparently a somewhat delayed reaction to events during World War II. While the studies in each case showed that only a very small proportion, from 2 to 3 per cent, of the respondents could describe the duties of a nuclear scientist, there was undoubtedly a realization that this scientist was the man responsible for nuclear weapons. It may be also that the nuclear scientist has become the most articulate representative of the scientific community.

At any rate, the post-World War II years have seen not only a change in the public image of the scientist, but an equally profound change in the scientist's attitude toward himself and his work. The soul-searching among scientists of the Manhattan Project, as that endeavor reached its culmination, is too well evidenced in subsequent events to need retelling. There is hardly a one of the major contributors to the project who has not felt the need to participate in developing public policy with regard to the control of nuclear energy in its military and peaceful uses. There has been a marked willingness of scientists in recent years to enter the public service, sometimes even at a considerable personal sacrifice.

What is now emerging, I believe, is an era in which the scientist will achieve increasing stature as a human being because he is willing to look beyond the immediate results of his scientific endeavors to their social consequences. He recognizes that even though he cannot presume to advise mankind with finality on the values that are most acceptable for our world, at least he may be able to help point out the probable consequences of pursuing alternative courses according to one or another set of values. And he realizes that he, in common with men generally, will be deeply affected by the course that is chosen.

I believe further that non-scientists are coming to accept the scientist and his science on these terms. They do so realizing that only by cooperative efforts involving scientist and layman, deeply committed to the need for resolving the extremely complex problems we face in this age, will we be able to achieve the higher levels of human behavior toward which we are all striving.

The New Imperatives (or, How to Succeed by Really Trying)

Commencement address at the American River Junior College, Sacramento, California, June 11, 1965.

I believe it was the American newspaperman Henry V. Wade who said, "Youth must be strong, unafraid, and a better taxpayer than its father." Now some of you may think this is a fine way to begin a commencement talk — by worrying the graduates about such things as future taxes when they are still anxiously waiting to receive their diplomas. Actually, when I selected that quotation to begin my remarks this evening, I did not think of it as an expression to cause concern but rather as one which might lead to a bit of optimistic thinking about the future. For I believe it is essential that young people everywhere today look at the future not as a series of insurmountable problems, but rather as a series of challenges which they can meet and overcome, given their heritage of knowledge and the gift of energy typical of their years.

Of course in being optimistic about the future one must also be realistic. It would be foolish to lead you to believe that a great deal of very hard work is not ahead of you — ahead of all of us. It *is*, and I think most of us realize this. For only by expending enormous energy and pursuing excellence will you be able to build that better world about which we hear so much these days; and thoughtful people today know that in reality there is no alternative to building it. Nor are there alternatives to those attributes of youth — vitality, creativity and daring — which will be necessary to do the job. This future world I speak of will still occupy, for the most part, the same earth we occupy today, except that by the end of this century it will need to support almost twice the number of people we have on earth today at a far higher average standard of living.

But before we discuss some of the implications of this statement, or examine some of the possibilities and problems of the future, perhaps I should say that I realize it may be difficult for many of you to be concerned with the world's future problems when you may have some closer, more immediate problems of your own. Of course there is nothing wrong with focusing on one problem at a time either, as did the little boy whose father took him to the zoo for the first time. When they visited the lion house, the father went into some detail on just how ferocious a beast the lion was. Following a rather vivid description of what the lion might do if he escaped from his cage, he concluded, "And if that lion should get out of his cage, he would not only tear me all apart, he would also eat me all up." Then he asked the terrified youngster if he had any questions. The little boy did. "Daddy," he said, "if that bad lion should get out of his cage, tear you all apart and eat you all up — what bus should I take home?"

That may be the ultimate in focusing on one problem at a time, but, assuming that most of you know "which bus to take home" at this point, let us take a broader look at the world before us — at some of the prospects of the future and the role you may be playing in it.

In discussing the future there is no better subject with which to start than education — and perhaps no better time or place to discuss it than at a commencement. For, as the word implies, a "commencement" is a beginning.

Let me put into perspective for you a few thoughts on the important role of education today and for the days ahead. There is no doubt that we are at a major crossroad of the history of man. The energy of science and technology has been propelling us at a breathtaking and accelerating speed in recent decades. In no comparable period has man acquired so much knowledge, put it to use so rapidly, nor moved so swiftly in his ability to produce goods, to change his environment, or to destroy himself. Here in California, across the nation and all over the world, new industries rise everyday. Our economy expands and changes. New skills as well as new tools quickly become obsolete. Along with the scientific–technological surge has come social and economic revolution the world over. The cry of men in want of dignity and of material satisfaction is rising in crescendo.

In the background are the terrible weapons of modern war. And above all hangs the lack of understanding of men for each other.

If one were to hazard a choice as to the world's greatest need, it would be, I think, understanding: understanding and adaptability, a capacity for tolerance and for the rapid change and evolution that men in our world must bear. Our best hope for achieving such a capacity lies in more and better education. It must be the goal everywhere to provide the opportunity and the encouragement for each individual to achieve the maximum education his talent will allow. We must have, more than ever, this maximum education of the individual which sometimes has been called the democratization of education.

Fortunately, this concept has guided California education over the years. And in California, the junior college, more than any other segment of the educational system, has represented this kind of democratization. The California junior college system stands as a model for our nation, and your Superintendent, Bill Priest, is one of the most outstanding leaders in the system.

The junior college serves many important functions in terms of our needs in modern education. It allows for an immediate extension of education and vital skills to many who would not otherwise have the opportunity to acquire them. It makes it possible to prepare many more students to go on with further education and offers them important guidance for the future at a time in their lives when this is so vital.

You are indeed fortunate to be a part of an educational system which is providing our country with a more literate, a more highly skilled, citizenry better suited to carry on in this increasingly complex society of ours and better equipped to face the future.

But let me quickly add that in today's rapidly changing world educational requirements do not stand still. They move forward and expand — at a rate as fast, or faster, than most aspects of society. Today, more than ever before, education must be an ever-continuing process. The explosive growth of information, the rapidity of communication and transportation, the shrinking of boundaries, and the breaking down of barriers — economic, social, and physical — all tend to render us, our knowledge, skills, and tools, obsolete unless we make an intensive effort at continuing education. This will be especially true for those entering scientific and technological fields. The knowledge acquired up to their final degree will be serving, in many cases, only as background for a continuing fund of knowledge. Because of this, constant self-education will be essential to professional survival.

The daily growth of information resulting from scientific research and development is truly phenomenal. We are told that the amount of scientific information published around the world every 24 hours would fill seven complete 24-volume sets of the Encyclopedia Britannica. This means that, reading around the clock from day to day, one man would need about 1000 years to cover one year's output. No doubt such an assignment would cause considerable eyestrain — even for one having completed a rapid-reading course.

However, the absorption of new scientific and technical knowledge is essential not only to the scientist and engineer, it is important to other people in other fields. To an increasing degree, people in other professions, and especially our social and political leaders, also find it necessary to become better acquainted with, and keep up with, science and its implications, as do our leaders in business and industry.

Not only is the specialist affected by this influx of information, but so is the general public becoming more and more involved and concerned in the results of technological changes as these changes more and more affect all phases of our lives. The average citizen is called upon to make vital decisions in his daily life on a wide variety of personal and family matters where scientific knowledge may be the most important single factor in his decision. The need for such knowledge to make sound judgments and decisions is bound to increase in the years ahead.

But the need for extensive and continuing education by the maximum number of people will not be limited to scientific and technical knowledge. If we are to have an enlightened citizenry — one capable of maintaining a democracy in the complex civilization of the future — extensive education in the humanities and the arts also will be a necessity. Never before have the humanities been more important — nor will their importance diminish in the coming years. Together with the arts, their study will serve to give our lives purpose and direction and to help make our increasing hours of leisure richer and more worthwhile. There is also a growing urgency for people educated in the social sciences, if society is going to be able to digest our scientific advances, to recognize their potential and be able to put them to maximum use.

Up to this point I have alluded to the future rather vaguely. Let me be more specific now and try to paint for you a picture of the future — that perhaps ten, twenty, and thirty years ahead. This is the way some of the leading scientists and thinkers of our day see the world in which you will be living and playing a leading role. For those of you interested in doing further reading or research into these projections of future wonders, there are many published articles and books. A recent compendium I would recommend to you is *The World of 1984*.

There is little doubt that our most immediate progress in the future will be made through the advances we are already experiencing in the field of electronics and its relationship to communications and the computer. The possibilities growing out of this combination seem fantastic. Among their most promising potentials are worldwide live television for entertainment, news, education, and private and commercial communication. This will be made possible by the use of nuclear-powered synchronous satellites orbiting the earth at a distance of 22,000 miles. These satellites, which in effect will be hovering over fixed points above the rotating earth, will not only make it possible for someone in New York, Nashville, or Nogales to see a live performance of the Bolshoi Ballet in Moscow or witness the Olympic games as they take place in Bangkok, but will allow a department store buyer in Dallas to see a fashion show in Rome and make direct purchases, an engineer in Schenectady to review plans and specifications in an office in New Delhi, or students in Seattle to attend a seminar at the Sorbonne in Paris.

When necessary, two-way TV systems will allow each participant to speak in his own language and have it automatically translated to the listener's tongue at the receiving point. This two-way communication will also be possible with the private portable radio-telephones that modern electronics may make available to almost everyone. Besides being automatically translated, your language might be electronically compressed in transmission and expanded upon reception to normal speaking speed. Speaking to your family, you might want the warmth and emotions of your own voice to come through, but, in a business conversation, it might be possible to electronically eliminate all emotion from your voice.

With your own private number then — perhaps your social security number — you will be able to reach anyone else having a similar instrument anywhere in the world. Of course, such marvelous systems do not preclude some possible complications. For example, how will you avoid getting a call from your mother-in-law, even though you may be visiting a sick friend in Tahiti or Tierra Del Fuego? To make matters worse, the chances are she will be calling collect! Or what will prevent someone from trying to sell you, who live in Placerville, a telephone subscription to a magazine published in Peking? And should you have a teenage daughter with a boyfriend in Melbourne or Manchester, or even in Archangel or Addis Ababa, your telephone bills may be astronomical!

Modern electronics will also make possible a world primarily controlled and operated through computers, data-processing equipment, and automated control systems 100 to 1,000 times faster than today and many times smaller and cheaper. These systems will affect almost every form of human endeavor. Let us look at just a few examples.

In medicine, information obtained from a physical examination, which may include blood and other body fluid analyses, X-rays, cardiographic and neurological tests, will be fed into a central medical center where the information will be rapidly analyzed and a diagnosis returned to the patient's physician. Such a diagnosis would be the equivalent of a consultation with, and diagnosis from, a dozen or more leading specialists, and the returning information might not only point out the disease but recommend a detailed course of treatment to cure it. An annual checkup of this type would be quite reassuring to a healthy person, but can you imagine the disappointment that such automated medicine would hold for the hypochondriac!

Since our world of the future will be filled with an unbelievable volume of information, electronic means will be essential for its collection, compilation, organization, and retrieval. Think what it will mean to have a worldwide library system available to you at your home or office. You will be able to read any work published anywhere in the world, in the original language or translated into your own, as it is projected one page at a time on your private viewing screen. Your daily newspaper or business reports could be brought into your living room or office in the same way. As a lawyer, you might subscribe to a special service through which you could obtain in minutes legal information which might ordinarily take many hours or days of research in a legal library. Similar systems could keep doctors, scientists, and educators informed and up to date in their respective professions. For those of us who are sport fans, we could really stay up to date on the scores and our favorite player and team standings.

Computerized and automated controls will not only help industry to give us better quality, less expensive products, but will also determine to a great degree how we use them. For example, in our cities, automobile traffic will be controlled by signals which can determine the flow of traffic at various cross-sections during different hours of the day and thereby regulate it. Of course, there is also the disturbing possibility that if we run through a red light or exceed the speed limit on what we think is an unpatrolled road, we might find at home a violation summons issued on the basis of information which radar, high-speed photography, and a computer in some central auto-registration office have supplied. In fact, rather than a summons, it might be just a notice that we have been fined — and that the fine has already been deducted from our bank account!

What will be the use of money — cash, that is — in such an electronically blessed age? Perhaps we will never handle any at all, and all "money" transactions will be carried out through the manipulation of credit. The world of business and industry has just about reached this point. How long will it be before each one of us conducts all his financial affairs with a single credit card and a single bank account? This may mean that even little Junior may have to be a member of something like the public school's Diners' Club just to buy his lunch milk, or perhaps carry his Dad's life's credit with him down to the candy store to purchase a single lollipop. Should he then learn of the tremendous financial power he controls, he is liable to come home with quite an upset stomach!

There is no doubt that automation also will have a growing effect on our home life — particularly the life of the lady of the house. Some day she may do most of her shopping by two-way television and have her groceries delivered to a portal through which they are automatically separated, checked, and stored. And, while the beds are being automatically made, she may sit at a console where she can "program" a meat loaf or a soufflé to be cooked by her automatic ingredients selectors, mixers, and ovens. Of course such kitchen equipment would always have to function with near perfect reliability in order to avoid some embarrassing moments and some strange meals. It is easy to imagine some of the unappetizing concoctions that might reach the dinner table should "all systems" not be "A-Okay" and "Go" at all times!

Although any look into the future is bound to reveal the importance of electronics, which I have been discussing up to now, let me explore some other aspects of life in the years ahead.

What changes will take place in travel in the coming decades? I have already mentioned something of auto travel. Other travel overland will be by high-speed trains, traveling hundreds of miles an hour on different types of rails, suspension systems, or through vast networks of tunnels. We will have other land transport vehicles not confined to tunnels, rails, or even roads. They will be the ground effects machines, or hovercrafts, which, traveling on cushions of air, will cover almost any type of terrain and skim over water as well as land. For sea cargo, there will be specially designed nuclear-powered ships, capable of years of travel without refueling and remotely controlled to deliver their important cargo to foreign ports.

Air travel of the future will not only be faster and safer but of greater variety. Supersonic aircraft of many sizes will be available, as will vertical take-off planes and very versatile personal, individual flying machines for short trips. Air traffic control will be completely automated. Much work is already being done in this field today. It has also been predicted that the end of this century might see global ballistic transportation — through which we would literally be "shot" to another place on earth in a matter of minutes. However, with the advent of worldwide instant television communication, global travel of this sort might not be so attractive. As a matter of fact, I know quite a few people who might find traveling in a ballistic missile unattractive today — even before the advent of global TV!

Progress in space activities today already forecasts some exciting events in this type of future travel. In the decades ahead, trips to the moon and activities on its surface may be commonplace. Reliable manned and unmanned missions to the planets will probably be under way. These will be made possible by nuclear-powered rockets capable of propelling heavy payloads for journeys taking years. Nuclear energy will also make it possible to establish substantial bases on the moon and planets — supplying extensive power for communications and life-support systems — *our* life-support, that is. Who knows what other life we may have discovered in the universe by then?

Here on earth we will learn a great deal more about human life — about all life — in the decades to come. A more complete understanding of the cell mechanism may help lead us to a cure for cancer and perhaps its prevention by immunization. Work in biochemistry may lead to the creation of simple forms of life in the laboratory. And there is a possibility that, before the end of this century, the elimination of hereditary defects by work directly altering the genes may be accomplished. Even greater control of human characteristics through control of heredity may become possible, leading to very serious questions of human planning.

While I have discussed some advances in medical diagnosis which will be made possible by electronics, I have not covered the use electronics may have within the human body. A field known as bioengineering is beginning to emerge which will have a tremendous effect in the future. Having successfully transplanted some human tissue and organs from one person to another, and also having used synthetic materials successfully in the body, surgeons are now considering the possibility of replacing entire natural organs, damaged or diseased, with artificial counterparts. It is already possible to regulate internally the beat of an abnormal human heart with an electronic device known as a "pacemaker." At present the patient carries the pacemaker itself outside his body. Placing such a device — powered by a tiny radioisotope — within the body is now under consideration. There is also serious thought being given to constructing a totally artificial heart — an electronically controlled organ impervious to disease — which could be placed inside a person's body and serve more reliably the same function as the human heart which was failing him.

What other possibilities are being considered in connection with human life in the future? There is talk of using biochemicals to bring about the growth of new organs and limbs and of drugs to produce personality changes and to raise the level of human intelligence. There is even the speculation that it may some day be possible to enlarge man's intellect by programming information directly from a computer-like machine into his brain. I wonder what a final examination will be like under those circumstances?

Will we some day be able to extend our life span another 30, 40 or even 50 years? Some people seriously think so — and will be working toward such ends. Will we be able to induce in people long-duration comas to permit extended time travel to distant planets? This also may be a possibility for the future.

In the meantime, there will be important work taking place here on earth. By the end of the century the world's food production will have to be more than doubled. This will be accomplished by vast projects making more of the earth's land surface tillable and that which is already under cultivation more productive. Enormous undertakings like these will require vast irrigation projects, soil conservation, and fantastic amounts of chemical fertilizers. In addition to food from our land, great amounts of it will have to be taken from the sea, of which we will have obtained far greater knowledge in the future. We can also look forward some day to the production of completely synthetic food to supplement or possibly replace our natural sustenance.

In the future, we will see more control of our land and even regional weather control. Large-scale, safe, nuclear excavation will aid in the making of canals, tunnels, mountain passes, vast underground storage areas, and reservoirs and flood-control systems. Weather will be more accurately predicted through the use of a system of worldwide observation satellites. It has even been predicted that, with the aid of these satellites, we will be able to secure the information needed to prevent the formation of dangerous hurricanes.

As you can imagine, the demand for power, particularly in the form of electricity, will be enormous in the years ahead. Our consumption of electricity is already at a point where it is doubling every ten years. Meeting this demand will require increasing use of our fossil fuels — coal, gas, and oil — and of nuclear energy. Higher efficiencies of conversion to electrical power will probably be obtained by the use of such new systems as Magneto-Hydrodynamics, referred to as MHD. We will probably see better use made of solar energy, and perhaps we will be able to harness the tremendous power of controlled thermonuclear energy, making it possible to use as our fuel the virtually limitless supply of heavy hydrogen found in common sea water.

In the broadest of terms, what kind of a world will it be? Obviously it will be a highly industralized one and one calling for the utmost in human as well as mechanical reliability. As I have mentioned before, it will be a world in which education will of necessity play a more important role in all our lives. Universal literacy will be a goal, but in addition there will be the need for constant reeducating and training of people all their lives. It will be the only way of keeping up with a world which will be changing so rapidly in so many ways. In spite of all the electronic communications gadgetry with which we will be surrounded, human contact will still be vitally important, particularly in the field of education. Highly trained, good teachers will continue to be essential.

With all this, can we prevent a nuclear war with all its terrible consequences? A world of peace is absolutely necessary if we are to live to see the world of the future that I have just described. I believe this can be achieved with your help and that of your fellow graduates throughout the world. The achievement of peace will require better education for more people, better communications, understanding, and a universally higher standard of living. Our vast efforts toward world peace should include increasing efforts in the field of "peace research" and in the areas of arms limitation and disarmament.

Hopefully, we will continue to see the growth of man's cooperation with his fellow man, of the rise of a "world patriotism" over narrower nationalisms.

I have devoted quite a bit of time reviewing some of the prospects of the future. What I have talked about is not in the realm of science fiction but in the realm of reality. These things not only *can* happen, they may *well* happen. Furthermore, they will happen in a world you can live in, a world in which you will be a participant. But most importantly, these things will come about whether you, as an individual, do or do not contribute. The world before you holds much promise. If you fail to work in it, to take part in it, to fashion it toward what you desire, you will be passed by and eventually become an obsolescent observer — not a productive participant. It is imperative that you realize this. I think you do, and that you will do your part in making it the best of all possible worlds.

Science and Freedom: A Challenging Partnership

Remarks made at Miller & Rhoads Virginia Woman's Forum, Richmond, Virginia, November 4, 1965.

I am pleased to have this opportunity to talk to the Virginia Woman's Forum, a gathering of women who, I know, are vitally interested in the major issues of the day.

While the subject of the partnership of science and freedom does not make daily headlines as such, it does underlie much of the material of today's major discussions and debates. To understand why this is so, and the implications of what is involved, is a first step, and a very important one, in being able to some degree to shape our future and assure a future for our children. Today I would like to spend some time exploring the role of science in our democratic society and the influence of each on the other.

During the past few decades, and particularly since the end of World War II, we in America have been experiencing a revolution which has had, and will continue to have in far greater proportions, as important an effect as any revolution this country has experienced before. We have been through the American Revolution — our War of Independence — and we have undergone the Industrial Revolution of the 1800's. Both have altered the course of our history and produced changes in our way of life. But today we are undergoing a Third Revolution — a scientific and technological revolution — the significance of which will stand out in history with our previous revolutions.

If the War of Independence gave us our political freedom — our "freedom of spirit" — the Industrial Revolution, which took place in the 1800's, was the beginning of our great efforts toward more physical freedom. Human incentive, so easily nurtured in a democracy, quickly took up the tools of the Industrial Revolution — the power machines, the ideas of the factory system and mass production — to struggle for "freedom from want" and to fill the material needs of a rapidly growing country.

During this period of material growth science was not asleep. Scientific thinking and investigation were thriving and the body of scientific knowledge was growing, principally in Europe, but also in America. However, this scientific activity for the most part remained separated from technological achievement. The Industrial Revolution was primarily the age of the inventor and the machine. Science made progress, but it was "scientific progress" and still apart from the mainstream of that which was moving the nation forward economically and socially.

When did this situation change? When did the true marriage of science and technology take place? And when did it bring forth those industrially influential and seemingly inseparable twins, Research and Development? While there may have been some flirtations between science and industry as early as the 1880's, the practical economics of the business community acted as a chaperon which kept these two apart.

At the turn of the century, however, some changes began to take place which made the laboratory scientist seem more attractive to industry.

In the 1890's, a Belgium chemist, Leo H. Baekeland, was attracted to America. Shortly after his arrival, Baekeland scored two important industrial successes in this country which were to have a significant effect upon the chemical industry. First, he created a new process for reproducing photographs on paper without the use of sunlight. He sold this process, called Velox, to George Eastman for a substantial amount of money. Baekeland's second success was the creation of one of this country's first "synthetic" materials — a plastic which was somewhat of a sensation when it was introduced to the public. Some of you may recognize its name — "Bakelite."

This remarkable synthetic proved quite profitable and made industry take a new look at the laboratory chemist. But it was not until after World War I that the chemical industry turned to the scientist on a large scale. At this point industry, in need of new development, set up new laboratories and turning to the chemists enticed them away from the college campuses in large numbers. Before World War I an American chemist with a Ph. D. was rarely seen working outside a university, even though the American Chemical Society had 7,000 members in 1914. But after the war, chemical "engineering" became a widespread profession in America, and a Ph. D. was not much more than a minimum requirement for an industrial position. Furthermore, the chemists produced results for industry. From the DuPont Company, for example, there were the coal tar dyes and chemicals of 1918. The year 1920 saw viscose rayon yarn produced. In 1923 the public was introduced to a "miraculous" transparent substance called "cellophane." The next year synthetic ammonia was produced, and in 1927 an organic chemistry professor, Dr Wallace H. Carothers, hired away from Harvard by DuPont, began work which soon led to the creation of a new yarn called "nylon."

But while the scientist made inroads in the chemical industry, and produced results important to industry and the country, chemistry was not the only field of growing scientific and technological progress. Electricity, and later, electronics, began to open up an area of progress for the physicist. It was not until 1909, however, that a major industrial firm hired a physicist, set him up in a laboratory and permitted him to do "basic research." This was the year that the General Electric Company recruited Irving Langmuir, a promising young physicist of the day, later to become a Nobel Prize winner. In 1916 the Westinghouse Laboratory also hired a Ph. D. who was to become a world-renowned scientist, Arthur H. Compton. Other major companies in the electrical and chemical industries followed suit, set up laboratories for both applied and basic research, and hired scientists during the period between the two world wars.

But the 1920's and 30's provided only a hint of what was to come. It was during the Second World War that the enormous value of electronics was forecast by the use of radar, that the computer was born and adapted to practical use, and that the seeds of the space age were sown in the development of military rockets. It was also during this era when the greatest single example of scientific power — the release of nuclear energy — was demonstrated, an event which has since produced a profound change in man's relationship with nature and his fellowman.

These scientific innovations, their wide and rapid development and applications during the ensuing twenty years, and the general adoption of science by industry and government have propelled us into our Third Revolution — the Scientific Revolution. And make no mistake about it — it is a revolution in every sense of the word, one characterized by rapid and far-reaching change, by change which has had and will increasingly continue to produce widespread social upheaval in our country and around the world.

The events of the last 20 years have removed science from the periphery of society. In these two momentous decades, science has become not only a regular generative activity of our society, it is now the central one, and it is this fact that constitutes our Third Revolution. Today we have whole industries whose products derive from a foundation of discoveries made within the last twenty years. Transportation, agriculture and medicine have been drastically modified during the same period. The means for our defense are essentially new. We are delving into new environments — space and the ocean's depths — and seeking to deal with nature on a different and more daring scale as a result of our scientific work. The expansion of scientific knowledge spreads a pervasive influence through industry, our economy, our social structure and our political system. Ours is a new way of life based upon the explosion of scientific information. And there is no turning back from the Scientific Revolution. The survival and effective progress of this new way of life depend upon continuing enlargement of the reservoir of basic knowledge. They also depend upon how effectively and to what end we apply this newly gained knowledge.

But human progress is inevitably accompanied by new human problems, and the rate at which our Third Revolution is propelling us forward, creating rapid and widespread change, presents us with many serious challenges. Among the most serious of these are: the explosive growth of world population due partially to our lower infant mortality and increased lifespan as a result of medical progress; the threat of technological unemployment as a result of automation, and its accompanying problem of reeducating and training a vast segment of our population toward new employment (and eventually toward the satisfying and productive use of more leisure time); the control of our natural resources, with particular attention to preventing the pollution of our water and atmosphere resulting from the growth of industry, transportation and even the blessings of modern sewerage; and finally the question of whether we can prevent the tremendous new forces at man's disposal from being used destructively — whether we can create an era of world peace long enough to allow for the disparities between the living standards of the world's peoples to be corrected — one step which would probably lead to lasting peace and the continued well-being of man.

There is no doubting the enormity of these challenges, but there is also no reason to believe that we cannot face these problems and with time, patience and perseverance, accomplish our ends. The answer as to whether we can do all this, or even an essential part of it, depends on the answer to still another all-important question: Given the knowledge and power of today's science and technology, how well can we in a free society control and intelligently direct the forces of our Third Revolution toward the accomplishing of major goals? Is our democratic society compatible with the scientific age it has spawned, and to what extent can we maintain our freedom and individuality in the swift current of change and growth brought about by our new revolution? There are those within our own society, as well as those who challenge us from without, who have their doubts as to how much freedom we can maintain as our civilization grows more complex.

I believe that a scientific society and a democratic one are not only compatible, but that the partnership of science and freedom offers us the best possibility of achieving the kind of world which most men seek today. I also believe that if free men want to live in a modern scientific age, enjoy its benefits and control its destiny, they must grow with that age, they must educate themselves to higher and higher levels and they must develop their capacity, through education, for the highest rational, humane and ethical conduct. Of necessity, I think we will see the Third Revolution forging a new breed of men and women in the crucial years ahead.

Perhaps it is worthwhile at this point to take some time to examine briefly the relationship between freedom and science, in order to have a clearer understanding of what will be required of us to make a scientific democracy work and flourish.

To a great extent science thrives on freedom. Human curiosity, the search for new truths, and the applications of new ideas seem to develop best in an environment which is not restrictive and where there is incentive and reward for good work. But while the progress of science is often greatly spurred on by the monumental achievements of a few creative giants, it is also moved forward by the relentless efforts of a multitude of others who work with and build around the new theories and discoveries. Today, more than ever before, the interdependence of the scientific world is obvious. The tremendous growth of scientific organizations — numbering at this moment in the thousands — is a result of the need for the exchange of information within and among the growing number of scientific disciplines.

Although science today is still nurtured by the freedom of individual scientific thought, it is clear that the complexities of modern science and technology have created not just more individual scientists and engineers but a scientific community which is both responsive and responsible — responsive to its own needs and responsible to industry and the general public supplying the means to fulfill those needs.

In many respects the nature of freedom parallels the nature of science. With the growth of our complex civilization the interdependence of free citizens has become a more obvious necessity. One has only to pause and reflect for a moment on the number of individuals and organizations involved in supplying the necessities of life to one's self and one's community today to realize how interdependent modern civilization has made us. A little further reflection will make one also realize how well our individual freedom is supported by a complex structure of individual responsibilities. And here, of course, I am referring to not just our political freedom but our freedom from want, our freedom of movement, of information and of action in general. Such reflection makes it clear that freedom demands responsibility — responsible leadership and a responsible citizenry — intelligent, educated, farsighted and alert, and adaptive to change.

Today in particular, when technological and economic change is so rapid and far-reaching, when the growth of our population impinges upon us in so many ways, the characteristics of farsightedness and adaptability are of utmost importance. In a statement made this year, President Johnson pointed out how aware he has become, since entering the White House, of the importance of planning. As the nation's leading executive he recognizes the urgency of looking, thinking and working ahead, of trying to anticipate and plan for conditions which will inevitably arise. Because of our ability to gather and analyze information today, we are better able to predict and forecast events and changes in our population, our industry and our environment, and therefore to make plans and take some action in advance to control forces which otherwise might overwhelm us.

But while recognizing change and pointing out and anticipating problems is one thing, controlling the course of events and solving problems is another. This is the area wherein the combination of science and freedom meets its greatest challenge. At this point the two can either work together in partnership or at odds with each other. As citizens of a democracy caught up in the scientific revolution, we can either be swept along in a rapid tide of change eventually losing much of our freedom and individuality through technological changes which take place without our knowledge or direction — or we can take an active part in the revolution and to some extent direct the forces of modern science and technology, using them to shape our society along the lines we wish. Needless to say, the latter takes great effort.

The success of the scientific–democratic partnership depends primarily on education — education that can fulfill three major objectives: turn out students who can continue to expand the flow of new ideas and production necessary to meet the needs of a growing population and rising economy; create men and women of foresight and wisdom for leadership in government and private enterprise; and, in general, establish an enlightened citizenry who can understand most of what is taking place in their rapidly changing world, make intelligent decisions on important matters and participate in guiding their government in doing likewise.

The success of maintaining a democratic society with a maximum of individual freedom in the years to come will depend heavily on how well the last of these three objectives is fulfilled.

A major dilemma of our day is that modern science and technology is placing at man's disposal an enormous range of power. As this power grows, man's margin for error in using it diminishes. On of the best examples of this I can offer is our use of nuclear energy. Here is a source of tremendous power, the greatest force man has yet wrought from nature, which can be used to light our cities, warm or cool our homes, provide us with fresh water from the sea and a multitude of other benefits, or can be used to virtually destroy all life on earth.

Other vital decisions large and small are forthcoming which will affect our future directly and indirectly. They involve such matters as the education of our children, the kind of cities and housing we will have, our systems of transportation, our jobs, our health, the water we drink, the air we breathe, the beauty of our countryside, and in general the degree to which we can lead a happy and fulfilling life.

In our highly complex civilization we naturally delegate the responsibility and decision-making for a great many things to the "experts." But if we wish to heed the words of Thomas Jefferson — "There is no safe depository of the ultimate power of society but in the people themselves" — we must all strive to be more of the "expert" ourselves, at least to seek as broad a knowledge as possible of those scientific–technological events and practices which will affect us all.

The effects of the Third Revolution extend far beyond our own national borders. They are producing dramatic changes around the world — changes the likes of which history has never before seen. These changes and the social upheavals they produce in turn have a direct bearing on our lives here. Modern communication, transportation and trade have spread to all countries of the world the idea that the scientific age offers them the possibility of rapidly achieving a better life. The belief in the power of science and and the power of freedom — the desire to shape one's own destiny — have produced "the revolution of rising expectation" in the many new and developing nations of the world. We who have been through two revolutions and are in the midst of a third cannot ignore the progress of these evolving potential democracies. Offering them our understanding and support has become a social, political and moral necessity of our time. In assisting these emerging nations in combining the forces of democratic ideals and modern technology, we hasten the day when they can gain the standard of living and political stability necessary around the world for a lasting peace.

The challenge of putting science and freedom to work for all men of good will has been thrust upon us — the citizens of the free world — in this Third Revolution. I believe that we are up to that challenge, and that history will some day show that the partnership of these two forces did prevail in bringing a new era of well-being to those who were wise enough to strive for it in a peaceful and productive manner.

Science: Its Truth and Consequences

Remarks made at the Western Regional Meeting of the American Chemical Society, Los Angeles, California, November 19, 1965.

I hope that those of you here this evening who are expecting to hear me talk about some aspect of chemistry will not be too disappointed if I cover another subject — one which is frequently discussed today and probably will be for some time to come — namely, the growing and somewhat controversial role of the scientist in society. There has appeared in the last year or so a rash of books and articles relating to today's scientist and his place in our culture.

The titles of some of these works are: *Science and Society*; *Science and Culture*; *The New Priesthood*; *Where Science and Politics Meet*; *The Scientific Estate*; *Science, Technology and Human Values*; *The Rise of Science in Relation to Society*; *Science in the Cause of Man*; *Science, The Glorious Entertainment*; and there are so many more that it would take most of my time this evening just to mention the titles.

No doubt many of you have read some of these books, but for those of you who haven't, before you begin to feel too flattered about becoming subjects of the literary scene, let me warn you that unfortunately none of these books has yet made the best seller list and in many of them you are not exactly heroes. Nevertheless, this literature, which many refer to as "the social science of science," is timely and provocative, and I think it indicates that an abundance of serious thought is being given to the scientist's position in society at a time when he is playing his greatest and most effective role in human affairs.

The topics which these books and articles cover are, in part, the responsibility of the scientist in public affairs, his role in government, the conflict between his teaching and his research, how much of his research should be basic and how much applied, how much should be spent on either, and his impact on the economy, the military, the morality and even the art of the country. This may seem like a potpourri of subjects but they are all related to a major theme I hope to follow — that the consequences of science have become, and will continue to become, so enormous that new roles and new responsibilities will necessarily be thrust upon the scientist by society. Today I would like to project a few thoughts on this subject — perhaps raise a few questions and a few eyebrows, but hopefully no one's ire or blood pressure.

Let us begin with some fundamental thinking about the nature of science in order to understand why it has arrived at its present position of importance. If one looks back into history, one sees that science was initially an outgrowth of another high calling of man — philosophy. Philosophy — "the love of knowledge," "the search for truth" — as practised by the ancient Greeks, held the beginnings of Western Science. Although man had previously speculated for thousands of years on the nature of his world, it was the logic and natural philosophy of the Greeks beginning in the fifth century B. C., and more notably the later work of such men as Socrates, Aristotle, Plato — and perhaps we should include the mathematics of Euclid — which began the organization of Western man's thinking and laid the groundwork for what was to evolve into science as we know it today.

The philosopher of those times was held in esteem, but he believed that for the most part the "scientific truth" as he knew it had little direct bearing on the daily lives of the citizens, except to elevate the mind of the philosopher or the student for such time as he devoted it to philosophy. In fact, practical knowledge was looked down upon by most educated Greeks, who felt that the highest type of knowledge was that arrived at by abstract thinking. Perhaps we might say that one of the earliest — and one of the shortest — arguments concerning the value of basic research as opposed to applied research was the exchange between Plato and one of his students who inquired of a mathematical proposition, "What is the use of all this?" Plato had one of his slaves give the boy a coin, and then, saying, "Now you need not feel your instruction has been entirely to no purpose," expelled him.

I must say that I do not recommend that we follow such a procedure in our schools today in dealing with skeptical students. If we did, it would not only cost the school systems and the taxpayers a great deal of money, but we would probably lose a great many good students. Furthermore, with a few thousand years of hindsight, I am going to say that I feel Plato was quite wrong in his thinking on the lack of practical value of Greek philosophy. And today, I would like to present this line of thought: that almost all truth and the manner in which it is arrived at, no matter how abstract at the time, have a "practical" value and consequences, even if they are not immediately recognized or used. While it is true that the science of early Greece did not physically change Greek civilization or even have much direct effect on mankind for the next thousand years, it was not forgotten. This body of knowledge and way of thinking was revived in the 13th century to influence the work of Thomas Aquinas in theology and Roger Bacon in natural philosophy, and still later Copernicus and Galileo, who put Bacon's type of inductive reasoning to work and established the basis for our experimental sciences.

This is not to say that there was no applied science practiced in all the time between the ancient world and the time of Galileo. Some of our first sciences actually developed out of a practical need. Mathematics, for example, developed as far back as early Babylonian and Egyptian times, served in many areas of practical measurement, including surveying, construction and even the determination of planting seasons in agriculture. Early astronomy, limited as it was, had some use in navigation. However, in the history of science a far more significant use was to be made of both mathematics and the early theories of astronomy by Galileo as he put them to what we might call the "practical use" of contradicting the science of his predecessors. What I am driving at here is that perhaps the major consequences of Aristotle's science and of many of the theories which followed were that they provided Galileo not only with material for thought but also with "absolute truths" to *disprove*. To a great extent such was the course of science for many centuries, particularly in theoretical science, of which most science was comprised in those days. The work of Kepler, Newton, Descartes and many others followed along these lines. The "truth" derived from much of their profound theorizing was to affect — to stimulate and provoke — the thinking of other men of science and thus produce an important chain of consequences which, though it perhaps started in theory only, would eventually have an enormous physical impact on our lives.

But while the body of scientific theory grew and took on new validity over the centuries, it was the advent of experimental science which hastened the day when the average citizen, the man-on-the-street, would recognize the power of science in his daily life. Experimental science was also eventually to become closely allied with technology and thereby to produce applied science as we know it today.

Actually, in alchemy, the forerunner of chemistry, we had one of our first applied sciences. While some may point to the alchemist's rather unscientific approach to his work, or decry his basic motivation, it is true that alchemy did give us, in addition to a number of important chemical compounds, the discovery of a number of chemical elements, including arsenic, bismuth, antimony and phosphorus, all of which have since assumed their lawful places in Mendeleev's table. We are also indebted to the alchemists for their work in developing and improving some of the utensils and equipment, procedures and techniques that are still employed in laboratory practice.

If one is to look into and examine the history of science and technology, studying the development of theory and the course of applied science, an interesting conclusion can be drawn: that the time between the birth of a scientific theory and the physical application of some aspect of that theory is rapidly diminishing. We might say that as we are experiencing an explosion and amplification of knowledge, we are experiencing an "implosion" or compression of its consequences, in the sense that while knowledge is being disseminated more widely and more effectively, it is also being applied much more rapidly. (As examples of this effect I might cite briefly the time differential between the invention and widespread application of the steam engine and the transistor. The one took many decades to create the Industrial Revolution, the other produced its own revolution in a matter of years. This, I think, is indicative of the exponential growth and effect of modern science and technology.)

This also presents us with the significant point that when a scientist comes up with an important theory today he is more likely to see it either destroyed or used fruitfully within his lifetime. If it involves something which can have a beneficial effect on man he will probably witness the good he has produced and in all likelihood receive rewards for it. If it involves something which can result in good or ill for man he faces the possibility of seeing it used either way, even though his intention in discovering and revealing the knowledge was simply the finding and revelation of truth (for centuries the essential, and at times the only, purpose of science).

This brings us to an important question of the day — the one about which so much is being written and discussed: what is the role of science and the scientist in society today? There are many who argue — and effectively so — that science is concerned only with knowledge, not values, that its business is the creation of fact, and as the philosopher Hume has pointed out, "there is no path from fact to value." It follows then that science is neutral — its information neither good nor evil. But today's science, aligned with technology — and to some extent inseparable from it — represents power, power derived from our understanding of and mastery over the forces of nature.

Today's scientific truths and their almost immediately following consequences are surrounding us and pressing in on us from many sides. More and more we are beginning to see the world as a closed system where the amplification of our technological progress produces some startling and often unexpected reverberations. Taking a quick look at some of the situations that science, as allied with modern technology, is creating, what resulting problems are we faced with? There is the major one of the possible proliferation of nuclear weapons as a result of the extension of nuclear energy for peaceful purposes, although many important safeguards are in effect today to prevent this, and we in the nuclear field are hopeful that the peaceful benefits of nuclear energy will someday far overshadow its military implications. There is the concern over potential unemployment and the need for retraining in new skills which will result from the increased use of automation in industry. However, there are many who believe, and we hope they are correct, that automation will result in a far greater employment of people — although a higher level of education and training of these people seems inevitable. Man's domination of his environment and his extended use of chemicals — with beneficial intent and immediate beneficial gain — lead to many long-range ecological problems, some of which are just now being brought to our attention. The exploration and the control of space create military implications and the need for international space law. The investigation of new ocean bottom frontiers will present similar complications. Rapidly growing industrialization confronts us with a major challenge in controlling the pollution of the environment. (And as you here in Los Angeles well know, the increased use of the internal combustion engine adds to this problem.) Rising concentrations of population and expanding cities, indirect results of our scientific age, are creating new economic and social problem, perhaps faster than we can cope with them — employment, health, education, crime, and even water shortages where water was always considered to be abundant. In addition to these current problems we may soon face those involving the control of human heredity and personality, as the biological and chemical sciences bring us closer to those secrets which, once revealed, may allow us to alter and create life itself.

Basically, the findings of science may be value-free, but the impetus of science, particularly in its alliance with technology, is derived from human values as well as from man's basic curiosity, and while we can talk of science being neutral, it is impossible today to act as if it were such. *Most of today's scientific truths do have consequences — and they are consequences we must be prepared to face. Furthermore, as I have pointed out, most of today's scientists will be alive to see their theoretical and experimental work applied in some manner or other. And whether or not they claim neutrality, the world is going to turn to them for some understanding and guidance in using their work.* This will probably happen even if the individual scientist responsible for a discovery is not the best equipped to direct the use of his discovery.

There are many implications involved in this line of thinking. For example: does this mean that every scientist must weigh the possible consequences of his work before he undertakes it? This sort of suggestion has been made but we all know this is impossible. Such thinking would stifle scientific creativity, and furthermore no one can possibly know all the ramifications of a piece of work before he undertakes it. There is also the argument that science must be carefully watched, its work restricted and controlled so that it will not release any information that civilization is not yet prepared to assimilate without danger. Of course, there is secrecy and restriction in science allied with military technology, but we cannot subject all science to this procedure. Such thinking is at once ridiculous and dangerous. Science could not operate under such a system — even assuming that such controls could be set up. Besides, who would be the watchdog of science? Who would determine which scientific accomplishments would bode good or ill for man?

Another extreme in thinking about the role of the scientist is that scientists should not only be responsible for seeing that their work is applied properly but that the scientist himself is necessarily a "supercitizen" in this scientific age and therefore best equipped to direct *all* of society's activities. As flattering as this technocratic argument may be to those here today and to other scientific groups around the country, we all know some of the dangers and problems to which this argument could lead.

But somewhere between these extremes lies some sensible, and I think workable, solution to the question of the scientist's role in society. Much of this solution is in effect today. It has the scientist in the important role of advisor, providing information and counsel on scientific and quasi-scientific matters to the government and private sector of the country. The Executive Branch of the government in particular has seen a tremendous increase in its scientific advisory bodies since 1863 when Lincoln established the National Academy of Sciences. This esteemed body was consulted by the War Department on only five matters between 1863 and 1913. Those matters were: "On the Question of Tests for the Purity of Whiskey; On the Preservation of Paint on Army Knapsacks; On Galvanic Action from Association of Zinc and Iron; On Exploration of the Yellowstone; and On Questions of Meteorological Science and its Applications." Considering the role that science plays in the Department of Defense today, I think it would be safe to say that "times have changed." (For instance, we all know that the Army no longer tests the purity of its whiskey — or does it?)

Today, government science advisory groups include the National Academy of Sciences, the National Research Council, the National Science Foundation and National Science Board, the Office of Science and Technology, the President's Science Advisory Committee, the Federal Council for Science and Technology, the Defense Science Board, the AEC General Advisory Committee, the National Aeronautics and Space Council and a host of others too numerous to mention here. All these organizations, which together make up what has been referred to as the "Science Affairs Community" or even the "Scientific Establishment," perform valuable functions and contribute much toward fostering the growth of science and bringing its accomplishments into the mainstream of the nation's progress. Without them I think we could hardly function today as a scientific democracy. But there is also concern today that the work of these groups alone will not be enough to organize the forces of science to meet many of the large-scale problems looming head — some of which, as I have pointed out, are the result of scientific advances in individual fields.

Such concern has also led to talk of the need for more integrated scientific planning, of the need for more looking and working ahead so as to avoid having to deal with problems on a crisis-to-crisis basis. Too often today we are apt to find ourselves caught up in such situations — situations which may well multiply in the future beyond the point where we will be able to cope with them.

There is controversy over the "pluralism" in scientific affairs and a call for a more integrated approach in the federal science program and by the scientific community as a whole. In an editorial written for *Science* last August, Dael Wolfle went so far as to suggest that the time is now at hand for something akin to "an extensive systems analysis of the problems of saving planet Earth as a pleasant, productive home for mankind ..."

There is concern today over the increasing specialization in science (although its inevitability seems to be recognized) and over the effect that specialization is having in our system of higher education. There is also the vital question of resources — resources which are not unlimited and which present us with problems involving difficult decisions to be made in choosing programs and courses of action.

It is perhaps interesting to note that today our federal budget for research and development alone exceeds our entire governmental budget before Pearl Harbor. If our R and D expenditures were to continue to grow at their current rate, they would perhaps equal our gross national product by the turn of the century. But while funds are limited and we must make choices, we also must move on and we must choose wisely. I therefore agree strongly that better scientific planning, better integrated with our economic and social programs, will be a necessity for the balance of this century and probably for all time to come.

There is no doubt that the current state of science and its relationship to society poses some serious problems and challenges to our democratic way of life — as a matter of fact, to all life — and they are challenges with which we must sooner or later come to grips. Some we have already tackled, others will demand our attention very soon. I think that we have reached the time in history when the power of science — its truth and its consequences — will force a certain maturity on mankind, when there can be no more isolated treatment of facts and values, when we can no longer afford the tolerance of what used to be called — and perhaps still is — "the cultural lag." We must live with science and science must help us live — if we are to live at all as we have come to believe man can and should.

If you have agreed with me up to this point, I think you will agree further that all this places a growing burden and greater responsibility on the scientist — on all of those who work in science and scientific affairs, no matter what their field or capacity. If this seems like too sweeping a statement, let me spend the last few minutes of my talk clarifying it.

Writing in The Fund for the Republic Papers, "Science: Scientists and Politics," Lynn White, Jr. stated:

"Scientists must become increasingly aware of the complexity and intimacy of science's relationships to its total context."

"The cultural support that science enjoys today," White continues, "rests more on fear of foreign enemies and of disease than upon understanding, and fear may not be a healthy or lasting foundation. Science needs its statesmen, and statesmanship demands the long view."

I find that I agree with White on both these vital points, and I believe that accepting his view imposes two important burdens on all scientists if they are to accept the responsibilities thrust upon them by our scientific age. First, scientists must attempt to broaden their horizons both scientifically and in the affairs of society. I know this is easy to say, and I realize as much as any of you the time and effort that must be expended in pursuing a scientific speciality these days, but the intelligence and training, the skill and talent, of the scientific minds in this country must reach beyond the laboratory and research center and add their rationality and intellectual vigor to the wider efforts of the scientific community and the community of man. This is not to suggest that every scientist should neglect his specialty for broader fields. The state of science demands the diligent pursuit of highly specialized knowledge today more than ever before, but when the scientist who has pursued a single discipline leaves his work at the end of the day, even he must enter the interdisciplinary world in which we all live and try to contribute to it.

Secondly, the scientist today must become more articulate within the scientific community and as a citizen if he is to assume any leadership or influence the course of human events which could result from his scientific work. On many occasions I have spoken to lay audiences and urged them as citizens living in a scientific age to learn as much about science as they could, so that they might be better equipped to be active participants in a scientifically oriented society and thus help keep science and democracy compatible. Tonight the shoe is on the other foot, and on behalf of all those audiences I have urged to understand science, I call on you to make science more understandable to them. I think much can be done toward this end, and the significance of such an effort cannot be overestimated. No one knows better than the members of the Atomic Energy Commission the importance of public understanding and its subsequent counterpart — public support.

There is no doubt that much good work is being done to explain the workings of science to the layman. Organizations such as the Scientists' Institute for Public Information, the American Association for the Advancement of Science, and branches of the various scientific societies — particularly the American Chemical Society — deserve much credit for their effective programs to promote a better understanding of science. There are many capable science writers at work today who are turning out excellent books and articles successfully interpreting science to the man-on-the-street, the housewife and the teenager. The efforts of these writers are receiving acclaim through such distinguished awards as the International Kalinga Prize for the Popularization of Science awarded by UNESCO, the Ettinger Award sponsored by the Rockefeller Institute, the University of Pennsylvania and New York University, and, of course, the ACS's important James T. Grady Award, which this year went to Dr. Isaac Asimov.

Some excellent work is being done in education to teach science and scientific concepts even in the early primary grades, and good science films and educational television programs are being produced and shown every day. But more needs to be done, and as scientists who will benefit, and will see our country benefit, by a wider public understanding of our work, its requirements and its results, it behooves us to promote that understanding.

Today there is an increasing tendency on the part of many interpreting science to emphasize the spectacular. This is easily explained. In an age when our eyes and ears are constantly flooded with information designed to attract our attention, the spectacular becomes almost a necessary device for motivating an interest. But in bringing science to the public, I question whether or not some of the emphasis on the spectacular is due to the fact that many of us have lost the sense of wonder and beauty about many of the basic truths of science or cannot communicate much of the excitement or challenge in our own work, not to mention, our inability to explain to the layman, in a clear and understandable way, what our work is all about. If we are willing to make the effort, I think we can go a long way in correcting this situation and in educating the citizen in science. In doing this we will be contributing much toward that which will in turn help us to move science and society along in a more workable partnership.

Towards the beginning of this talk I mentioned the fact that I was going to project a few thoughts and raise a few questions. I hope that these have balanced each other out during the course of what I had to say this evening. But the chances are that there are a great many more questions to be asked and answered about the role of the scientist today. Many more books and articles are going to be written on this subject and many more discussions will take place. We all know that there is no one answer to the question of the scientist's role any more than there is a simple answer to the question of where science is taking us. If science has taught us anything throughout the centuries, it is that new answers bring up new questions and thus the search for truth goes on.

But perhaps in the march of science today there is one truth which we must recognize as a consequence of all science, simply that we have entered a new era — an era in which man can survive only if he realizes the power of science and the fact that in the light of "neutral, amoral science" we cannot remain neutral but are forced to a new level of rationality and morality. If science provides any one truth in the years, the decades, to come, it may well be whether or not men can learn to live together in peace and harmony with each other and with those forces of nature which men are beginning to understand and control. I believe they can, but the answer to such a question is too important to be left to chance. We must all work toward proving that one of the consequences of science is truly the best of all possible worlds.

The Legacy of Alfred Nobel

Remarks made to the Washington chapter of the American–Scandinavian Foundation, at the Smithsonian Institution, Washington, D.C., January 7, 1966.

As some of you here probably know, and as I am happy to tell the rest of you, I am quite proud of my Scandinavian ancestry and pleased at the opportunity to talk to members of the American–Scandinavian Foundation and their guests. I am also pleased that when you invited me to speak, it was suggested that the subject be Alfred Nobel. Nobel was such a remarkable person that the more one reads about the man, the more fascinated one becomes with the story of his life, his thoughts, his accomplishments and, ultimately, his influence on humanity.

Time will not permit me to go into great detail concerning Alfred Nobel's life — a life so filled with events that no single biography has yet been written to do it justice — but I would like to touch on a few highlights of his life before discussing some of his ideas and the legacy he left us.

Alfred Nobel was the third of four sons, of a self-trained Swedish architect and inventor, Immanuel Nobel, and a Swedish woman of hard-working peasant stock, Caroline Andriette Ahlsell. Born in 1833, at a time of family poverty due to serious business reverses, Alfred was sickly at birth and remained in poor health through most of his life, although he was to outlive his three robust brothers. As he was to later describe his infancy in an autobiographical poem — "My cradle looked a death-bed, and for years a mother watched with ever anxious care, so little chance,

Reprinted with permission from *Chemistry*,
Vol. 41, No. 1, pp. 22, 24–25, January, 1968.

to save the flickering light." But a mother's tender care in his childhood, and her continued devotion in later years, sustained Alfred through many trying times.

In order to avoid debtor's prison, Immanuel Nobel had to leave Sweden and his family for a number of years. It was not until 1842, when Alfred was nine, that he, with his mother and brother, was reunited with his father who had established himself in the capital of Czarist Russia, St. Petersburg, as part owner of an engineering workshop. Also it was at this time that Alfred began to receive his education, mostly from Russian and Swedish tutors, when he started to learn five languages in which he was to become fluent, when his interest in science — particularly chemistry — was born, and when he first became fascinated with an interest of his father's which was to shape his life and affect the lives of countless others. The interest to which I refer was in the highly dangerous oil — nitroglycerine. His father had become attracted to this powerful but unreliable explosive because of his work on military mines and sought to develop this material, which was introduced to him through the work of the Italian chemist Sobrero.

Only years later, in 1863, did Alfred develop a method of safely detonating nitroglycerine. By 1865 he had what he called his "Initial Ignition Principle" patented, which was to become the major technique for all future explosive detonation. By 1867 he had invented and was manufacturing dynamite and was on his way to becoming one of the richest men in Europe. By 1875 he had invented Blasting Gelatin, later called "the most perfect explosive known." By 1880 he had invented Ballistite, a smokeless explosive powder, and established an industrial empire with factories and major holdings in over a dozen countries around the world. At his death, in 1896, he held 355 patents in various countries, was one of the wealthiest men in the world, and had given mankind not only inventions of great importance but left a legacy devoted to inspiring and elevating man to far greater accomplishments and ideals than he felt he had achieved in his lifetime.

If I stopped at this point, we might have one of the shortest biographies of Alfred Nobel on record, but I have no intention of stopping here. There are far too many interesting facets of Nobel's life for me to explore a large fraction of them tonight, but let me examine some of the more striking ones.

First, the accomplishments for which he is best known and through which he made his fortune are those in the field of chemical explosives. As I mentioned before, Nobel early in life developed a fascination for the novel substance, nitroglycerine. At the time he was introduced to nitroglycerine, it was an almost useless explosive mixture since its nature was so poorly understood and it was so dangerous to handle. But with remarkable patience and a relentless perseverance, Nobel devoted over fifty years of his life to developing various useful forms of this substance and techniques for using them, and to manufacturing and distributing them all over the world. His first, and to some extent his most important, innovation was the development of a detonator which could reliably explode the "blasting oil," as nitroglycerine was called. This detonation technique not only made nitroglycerine practical to use but was to serve further as the means for setting off almost all future explosives. (I might add that the same basic principle has been adapted to systems which detonate nuclear explosives today.)

But though Nobel made nitroglycerine useful through the invention of his detonators, it still proved to be a highly dangerous material. A worldwide series of tragic accidents in its use, one killing Nobel's own younger brother, drove the chemist to develop a safer explosive. That explosive was Dynamite, which Nobel created by mixing the nitroglycerine with an absorbent claylike material, kieselguhr (diatomite). Dynamite was an immediate success; the demand for it was so great that Nobel was able to manufacture and sell it all over the world. He had developed it at an ideal time — the height of the Industrial Revolution — when railroads were spreading across the continents of Europe and America, necessitating the excavation of tunnels and mountain passes, when harbors, canals and riverways needed to be built, deepened and widened, and when mining was growing to provide fuels and materials for the new methods of transportation and new ways of life. Some of the monumental engineering projects for which Dynamite was used in those years were: the tunnels on the St. Gotthard line (through the Alps), the blasting out of underwater rock at Hellgate in New York's East River, the clearing of the Danube at the Iron Gates, and the 295-foot deep cut, nearly four miles long, of the Corinth Canal in Greece.

In addition to its many uses in excavation and mining operations in those early days, Dynamite was used in other ways — some of them bizarre, but oddly enough, apparently successful. In his biography of Alfred Nobel, Erik Bergengren recalls that, "In New York, a continued drought had caused a scarcity of water in the city supply from the Croton Aqueduct (sounds familiar, doesn't it?); a balloon was sent up with 200 pounds of dynamite which was exploded with tremendous force, the explosion being followed by a heavy downpour of rain." Bergengren points out that a similar experiment with similarly good results was also carried out in Texas. (I call these events to your attention only because they are interesting, not because I am recommending that Governors Rockefeller and Connally use these methods to solve any water problems they might have today.)

With the worldwide acceptance of Dynamite and with a great deal of business acumen, Alfred Nobel was able to establish factories and control the manufacture of explosives in Sweden, Finland, Norway, Germany, France, America, Australia, Hungary, Scotland, Switzerland, Italy, Spain, and Portugal. And all of this was accomplished by the time he was forty!

But even the success of Dynamite did not stop Nobel's fascination with nitroglycerine or his desire to use it even more effectively. His continued work with this material led to his development of Blasting Gelatin, a safe explosive more powerful than Dynamite. His further work in explosives brought forth even another product, Ballistite, a "smokeless progressive powder."

Although it is true that Alfred Nobel made his fortune in explosives, and probably devoted the major part of his life to them, it is important to note his interest and work with a variety of other scientific and technical ideas. He made considerable effort towards creating synthetic leather and rubber. He invented a method of producing the filament for an artificial silk, similar to the rayon developed years later. He worked on the development of rockets and, using his rockets and parachutes, conceived and tested probably the first system of aerial photography. As an outgrowth of his work with nitrocellulose, he developed important varnishes. He experimented on the improvement of the phonograph, the telephone, the electric battery and incandescent lamp parts. In trying to improve Edison's inventions, he was probably more successful than Edison in trying to improve nitroglycerine, since the story goes that Edison and a co-worker who became interested in Nobel's Blasting Oil were so alarmed by its explosive power that they disposed of their only bottle of it in the New York sewer system!

In addition to his own accomplishments as an inventor and industrialist, Alfred Nobel was instrumental in helping his brothers, Ludvig and Robert, to develop the Nobel Brothers' Naphtha Company at Baku, which at the turn of the century was producing more oil than any other place in the world.

Rather than dwell further on Nobel's accomplishments, let us turn to an examination of Nobel the man — his personality, his thoughts, and those ideals which led to his famous will.

It has been said that Alfred Nobel was one of the loneliest men on earth, one of the richest men with the fewest friends, and a man "that nobody knew." These claims are highly exaggerated, but there are reasons why they are made. His poor health early in life, his extreme intelligence and sensitivity and the long periods of separation from his family, did contribute to an introverted personality. For many years he traveled widely and knew no one place as home. The fact that he never married and raised a family of his own made him acutely aware of the joys of family life. Those who emphasize his loneliness usually refer to a letter he wrote to his sister-in-law after returning to his own empty house from a visit with his brother Ludvig's family. The letter, obviously written at a low emotional ebb, showed — as did all of Nobel's correspondence — great sensitivity and perception. In it he speaks of the contrast between their homes, the warmth, love, joy and bustle present in Ludvig's and missing in his own. He demeans himself as "a wreck on the sea of life" with no bright memories of past years and no false but beautiful illusions of the future.

But this isolated example of self-pity does not do justice to Nobel. Though he had moments of loneliness, he was far from being a recluse, and when he was not engrossed in his work in his laboratory, most of his life was a stream of activity and full of contact with all types of people, many of whom got to know him quite well, and a number of whom became his close associates and good friends.

Those who knew him closely tell of the personal warmth and humanity which emerged after his initial shyness and reticence had vanished. While he did not seek society and avoided personal publicity, he was not averse to entertaining his good friends in a royal manner. At these parties he was a brilliant, witty conversationalist and a perfect host.

If he expressed mistrust of people at times it was only because he had trusted too many too much in his early years and had been bitterly disappointed on several occasions. Though at times human folly made him doubt the wisdom of mankind, he was not a misanthrope. To the contrary, he was a great humanist and humanitarian — as was clearly indicated by his ideas covering religion, government, labor, literature, and war and peace.

We might briefly look at a few of these ideas in order to reveal more of Nobel, the man.

His religious beliefs centered more on ethics than on dogma, and he was not ashamed of this. He once wrote Nathan Soderblom, pastor of Swedish Church in Paris and later Archbishop of Sweden: "Our religious views differ possibly more in form than in substance, for we are both agreed that we should treat our neighbor as we would be treated by him." (This difference in religious views, incidentally, did not prevent his contributing most generously to Soderblom's church — far more generously than the pastor expected.)

Nobel's adherence to the Golden Rule and his basic humanism carried over into his social thinking. It was quite evident in his relationship with his workers. He treated them with respect and generosity. He praised their good workmanship, and he refused to propagandize them to counteract the agitation against industrialists which was taking place at the time. His actions spoke louder than any leaflets could. In his factories he had introduced workers' bonuses, free doctors and medicine, and a retirement system. Although he ruled a vast industrial empire, no strike ever took place in any of the plants for which he was responsible.

It is interesting to note Alfred Nobel's prophecy of future social benefits. In 1879 he wrote: "Some day, when the world is really civilized, those who cannot yet work — the children — and those who cannot work any longer — the old — will no doubt be entitled to a general state pension. This would only be fair, and to carry the scheme into effect would surely be easier than people think."

Alfred Nobel saw great hope for the world — but not without hard work. In this vein he emphasized the need for education and knowledge as means of achieving peace and prosperity. "To disseminate enlightenment," he wrote, "is to disseminate prosperity — general prosperity and not individual riches, and," he continues, "with prosperity disappears the greater part of the evil which is our heritage from darker times." The same writing goes on to point out his belief that scientific research would play a most important role in man's future. (And here I must remind you that Nobel's emphasis on science and his legacy to science by both his own work, and the prizes I will describe later, were made in an era well before the beginning of the scientific explosion we have been experiencing in the past few decades.)

The interest and dedication which Nobel devoted to science and his business enterprises did not prevent him from having a lively interest in the arts. He had a great love of literature and read and studied it profusely. He liked the "idealistic" writers — the French writers Victor Hugo, Maupassant, Balzac, and Lamartine; the Scandinavians Ibsen, Björnson, Jonas Lie, Garborg, Kielland, Victor Rydberg, and Selma Lagerlöf; and the Russians Gogol, Dostoyevsky, Tolstoy, and Turgenieff. He was particularly fond of poetry — especially the poets Byron and Shelley, the latter greatly influencing his own thoughts and writing.

One of Nobel's accomplishments which affected his interests and work was his mastery of languages. Fluent in Swedish, Russian, French, German, and English, he spoke, wrote correspondence and even translated literature in these languages. As a diversion he translated Voltaire into Swedish and then back into French. Since he could never find a personal secretary who could handle his daily correspondence in the language of the incoming letter, he wrote his own replies, as many as forty handwritten letters a day — in five different languages.

Besides the writing of correspondence, he was for many years engaged in his own literary efforts. He wrote poetry, drafted novels and, in his later years, finished a play based on a classic tragedy. The ideas he presented were superior to his writing talents, although he did show flashes of literary creativity, particularly of wit and satire. The scope of his ideas and thoughts towards the end of his life can best be realized by a consideration of the list of subjects on which he was still planning to write at the time of his death. Under his general heading of "Philosophic Reflections To Be Written" were such titles as "The Interacting Atoms" (he was certainly ahead of his time in this area); "The Function of The Brain, Thought and Memory" (he always had a great interest in biology and medicine); "Penetration of the Various Religions"; "Government Systems Based on New Ideas" and "The Philosophy of Cells and the Cosmos."

Among his most interesting ideas are his thoughts on world peace. He was deeply concerned with this subject. There are some who try to pinpoint his interest in peace to certain events in his lifetime, or look upon it solely as personal reaction to the fact that he had invented a material which could be used destructively, although he had not created it with that in mind. No doubt his long friendship with Baroness (Bertha) von Suttner, who became a leading peace advocate of the time, stimulated his thinking on this subject. They had many discussions and carried on a lively correspondence about the possibilities of abolishing war. But most of Nobel's notions on how this could be done were his own, and he disagreed with the Baroness on many ideas. It is remarkable how many of his thoughts on the subjects of disarmament, peace-keeping, and the worldwide rule of law are applicable today and currently among our best hopes of achieving peace.

What were some of these beliefs? Basically, he believed that it would be some time before the entire human race was mature enough or shared enough material prosperity to live together without any conflict. Meanwhile, he believed that the growing terror of the weapons of war might act as a deterrent which would force men to live together in relative peace until such time as they could create the basis for a real and lasting peace. In this respect he hoped that his explosives — awesome as they were at the time — would be an aid in the cause of peace.

Nobel believed strongly in a worldwide recognition of the rule of law, of the need for a world body to settle differences between nations. He felt, however, that such a world body could only be effective if it had an international peace-keeping force strong enough to enforce the rule of majority. But he also expressed some concern over the control of that "police force," concern over the possibility of its falling into despotic hands.

He seriously advocated a "cooling off" period of one year for nations about to go to war, hoping that during the period they might reach some reasonable agreement over their dispute. And if that one year was not enough to solve all problems, why not a second year of peaceful effort in the right direction? Time and human progress, these were what he hoped might buy peace. Such was the thinking of a man who in his own words "avoided dispute like the plague" and called war "the horror of horrors and the greatest of all crimes."

Having taken a look at Nobel the man, let me turn now to some thoughts of his legacy to us — particularly to the Nobel Prize.

Nobel dedicated his vast fortune to the ideals he cherished. In a simple, powerful passage in his will he wrote:

"The whole of my remaining realizable estate shall be dealt with in the following way:

"The capital shall be invested by my executors in safe securities and shall constitute a fund, the interest on which shall be annually distributed in the form of prizes to those who, during the preceding year, shall have conferred the greatest benefit on mankind. The said interest shall be divided into five equal parts, which shall be apportioned as follows: one part to the person who shall have made the most important discovery or invention within the field of physics; one part to the person who shall have made the most important chemical discovery or improvement; one part to the person who shall have made the most important discovery within the domain of physiology and medicine; one part to the person who shall have produced in the field of literature the most outstanding work of an idealistic tendency; and one part to the person who shall have done the most or the best work for fraternity among nations, for the abolition or reduction of standing armies and for the holding and promotion of peace congresses."

These were the words which were to establish the Nobel Prizes — among the world's most important awards and a continuing memorial to their remarkable founder. The areas selected for reward are certainly fundamental to the well being of mankind, and it is a significant commentary on Nobel's intellectual stature that he was a competent participant or deeply interested in all of these fields. There are the two physical sciences, which feed the technologies upon which material improvement is based; physiology and medicine, the sources of better health and conquest of disease; literature, the mother of learning and culture; and peace, the aspiration of all enlightened men.

Nobel's legacy was idealistic and, in the best sense of the word, visionary. If it were to have meaning, it must be carried out with integrity and devoted effort. Political, national and petty considerations — always a risk in such matters — must be eliminated from the selections. Integrity commensurate with the concept was found in Swedish and Norwegian institutions. The conscientious administration of the trust has been, aside from the conception itself, the foremost reason why the Prizes have had such an important impact on our times.

It is not possible to estimate quantitatively the effect of the Nobel Prize on science and the general welfare of man since the turn of the century. But one can identify some of the ways in which Nobel's legacy exercised definite influence. At the turn of the century, science was on the threshold of exciting new developments, especially in physics, in chemistry, and in biology — the fields selected by Nobel. For example, the investigations that gave rise to nuclear science were just beginning. Certainly, Alfred Nobel's bequest came at a propitious time in scientific history.

Above all else, I believe the Prizes are an encouragement to excellence. Ideally, of course, a scientist strives for excellence in all his work. By the nature of science — the requirement of repeatable, measurable, precise evidence — I believe scientists as a group are able in their work to come as close to the ideal of integrity as any class of men. Scientists, for the most part, are quite human and cherish the recognition of their peers. Few of the members of three generations of scientists have been unaware that the crowning recognition of their peers, the world, and probably of history comes through the award of the Nobel Prize. I should hesitate to guess how much insignificant work has been passed by, how many papers remained unpublished, how many experiments repeated — because of the influence of the standards of excellence of the Nobel selection process.

The Nobel Prizes have enriched scientific research in other ways. They have dramatized the potential of, and stimulated research in, essentially new fields. They have been the instrument of financial and moral support for important new research centers built around Nobel Laureates and populated by brilliant students. The annual selection process forces leading scientists to evaluate the world's work in science, and in this respect the process constitutes an important form of self-examination by the world scientific community. The Nobel Prizes have created heroes as models for young people, and have been an important source of public education in the progress of science. By the impartiality and objectivity with which the Nobel Foundation has acted, the Prizes have promoted the international values of science, stressing the universality of interest, achievement and benefit inherent in science for all men. The Nobel legacy has helped to make Sweden a center of scientific excellence, and brought worldwide honor and admiration for Swedish institutions and the Swedish people. The Nobel Prize then, has influenced, and will continue to influence, the scientific revolution of our times.

While I have been emphasizing the effect of the Nobel Prize on science, we must not overlook its important role in the two other areas in which the prize is offered — literature and peace. For without the human values represented in these areas — without the beauty and truth of art and the hope that men can learn to live together in harmony, trying to alleviate human misery and build a better world for all — without these things what is the purpose of science and where can it lead us? Nobel, a man of science, realized this, and his life and his legacy bore proof of this realization.

We are reminded of the emphasis which Nobel placed on values by his love of "idealistic" literature — writing which would reveal the nobler side of life and uplift the human spirit. In the light of this, he would have been pleased had he been present at the awarding of his prize in 1950 to William Faulkner, when Faulkner in accepting the prize, during a trying era of our time, exhorted young authors to write not out of fear but to return to "the old verities and truths of the heart — the universal truths ..."

In his short but brilliant speech, Faulkner concluded: "I decline to accept the end of man ... I believe that man will not merely endure: He will prevail. He is immortal, not because he alone among creatures has an inexhaustible voice, but because he has a soul, a spirit capable of compassion, sacrifice and endurance. The poet's, the writer's duty is to write about these things. It is his privilege to help man endure by lifting his heart, by reminding him of the courage and honor and hope and pride and compassion and pity and sacrifice which have been the glory of his past. The poet's voice need not merely be the record of man; it can be one of the props, the pillars, to help him endure and prevail." I believe that Alfred Nobel would have agreed with these thoughts.

Finally, I believe that Nobel would have approved wholeheartedly of the decision of the Nobel Foundation to award last year's Peace Prize to UNICEF — the United Nation's Children's Fund. As a man who had no children of his own and so missed, as he often expressed, the joy of their laughter, their warmth, trusting innocence, and their abiding hope in the future, he would have been pleased to aid and honor an organization which is working so hard to better the life of children all over the world — children in whose hands will soon lie the future of mankind.

It has been a privilege to speak to you this evening and to try to convey to you some thoughts on the life of Alfred Nobel, his accomplishments, his ideas, and the heritage he has left us. We of Scandinavian ancestry can take great pride in this man, and I believe that the entire world can be grateful for his life and the legacy which he has left to us — a legacy he hoped would in some small measure make our world a better world, a more fruitful world, and one in which, hopefully, we all could learn to live together in peace and understanding.

Science in a World of Widening Horizons

A lecture delivered under the sponsorship of the U. S. National Commission for UNESCO and the Joint Graduate Consortium of Washington, D. C. Universities at Georgetown University, Washington, D. C., January 17, 1966.

I am pleased to be here with you this evening and to have the opportunity to present one of the five "UNESCO Lectures" being given at our five Washington, area universities. This past year we celebrated the twentieth anniversary of the founding of the United Nations and of its Educational, Scientific and Cultural Organization — UNESCO — created at the same time as its parent organization. The first UN delegates realized, and wisely so, that their political and social goals could best be attained in a world which had established better international contacts in education, science and culture. UNESCO, one of the largest members of the United Nations family, has a roll of 117 nations and two associate members, and a staff made up of nationals from over 90 countries.

While I am not going to speak about the work of UNESCO directly, the topic I am going to cover, "Science in A World of Widening Horizons," makes clear the importance of having a world organization such as UNESCO and of the need to support and strengthen its work. I hope to present several different ideas, but summed up they might lead to the following conclusion: that man may well have reached that point in his history, that stage of his development, where he has not only been made master of his fate, but where his technology and his morality have come face to face, where he can scarcely treat fact and value separately, and where he may see principles as diverse as the Second Law of Thermodynamics and the Golden Rule being considered side by side in the making of decisions which determine his future. I believe — as I am sure many of you do also — that the major forces that have shaped our lives in the last twenty years — since the birth of UNESCO — are not only widening our horizons but forcing us to shape up for a rigorous journey toward those horizons. We are headed for some new undiscovered shores, and after a million years of headway — of struggle and growth, of conquest and accomplishment — during which time we believed we were all sailing separately, I think we have reached the point where we now realize we are essentially all in the same boat together. I think it is science which has made this so. And it is the power derived from science, and how we use or misuse it, which will determine our future.

Of course, we all realize that science has been with us for longer than the past twenty years. As an organized way of thinking and acting for man, it has been with us in many respects since ancient times. Experimental science, the foundation of today's science, had its beginnings a few centuries ago. The forging of the bond between science and technology dates back to the last century, and the adoption of both by industry took place at the beginning of this century. But it has only been within the last few decades that the impact of scientific theory, transformed with increasing rapidity into technology — into tangible products and power — has been felt by society as a whole.

Before I go into some of the results of this impact and the challenges it has placed before us, let me review, in broad terms, some of the recent progress of science and the ways in which it has widened our horizons.

Endowed with an intellect that allows him to reason, and to remember, record, and transmit his ideas, man has been responsible to a large degree for his own creative evolution. In speed and scope it far exceeds his own natural evolution. Physically we are not that much different today than, our forebears of the Stone Age. We exist on a pinpoint of time, temperature, and chemical balance, individually weak and ineffectual, and our only contact with the real world is through our five meager senses. But through the highly organized intelligence we call "Science" and the highly organized group we call "Society," we have been able to extend our senses and amplify our energy to an incredible degree. We have been able to learn much about ourselves and our environment, and we have developed ways to adapt ourselves to environments not only unsuited to man but to those where no other life has ever existed.

Our sense of vision has always been our primary method of knowing about our world. Yet marvelous as our eyes are, they allow us only a very limited picture of our immediate world. Today, through scientific innovations which have been developed or refined mainly within the last twenty years, we have extended the limits of our sight by orders of such magnitude and in such terms that the very expression "sight" needs redefining. We are now able to witness indirectly events taking place almost anywhere on the surface of the earth at the time they happen. We are able to "see" the existence of stars far beyond those at what we believed was the "outer edge" of space, and to film cosmic events which happened billions of light years ago; we are able to visualize the structure of an atom and to track the path of its disintegrating particles, though their lifetimes be far less than a billionth of a second. We are able to observe the microcosmic world of life and see the detailed structure of such simple self-reproducing things as viruses, some of which are so small that a trillion of them could comfortably live on the head of a pin. This same extension of sight toward the inner world allows us to study in detail the basic structure of our own living cells and the chemical basis of our heredity.

Our sense of hearing has also been enormously extended by technologies which transform understandable sounds into various forms of energy, carry them to us over vast distances, then transform them back into familiar patterns and the wavelengths our auditory nerves can handle. From such diverse places on earth as Sugar Grove, West Virginia, and Jodrell Bank, England, we cock those gigantic "ears" — our radiotelescopes — toward the farthest galaxies and beyond and pick up sounds from distant worlds, listening for (and hoping for) patterns which might indicate the existence of intelligent, communicative life somewhere in those remote cosmic regions. We can "hear" the sounds of moderate-sized earthquakes anywhere in the world, determine their intensity, and often their exact location. We have extended our hearing (as well as our sight) beneath the sea, and in doing so have learned not only how to detect animate and inanimate objects in that almost lightless world, but that, dark as its depths may be, the sea is not a silent world, and that the voices of the fish and the other marine life in it make it a veritable screeching, croaking jungle of sound.

Fortunately, we have created new senses as well as extended the range of our natural senses. We have no natural sense organs designed to detect magnetic fields, or the greater range of radiation — that not in the limited areas of the spectrum which produce light, sound or heat — yet we can detect almost all forms of radiation in the smallest amounts, whether in the confines of a laboratory or in orbit around the earth or sun.

Miraculous as these extensions of our natural senses may be, they would all be for naught if men were not endowed with the ability to remember, to record and transmit these impressions and the ideas he develops from them. It is only through memory and communication that the collective intelligence of mankind grows, and that he is able to make any use of the impressions gathered by his senses. It is, therefore, most fortunate for us that the same scientific revolution which has magnified our knowledge-gathering ability has given us the electronic tools to handle this great new accumulation of data, to sort it, store it, recall it when we need it, to solve problems with it, and to some extent put the answers to work for us. If it were not for these tools, how would we begin to handle the incredible input of our information explosion, one which in science alone sees the publishing of 100,000 journals a year in more than 60 languages, a figure doubling every 15 years?

If the past twenty years has seen a modification of man through a vast technological extension of his senses and collective intellect, certainly it has also seen an important, though perhaps less spectacular, modification of man himself. The understanding we have gained of the life processes and of the many biological forces at work in ourselves and our environment has extended our life span, has allowed us to reduce disease and suffering and to enjoy longer, healthier lives. But this progress, important as it may be and as grateful as we are for it, may soon seem insignificant compared to that which appears to be just ahead of us. The possibilities of new and rather dramatic changes in man loom large, some of them in the not too distant future. They may stem from advances in a number of diverse fields — from new found knowledge of the body's chemistry, of the electro-chemical nature of the brain and nervous system, of our genetic make-up, and from advances by the chemical engineer, the surgeon, the electronics expert, and the biochemist. Man has been modifying himself for centuries by such simple devices as external braces, artificial limbs, hearing aids, and even dentures. Now we are beginning to think and act in terms of internal modifications, of replacing parts of organs and eventually even entire organs with synthetic, man-made substitutes. The replacement of an artery with a dacron tube is now commonplace. Externally used artificial kidneys, heart-regulating pacemakers and heart–lung machines perform life-saving functions today. Serious effort is now being devoted to developing artificial kidneys, pacemakers, and even mechanical hearts which can be carried internally by the patient and in effect become a part of his natural body. Efforts are also being made to repair man's damaged or diseased parts with the healthy parts of other men and even animals, and just in recent years some remarkable new progress has been made in the understanding and technique of the transplanting of human organs.

I do not have time to go into the great strides that have been made by our biochemists in recent years. The extent of their progress might best be expressed by the fact that work in an understanding of the life processes has reached the point where it was seriously suggested this past year by the president of the American Chemical Society, Dr. Charles C. Price, that a national crash program, similar to that taken by the government in nuclear fission in the 1940's, would probably lead to man's ability to create life. Even without such a crash program, we may see this accomplished before the end of this century.

Paralleling our growing understanding of the biochemical basis of life has been a growing interest in, and knowledge of, the basic mechanism of the brain and nervous system. At the recent annual meeting of the American Association for the Advancement of Science, there was much discussion of the work being done now in the chemistry of the brain and central nervous system. If current research in this field — including the use of chemical compounds to influence memory and learning — offers any indication of the possibilities ahead, we may see some revolutionary advances in the treatment of mental illness. We may also see some other startling uses of chemicals in relation to man's mental and physical processes which could have very serious social ramifications and call for the utmost consideration and control in their uses.

The electrical nature of the brain's activity and its comparison with the computer has become the basis of the relatively new science of cybernetics. Cybernetics, together with another new discipline known as bionics, give us cause to speculate on how far man can go in combining his naturally endowed self with the mechanical products of his creative evolution. Those involved in these new fields suggest that some day in the future it may be possible to "program" information directly into the human brain, or develop cybernetic control systems through which a man's thoughts alone can remotely control a machine. Work in this area of so intimately combining man and machine has already led to the coining of a new word — cyborg — standing for cybernetic organism. And to a great extent Man — The Cyborg — is with us today and making valuable contributions to our programs in space, nuclear energy, and oceanography — in those areas where man must operate in a hostile, hazardous environment.

In a literal and figurative sense man's horizons have been widened most by his ability to venture into, adapt to, and work in new environments. No discernible amount of natural evolution has allowed him to do this, but his creative evolution, through science and technology, has allowed him in recent years to take giant strides in this direction.

Our adaptation to new environments has taken many forms. In some cases we work indirectly in a hazardous environment. A good example of this might be that of a laboratory worker in a nuclear facility who, from behind thick radiation shielding, is able to use the master–slave manipulators, or "magic hands" as they are sometimes called, to work with highly radioactive material. These extremely sensitive mechanical extensions of his own hands permit him to work in an environment he cannot otherwise enter.

When we enter an environment we could not normally survive in, we take the necessities of our own environment with us. These might take the form of an aqualung for limited underwater work, or some type of submarine vehicle should we need warmth, light, oxygen, greater mobility, and protection from great pressures. The fact that we have been able to descend almost to the bottom of the deepest trench in the oceans — a depth of over 35,000 feet — seems remarkable, and yet we are just beginning major efforts to explore this environment of "inner space."

There is no need to go into detail on the efforts we are making to explore our horizons of outer space or the progress we are making in this endeavor. In recent years a good part of the world's attention has been focused on our entrance into this new environment. Here again man is bringing his own environment with him on his journey, and he will have to bring a great deal of it great distances if he is to go far or accomplish much in the incredibly vast areas he hopes to explore and possibly inhabit.

What is perhaps most remarkable about our space program, and perhaps not realized by many, is that the fulfillment of this desire on the part of man to leave his mother planet requires an effort involving almost every one of the scientific disciplines known to man. There is scarcely an area of human knowledge or technology that will not come into play in putting a man on the Moon and eventually on the planets.

No review of how science and technology has widened our horizons would be complete without some discussion of the role of energy. For it is man's extraction, conversion and control of energy which has made all that I have mentioned to this point possible. And it is our relationship to energy — how we use it and what we do with it — which offers us both our greatest promises and our greatest challenges.

Less than two months after the birth of the United Nations and UNESCO — and very influential in the early development of these organizations — was the revelation that man could release the enormous energy of nuclear fission. With this discovery came the realization that it would be only a matter of time — a relatively short time — before we would have at our disposal enough energy, and the technologies for using it, either to completely destroy our civilization or to lift it to great new standards. But the energy of the atom is remarkable not only because of its total power. The range and usefulness of nuclear energy, from the standpoint of both fission and radiation, opens to us whole new energy horizons. With our nuclear science and technologies, we will be able to convert, through the man-made element plutonium and a man-made isotope of uranium, almost all of the energy bound up in the natural elements uranium and thorium into a nearly inexhaustible supply of power. And we will be able to do this just at a time when the demands of our high-energy civilization are growing so fast that a new energy source to supplement our fossil fuels will soon be needed. Through our nuclear science and technologies we will also be able to produce enough compact power to make possible our forthcoming adventures in space — to take us, and the equipment we will need, to the planets. This will be possible through nuclear powered rockets and systems for nuclear auxiliary power, all of which are under development, and some of which have already been used successfully in various ways.

I mentioned before the range and usefulness of nuclear energy, and perhaps this versatility of the atom is one reason why it is so valuable to man. To give you an example of what I mean, let me contrast some of the work of two of our divisions at the Atomic Energy Commission, both involved in entirely different aspects of the peaceful uses of the atom. At one end of the scale is the Division of Peaceful Nuclear Explosives and at the other is the Division of Isotopes Development. I place them at opposite ends of this arbitrary energy scale because while the nuclear explosives group deals mostly in energies measured in kilotons and megatons — equivalents of thousands and millions of tons of TNT — the isotope group is more apt to be using such power terms as watts or microwatts, or such terms as curies or millicuries, and at times even microcuries or picocuries. The Division of Peaceful Nuclear Explosives administers the AEC's "Plowshare" program — the term "Plowshare" derived from the biblical expression concerning the beating of swords into plowshares. This program involves the development of nuclear explosives, and related technologies, for large-scale excavation such as the digging of canals, mountain passes, harbors and underground storage areas. To give you some idea of the energy involved in such work, or more accurately, the work done with such energy, I might point out that in an early Plowshare cratering experiment — in 1962 — a single 100 kiloton thermonuclear device was able to lift 6.6 million cubic yards or 8.4 million tons of earth in a matter of seconds. Contrast this with the energy of the radioisotope, perhaps the ten to fifty microcuries of iodine-131 in a glass of water, which a person might drink in the course of a routine diagnostic test of his thyroid gland. Such an isotope, finding its way to the thyroid, would be emitting a constant signal which only a highly sensitive electronic scanner could detect. And, of course, even this represents an energy level far greater than many used in other nuclear work.

The energy revolution of the past few decades has been more than one of a growth of sheer power. It has been one of tremendous refinement in the production and control of power, and of heat, light, and various forms of radiation. It has been one of incredible speed, bringing into almost common use in some fields such previously unused divisions of time as the nanosecond (a billionth of a second). And it has been one inseparable from our revolutions in transportation, communication, and the growth of human knowledge.

Review of the widening of our horizons over the past twenty years may leave us very impressed with our collective accomplishments. And such a review with some contemplation of what the future holds is in many respects overwhelming; to some, it seems almost frightening. But the facts and figures alone do not tell the most important part of the story. Our scientific progress must not be considered apart from our social progress, nor our technology talked of in terms not relating to human values. Wherever we are headed on our new wide horizons, we are traveling there as the family of man. As such, what has this journey done for us? And what future does it hold? I would like to devote the remaining few minutes of this talk to some thoughts on these matters.

Certainly while science has widened our horizons, it has brought us physically closer together. I believe it has done the same in spirit — in engendering broader human cooperation and understanding. Science comes to us from an international heritage and it promotes a growing internationality. In spite of the scientific ascendancy of some countries, no nation has a monopoly on scientific talents. One does not have to go far back in history to see an outstanding example of the contributions of scientists of many national backgrounds to a single scientific field. Within the past three decades, the nuclear age was born of the labor of such an international group of scientists as Hahn, Fermi, the Joliot-Curies, Rutherford, Bohr, Einstein, Szilard, Lawrence, Compton, Oppenheimer, de Hevesy, and a host of many other multinational names too numerous to mention at this time.

Today, the sharing of scientific brainpower is an important inducement toward international cooperation. In this regard, it is interesting to note that last year representatives of the United States attended about 600 international, intergovernmental conferences, mostly on technical subjects. We belong to 53 international organizations, again, mostly scientific and technical. These are organizations which exist because they are needed, because they serve the purposes of many nations and many people. It is a fact of twentieth century life that, while nations may stress national sovereignties and avow national purposes, science has created a functional international society. It exists, and it works, in spite of all who doubt it.

Modern science is such a dynamic force that its success — its very existence — demands growing cooperation from all quarters. The explosion of scientific and technical knowledge, from a multiplicity of specialized fields and diverse sources, calls for tremendous cooperative efforts in the gathering and exchanging of information. Expanding international communications are vital to scientific growth. We are just now beginning to learn how to efficiently handle the overwhelming flow of material being recorded and transmitted throughout the world each day.

Cooperation is again a factor in science in the use of scientific equipment. The staggering expense of some of today's scientific tools makes the sharing of large facilities and laboratories mandatory. I think that the future will see the development of many cooperative international facilities such as the one operated by CERN, the European Center for Nuclear Research.

Aside from the benefits which science itself gains from cooperation, the social applications and implications of science demand international cooperation. The consequences of decisions in scientific matters and the benefits of scientific progress are inexorably connected to the well-being of all men. They affect life itself — birth, health and death; they affect living conditions — food, shelter and environment; they affect human contact — communications and transportation; and they affect the general state of the world — war or peace. In this regard — because of the consequences of science — we have a vital stake in what is being done abroad and an obligation to disseminate the beneficial knowledge we gain here.

In many ways the physical scope of scientific investigation today transcends national boundaries. Global science demands global cooperation. The fields of Meteorology and Climatology, Oceanography and Geology, Astronomy and Space Exploration need as their laboratories all the Earth's oceans, its lands, its atmosphere, and the boundless space of the universe. These world-straddling sciences have been responsible for a surprising degree of worldwide cooperation, even among countries harboring great political differences. In the field of Meteorology, our current program involves the cooperation of more than 100 other countries. In our space program we have the cooperation of more than 50 countries and are beginning to cooperate directly with the Soviet Union in this field. In addition, we have had much success with the Soviets in exchanging information and ideas in the nuclear energy field.

I think that all this has brought us face-to-face with the facts that we can no longer live in a world where only a small percentage of the people enjoy the benefits of twentieth century life. Each year brings us closer to the realization that it is a technical, social, political, and moral necessity that we learn to live together in peace and understanding, sharing more and more of the knowledge we gain and the abundance that knowledge brings, until such time when all have the ability to share the workload evenly and all can enjoy the benefits which only some of us enjoy today. We no longer are a world of local or regional civilizations which can flourish or wither away by themselves. Our civilization is now global and absolute, and the actions of each nation, and to some degree, of each individual, strongly affect our collective destiny.

There are many people who fear the current explosion of science and technology, who see in it a dehumanizing effort, who believe that it is running away with us, and that we are becoming its slaves instead of remaining its masters. I cannot agree with these thoughts. A comparison of the conditions of people in any age with those of today will show that we are far better off. And I do not mean just in creature comforts, suffering less from disease, hunger, and the whims of our environment. More of us share a larger, richer culture — one international in scope, containing a vast heritage from each other's lands, and from the past — the art, music, and literature of bygone days preserved, reproduced, and recreated for our enrichment. And these are in addition to our own cultural contributions which are substantial by the standards of any age. We have more leisure to enjoy all this. If there is a certain framework of superficial conformity as a result of our methods of mass production, there is also within that framework a larger degree of freedom and individuality. Today we have more freedom of choice than any of our ancestors ever had. We have more freedom of expression. We have more freedom *from* ignorance, superstition, and iron-clad tradition, and, as a result, more freedom to change — to control and direct our future, our creative evolution.

I believe we *can be* masters of our fate. We *need not* be trapped by the onrush of our technology — simply because we recognize it. We realize (and not too late, as some would have us believe) its speed, its effects, and their cumulative impact. And in the light of this we can predict, plan, and work to shape our future.

Most of you are aware of the massive challenges facing mankind today. You read about them in your newspapers and magazines, see and hear about them on television and radio — environmental pollution, hunger, water shortages, the population explosion, the problems associated with urban growth, and, above all, that plague which mankind has not yet eliminated, the threat of war, compounded today by the overwhelming destructiveness of nuclear weapons.

Yet, I believe that none of these problems are insoluble if we face them without delay and work hard toward their solution. There is an enormous amount of national and international effort already going into the meeting of every one of these problems. Unfortunately, the mass media do not direct as much attention to our work in meeting these challenges as they do in dramatizing the problems and some of our failures in dealing with them.

Explaining some of the ways we are attacking these problems would take another talk — in fact, a whole series of them. But let me drop a few national and international names, and give a few random examples of the efforts being made to meet these challenges.

Essential to the meeting of all the challenges ahead of us is education — education on all levels. A technologically developing world and an uneducated one are incompatible. Stating its support for the goal of universal education, UNESCO's Advisory Committee on Science and Technology has said: "Governments of many developing countries must find the quickest formulae to teach children of largely illiterate people how to read and write and, at the same time, to train engineers and scholars drawn from among the same people. This is a phenomenon of the present century in which supersonic aircraft will be used as a means of transportation, while inside these same countries, traffic is still carried on donkeys."

In support of its strong belief in the importance of education, UNESCO will this year launch its five-year Experimental World Literacy Program, which will also include the vocational and technical training of millions of now illiterate workers.

With regard to what is being done about the problems of pollution, I call your attention to our own battle (in the U. S.) against environmental pollution and to the recently released report to the President by the Environmental Pollution Panel of the President's Science Advisory Committee. This report, the work of some forty experts and their staffs, carefully outlines the problems we face and some possible approaches to solving them. It is an important first step in our national campaign in — as the report is titled — "Restoring The Quality of Our Environment."

The problem of hunger, still with us in so many areas of the world, is being attacked on an international scale by the work of many UN organizations. Under the leadership of the UN's Food and Agricultural Organization, a "Freedom From Hunger Campaign" has been launched which involves over 80 countries working with UNESCO, UNICEF, the

World Health Organization, the International Atomic Energy Agency, the International Labor Organization, and the World Meteorological Organization. This formidable combination of international scientific and educational groups is exploring every natural, technological, economic, and social means of providing basic food staples for all human beings everywhere.

This new year — 1966 — has been designated "International Rice Year" by a joint FAO–UNESCO drive "to focus world attention on the role that rice could play in furthering the aims of the Freedom From Hunger Campaign; and to improve international understanding of the rice economy."

Related to the solving of the world's water problems — of which we have become more aware because of our own droughts — far more than a year has been designated. Under the sponsorship of UNESCO, FAO and the World Meteorological Organization, an "International Hydrological *Decade*" has been proclaimed, during which a massive program will be undertaken to find ways of averting the water crisis that so many nations face. During the IHD, a sustained and coordinated program of scientific research will be carried on to find out, among many other things, where the sources of fresh water lie, how we can recover them, what the best methods are for desalting water, what can be done to salvage the water that evaporates or runs into the sea — and what all of these operations will cost. With all this knowledge, it is believed we can make substantial progress in solving our water problems for many decades.

Finally, I call your attention to the work of the International Atomic Energy Agency, mentioned briefly before because of its cooperation in the Freedom From Hunger Campaign. The IAEA is not only making an important contribution to the world's efforts in developing the many peaceful benefits of nuclear energy, but it is also playing a leading role in seeing that the atom remains peaceful, in controlling the proliferation of nuclear weapons. This organization, with its 95 member nations, oversees and administers an international "safeguards" system, a system of inspection and control designed to assure that the nuclear materials, facilities, and technology furnished for peaceful uses are not diverted to military applications. Right now — today — the IAEA "safeguards" inspection system is at work. It is the first program of international inspection in the arms limitation field to be put into operation and it is operating with the blessing and encouragement of both the United States and the Soviet Union. Such an organization truly deserves our strongest praise and support.

We look to the IAEA, to UNESCO and to all the organizations of the UN to help us achieve world peace — true peace, not just stalemates of political and military power, but lasting peace based on reason and understanding, on world stability, and on the well-being of all men sharing a newly created abundance on this earth.

I believe that it is possible, through science and technology, to achieve these long unfulfilled dreams of mankind. Our progress of the past few decades has, indeed, widened our horizons. It has given us much new knowledge and many remarkable new tools. It has also given us a glimmer of a better new world which can be reached with these accomplishments as a beginning. Let us not sit back then and revel in what we have done. Let us continue our progress. Let us continue to grow in knowledge, to apply that knowledge, and to work hard with our new tools so that we can continue to move toward those wide horizons and the greater promise they hold for all mankind.

Science, the Humanities and the Federal Government — Partners in Progress

Remarks made to the American Philosophical Society, Philadelphia, PA, April 22, 1966.

Speaking before a U. S. congressional panel earlier this year, C. P. Snow, the eminent British writer and scientist, stated in a discussion of the role of science in society: "I would far rather have choices made by wise men who are not scientifically educated than by unwise men who are." I am not sure into which category most of us here tonight fall, but I would think, or at least hope, that the majority of us could claim both some degree of wisdom and some scientific knowledge, because Lord Snow also went on to point out in his talk the importance of being scientifically informed and of having scientifically informed people in public life.

I bring up C. P. Snow at this time because the arguments resulting from the thesis he so effectively presented in his essays "Two Cultures" many years ago are still continuing. In fact, discussion of the roles of science and the humanities and arts and their degree of influence in our country have been somewhat intensified recently because of the role that the federal government is now playing in these areas.

I would like to spend some time discussing the relationship of science and the humanities and arts, their influences on each other and society,

Reprinted with permission from the *Proceedings of the American Philosophical Society*, 110, 318 (1966).

and what bearing government interest and support have upon all of them. This may seem like a lot of ground to cover in one talk, but I think there is a focal point we can use. It is the quality of life we expect our civilization to provide in the years ahead.

In pursuing this theme I would like to make and expand on two major points: The first is that man's growing domination of his environment today demands that we take a new and broader outlook toward our future development. And secondly, in dealing with mankind's physical growth, with the mass society in a world of growing population, we must not neglect the individual human being and his basic need for a more satisfying, purposeful life.

The Scientific Revolution

One of the most significant phenomena taking place in our society today is our recognition of change — the extent and the rate of change occurring in this scientific–technological era, this Nuclear–Space Age. This Age is a relatively new era in history. It was born some twenty-odd years ago and has been growing rapidly ever since in its scope and impact. But it is only within the past few years that there seems to have appeared an overwhelming awareness of the implications of the ascendancy of science and technology, of their present effects and those possible in the future. Now it seems that nearly everyone is rushing off to conferences on technological change, the whole relationship between science and society is a subject of symposia and seminars, and a flood of books and articles is appearing on the topic, almost all of which are in essence asking the same questions — "Where are our science and technology taking us?", "What does this mean?", "What can we do about it?" — and apprehensively recognizing the rate of change, the word "explosion" is becoming one of the most overworked in our vocabulary.

I think this has all come about because more and more we are feeling, or being made aware of, some of the problems growing out of what in the past many of us have viewed as unmixed blessings. Improved medical science has brought better health and a longer life span but also an expanded population growth. More automobiles and better roads have brought speed and convenience but also traffic-congested streets and an accident rate of epidemic proportions. More efficient agriculture has allowed the increased production of food by fewer farmers, but as rural population has declined, urban areas have grown rapidly, causing a multitude of economic, social, moral, and political changes in our cities. New technologies in business and industry provide greater efficiency of operation, but demand the reeducation and retraining of workers as certain human tasks and skills become obsolete. And as our industries have produced more and better products, so also have many of them — both producers and their products — contributed to the alarming pollution of our environment, a condition which has made us realize that the control and recycling of waste may be a major human problem from now on.

To add to these problems and pressures rising from "normal" growth and progress, we face other mixed blessings of potentially enormous consequences — potentially good or bad — rising from the development of new technologies whose full influence in either direction we have yet to feel. In this case, I refer to the use of nuclear energy and the technology of space. Naturally, we hope and we work toward making these outgrowths of our scientific age serve us beneficially (and I think we are making great progress in that direction); nevertheless, they add to the total picture of change and dramatically call our attention to it.

Another factor of which we have become increasingly aware is the totality of change and the influence and interrelationship of all the scientific and sociological forces producing change today. We see quite vividly, in our modern world, not only the interdependence of man as brought about by the web of specializations in a global civilization — one we have created by modern communication and transportation — but also man's ascendancy over his environment, now reaching the point where certain changes by man can have widespread, perhaps devastating, effects on our planet.

Having recognized the problems and complexities of the rapid change of our time, we realize that planned action is called for in order to control and direct it toward our needs and benefit. We are intrigued with our new-found power but we are also aware of its possible consequences and pressed by a growing sense of responsibility. We realize that mistakes made today are apt to be big ones and costly ones, and that even apparently small mistakes may have large and important consequences in terms of future generations. But at the same time the demands and dynamics of the times call for decisions. In most cases, inaction means retrogression — and a type of retrogression which, because of the pressures of growth, may be catastrophic.

All these factors — a recognition of the rate of change, its totality, and the fact that we can exercise some control over it through planning — lead to some of the most important questions facing the American people today. And they are questions not alone concerned with controlling physical change. For we have come to realize that in planning and working for the future, we must also reevaluate our goals. We must temper, adapt and redirect the enormous technological forces we have created to the end of serving the individual human being — an end for which the forces were created but which has begun to elude us as we have become more and more taken with the power, the momentum and the "open sesame" fascination that our modern science and technology have provided.

In a democratic society, the underlying quality of life for which we strive is represented not only by total economic and social change but by the welfare — the well-being — of the individual in society. Maintaining a maximum of freedom and opportunity for growth for the individual in a world of overwhelming technological forces and growing social interdependence and responsibility is a major challenge of our times. And it is a challenge which many feel cannot be met fully without a more meaningful interplay between the sciences, the humanities and arts, and government — forces which up to now have acted to a great degree separately, and sometimes at odds, in our society.

In bringing up the influences and interplay of these forces in a nation, and to gain some insight on the growth of the challenge facing us, we must take into account several factors which had an important bearing on our early development. The pragmatic spirit of our early settlers and our first citizens was of prime importance in shaping our entire destiny. From its birth ours was a nation of opportunity, of action and of mobility. The need for movement and growth was characteristic of America from its beginning. Fortunately, we were blessed with a land of vast natural resources, a growing population of vigorous people from a variety of cultural backgrounds, and, most fortunately, our early years coincided roughly with the beginning of the Industrial Revolution. All these circumstances combined to allow us to grow and prosper rapidly.

From our earliest times it was the federal government's responsibility to minister to the needs of the people, but to do so only when those needs were considered national and federal action did not conflict with individual and states' rights. This was evident in the federal government's early role in science and technology — in its support of such a scientific venture as the Lewis and Clark expedition and its establishment of our first federal scientific agencies, the Geological Survey, the Naval Observatory, and the Coast and Geodetic Survey. Other federal organizations of a scientific and technological nature were ushered into being as their need became apparent. But it was not until the 1990's, the period after the first World War, and particularly after World War II, that large-scale federal involvement in science evolved. It was during this era when the chemical industry saw the beginning of rapid growth based on new discovery and innovation, and when in physics, as Robert Oppenheimer characterized it, we shook the tree of earlier theoretical science and applied the knowledge we gathered to help us develop new and potent technologies to help us foster the scientific revolution, the tremendous effects of which we are experiencing today.

This scientific revolution, since the birth of the Atomic Age — and particularly since the challenge of "Sputnik" — has had a pervasive influence throughout our society. It has led many people to "discover" science; that is, to realize its power to produce substantial material gains when its methods and knowledge are put to use by modern technology.

But what has been most notable about this scientific revolution — and probably what characterizes it as a revolution — has been the increasing speed with which theoretical discovery has found its way to practical application. And today as the time gap between discovery and application continues to narrow, so does the social impact of science grow with an almost exponential force.

There is no need to go into detail about the government's role in this revolution and the combined effect on our society. Its overwhelming influence in most aspects of our national life is, of course, obvious. It has continued to shape our national personality along the pragmatic lines of our formative years — and I do not say this disparagingly, as we have accomplished a great deal in bringing a better standard of living to our people in our relatively short history.

Have We Neglected the Humanities?

But what of the other side of the coin? As opposed to our technological know-how, how have we developed "value-wise," "culture-wise," or even "wise-wise"? If we view America in terms of its intellectual and cultural growth, we see that we arrive at a problem when we compare this with our "scientific explosion" of recent decades.

As I mentioned last year while testifying before a Senate Subcommittee in support of a National Humanities and Arts Foundation, I feel we are developing an imbalance in our national personality, one which is arising from a well-intentioned emphasis on technological gain but one which is weakening our framework of values nevertheless. Among the elements which characterize this imbalance is a strong desire to satisfy material needs — some of which we create artificially — at the expense of satisfying some very important psychological and spiritual needs.

Today our concern with efficiency and economy, with power and production — with our overall desire to "do the job" — dominates our reasons for doing the job. In our pre-occupation with "means" we have lost sight of many of the "ends." We set more production goals but fewer human goals. We think more of developing "systems" and less of developing individuals. We chart the course of our businesses and our economy but not the measure of our success or failure in providing human fulfillment. Granted there is no equivalent to the Gross National Product in the fulfillment of human values, we seem to give the former far more thought and concern than the latter.

Being a nation founded on concern for the individual — his freedom, his welfare, and his pursuit of happiness — it is all the more ironic that today our concern with the techniques for achieving many of these goals has somehow pushed the individual into the background. At times he feels sorely neglected as a human being. This is becoming more evident on our college campuses, in our big cities, and in our industries, where people, living and working together in large numbers, are reacting to the strange combination of the pressure of competition and yet the lack of meaningful human contact. It is evident in much of our current social behavior — in the pursuit of some of our more exotic fads and cults, in an excess of hero worship, in some rebellion only for its own sake, and even in an abundance of conformity to non-conformity.

But most of this behavior has some constructive meaning. It acts as a form of release and fills certain superficial needs of the times. More importantly though, it calls our attention to needs which are being overshadowed in today's technological society — the basic needs of a person to have an individual identity, a sense of purpose and accomplishment in his life, and outlets for his physical and creative energy.

In today's restlessness and rebellion, once again we are made aware of change and of the need for action to deal with it. Within the context of the total change I spoke of before, I believe we can deal effectively with our social and psychological needs — as well as our technological and economic ones. I believe we can deal with them by emphasizing in our plans and work the human values which we have had for ages, but many of which in recent times have been overshadowed, overlooked or inadvertently bulldozed aside by the momentum of a benign but often bumbling technology.

In the light of recent events I think we have begun this reemphasis on man's role in the scientific age, or at least are at the point of beginning it. I feel that perhaps we are now at a time in our history where, breathless from a headlong pursuit of material gains, we have paused to look around and reflect, and this has engendered in us a new awareness of the total condition of our lives and a broader outlook of the future. During our reflection we have arrived at some badly needed insight into significant needs. And, among other things, this is leading to some important national planning toward preserving the natural beauty and resources of our land, toward "beautification," toward fulfilling some of the obligations of our heritage which we have bypassed in our rapid growth. Who can doubt that these also are major goals of a truly Great Society, the pursuit of which deserves our firm support?

Today with our new awareness of change and our realization that we can and must be the master of that change, we are also taking a more positive attitude toward emphasizing and cultivating human and aesthetic values. Most recently this attitude has resulted in some important new action taken by the government which has long remained at some distance from any direct interest in or support of our humanistic and cultural growth, although it may have realized the importance of those values to us as a nation. There have been many reasons for federal inactivity in these areas.

The primary one parallels what has taken place in our scientific development and also grows out of our basic philosophy of a minimum of government interference in all segments of life. As I pointed out in discussing the relationship of science and government in our history, the government's role developed basically along the needs of the country. Unfortunately — or, in some respects, perhaps fortunately — wars and national crises have not demanded government support of the humanities and cultural pursuits. With the exception of some federal support of the arts by WPA projects in the depression years, scholars and artists have continued on their own during these periods when vast sums of money went to support science and technology.

Also, it has been traditional in our society to view scholars and artists as people who need no public support, who work best on their own; who, in fact, derive some kind of inspiration from their isolation if not their poverty. Considering the arts, most people who have this attitude conveniently overlook the fact that many of the world's great masterpieces were commissioned — masterpieces which may never have been created but for the support of a patron. America has had its great sponsors of intellectual growth and patrons of the arts in the past, and today we have numerous individuals, foundations, and business organizations that are helping to support a substantial segment of the scholastic and cultural community. But private support only begins to meet the needs being generated today.

Finally, there has been the attitude in America that government support of any educational or creative process would lead to its control by that government and to dangerous interferences with our most basic freedoms. Many have used as an example of this sort of culture-control the book-burning tyranny of Nazi Germany or the directing of the arts toward propaganda in Communist countries. But it does not follow that governmental support of creative or scholarly pursuits necessarily leads to controls. I think that what transpires in this relationship between support and control depends on the basic philosophy of the nation. And within the American framework of thought and action it seems highly unlikely that government control of the arts would evolve here.

It is of greater concern today to find ways to support the humanities and arts so that they can have the economic freedom to reflect some of the truths that a nation needs to have revealed to itself. Through this process we can maintain and develop further the values and outlook that we so crucially need to guide our technological society, to assure its growth and development as a human society as well.

Bills for Cultural Development

Today, recognizing the desire of our people for a more meaningful life — for a fulfillment beyond that of the materialistic life — has come a response from government. The response is taking place on local, state and federal levels, and in many forms. There is not time to go into all of the work being done or planned, but for a few minutes I would like to touch on some of it being carried out by the federal government, particularly by the newly created National Foundation on the Arts and Humanities.

This foundation, created by a bill signed into law last September, is comprised of a Federal Council on the Arts and Humanities, a National Endowment for the Arts, a National Endowment for the Humanities, and National Councils on the Arts and the Humanities. Its purpose is to help support and encourage literary and scholarly pursuits and creative arts of the highest caliber. In the Arts its interests include (but are not limited to) music, dance, drama, architecture, painting, sculpture, photography, graphic and crafts arts, industrial design, fashion design, motion pictures, television, radio, tape recording and all the other arts related to the preservation, performance, execution and exhibition of these major art forms. In the humanities it is concerned with our knowledge and understanding of literature, language, archaeology, history, the classics, religion, philosophy, and the preservation of our heritage in all these fields. This is indeed a lot of ground to cover, but the range of interests indicates, I think, the thoroughness of consideration that went into the planning of the foundation and its activities. It has carefully considered the complex needs of our creative and intellectual life. What are some of the ways it plans to help fulfill these needs?

As many of you probably know, Roger L. Stevens, who was formerly the President's Special Assistant on the Arts, was designated as Chairman of the National Endowment for the Arts, and the National Council on the Arts which advises the endowment. Mr. Stevens, who has combined a highly successful career in business and finance with an equally successful one in the world of art, recently elaborated on some of the plans of the Arts Council. And I would like to mention a few of them now just to bring out some of the areas of work involved in supporting our cultural growth. I think you will find many of them as surprising as I did, particularly in the very practical and far-reaching way they will help the arts — and without regimenting or controlling them. Not all of the plans involve direct financial support, but some, because of their widespread effect, may be more significant than direct financial support.

For example, Stevens points out that there is already an Executive Order in effect that calls for the expenditure of one percent of total construction costs for art in all federal buildings, but that this is the first item cut in a building's budget when the money starts to run out. The average expenditure for government buildings runs about $600 million to a billion dollars a year. Stevens argues that if the one percent figure could be made mandatory it could create a market for the purchase of art of about five to ten million dollars annually.

A similar situation exists under FHA regulations which, in view of FHA's annual expenditure of about $5 billion, might account for another forty to fifty million dollars worth of sculpture and painting. Aside from the support this could lend to American artists, think of what it would mean to the beauty of the buildings and the enrichment of the lives of the people working and living in them.

The Arts Council further suggests that there is no reason why the immense number of government-owned buildings throughout the country cannot to some extent be used for art exhibits, particularly in cities that do not have museums.

Paralleling these suggestions, Stevens points out that through National Park Service receipts controlled by the Department of the Interior it may be possible to support many cultural activities such as the theater, concerts, and museums in our national parks, thus providing an excellent medium for the work of actors and musicians.

Another very practical recommendation involves making available government surplus property construction materials to museums, theaters, and cultural centers for building or adding to existing facilities.

The Council recommended that direct aid to artists be made not only in the form of grants and commissions but in such minor but important matters as providing money for crating and shipping works to an exhibition — a matter perhaps trivial to us but quite important to the artist.

Another problem artists face is finding a place where they can afford to live and work, especially in large cities where building costs are high. Many are willing to fix up older buildings, and the Arts Endowment has arranged for a number of FHA loans for this purpose. It has also established a revolving fund of $100,000 for an artists' housing project, with studios and living quarters for painters and sculptors.

In the area of the performing arts, the Arts Council has been even more active than in the creative arts. Among the many things they have done is to recommend the setting up of four Youth Symphony Orchestras in four sections of the country where there is no strong local symphony. The Arts Endowment would sponsor these with the idea that eventually they would support a National Youth Symphony made up of the best instrumentalists.

The Council is also interested in setting up repertory theaters, "of the highest professional quality," in different parts of the country, and the Arts Endowment has set aside $500,000 to help found three such companies. These funds would be augmented by local school authorities and the U. S. Office of Education. In addition to these, the Endowment plans to underwrite a playwrights' experimental theater for $250,000, the purpose of which would be primarily to present plays "that cannot be produced under economic conditions facing Broadway today." Many will argue that all these funds and plans represent an infinitesimal amount of financing and effort as compared to government support of science and technology. Nevertheless, this action is an important first step in direct government participation in our cultural life, participation which certainly will be increased as it proves successful.

Up to this point I have been speaking of the plans of the Arts Council and Arts Endowment, but what of the work of the National Council on the Humanities and National Endowment for the Humanities, also created by the same bill? The Chairman of these new federal organizations happens to be the President of our American Philosophical Society, Dr. Henry Allen Moe. It is in the fortunate interest of the government to have someone of Dr. Moe's standing serving in all these capacities. Dr. Moe recently reviewed some of the plans of the Humanities Endowment, and once again it is interesting to see the extent and the scope of work anticipated.*

---

*The current Chairman of the Humanities Council and Humanities Endowment is Dr. Barnaby C. Keeney.

To further individual work in the humanities, the Endowment will support scholars with several types of grants. Among these are Summer Fellowships to teacher-scholars, Full Fellowships to support significant humanistic scholarly efforts, and Advanced Study Grants for those who, working within the university or outside, "are on the edge of seizing truth — historic truth, philosophic truth, aesthetic truth."

The Humanities Endowment plans to aid archaeological–historical studies by supporting a number of excavations in both domestic and foreign territories, and such excavations and studies will reveal important new knowledge for scholars and historians.

Support will also be given by the Endowment to local archaeological and historical museums — to their staffing, cataloguing, exhibits, and publications.

A most interesting area, which brings out the valuable partnership between science and the humanities, is the field of computer-oriented humanistic research. And this is an area in which the Endowment has great interest.

In addition to those activities mentioned, the Humanities Endowment will also be active in and lend assistance to such interests as the study of classical languages and cultures, the support and training of critics of the arts and humanities, the study of the educational uses of radio and television, the recording of literature, the support of American historical studies, and the rediscovery and preservation of American literature.

I could go on to mention other plans of the Arts and Humanities Endowments which will lend support and encouragement to ventures ranging from performances of the classics to the creating of avant-garde films and electronic music, but I think you will see at this point that a new source of support for our cultural life has come into being, and it is one which will be taking encouraging and effective action. Of course sufficient financing of these efforts is essential but has not yet been forthcoming. If Congress is as generous in providing funds for these Endowments as it was wise in legislating them into existence, I believe we can look forward to seeing some interesting and important results from this new venture of the federal government into the neglected one of the Two Cultures.

Foremost among these results will be refocusing on those values which give our lives purpose and direction. As I indicated before, we have entered an era in which mankind — and particularly individual man — will have a growing need for an inner strength, for a feeling of worth and fulfillment, and for an appreciation of the things in life which, shared or experienced alone, make life worth living. The arts and the humanities offer many of these more evasive necessities of life, supplementing those others supplied more abundantly today by science and technology.

Two Cultures Join

The new support of the arts and humanities by the federal government should help break down some of the artificial barriers we have created between the world of science and technology and that of the arts and humanities. I call them artificial because I think we have created them to some extent only by our words and deeds and can remove them by new outlooks and attitudes.

I believe these barriers are partly the result of our desire to conveniently categorize and catalogue our ideas and activities. But though science may have a pervading and ascending influence in our lives today, there cannot be any clear-cut division between science and non-science in interdisciplinary civilization which a liveable future world will require. The growing overlapping of cultures becomes obvious to those who care to pause and look for it. Some people, apprehensive over the rate and degree of change brought about by our applications of science, talk of science as if it were a force apart from man. What they tend to overlook is the simple fact that science is after all a human endeavor and that it does not exist independent of man. We must not forget that in overall history science has, so far, done more to "humanize" than to "dehumanize" man.

In trying to separate science from other aspects of our lives many tend to overlook the large areas of direct interplay between science and technology and the humanities and arts. As most of you know, science has become a valuable tool of the humanities. In the fields of history and archaeology it has provided remarkable means in helping us to discover and analyze many traces of our past. Such techniques as carbon-14 dating, the use of the proton magnetometer and neutron activation analysis are today helping to reveal much new knowledge about our ancestors and their various civilizations.

A unique new example of symbiosis: science and the humanities may bring together, in a most unusual endeavor, the most unlikely partners, high energy physics and Egyptology. This proposed project, in which high energy physics (a field not yet having many direct practical applications) may help archaeologists, is one involving the search for a possible hidden burial chamber in Chephren's Pyramid, one of the great Egyptian pyramids built about 4500 years ago. In addition to bringing together widely divergent disciplines, the project would be one of international cooperation bringing together physicists and archaeologists of the United States and the United Arab Republic. Those who would be involved are Dr. Luis Alvarez and his staff at the University of California Lawrence Radiation Laboratory at Berkeley, Dr. F. El Bedewi, nuclear physicist of Ein Shams University in Cairo, and Dr. A. Fakry, archaeologist and well-known authority on the pyramids from the UAR Department of Antiquities.

In this high energy physics approach to archaeology, originally proposed by Dr. Alvarez, the plan is to use cosmic ray "muons," occurring naturally from space, and high energy particle detection equipment to literally "x-ray" the great pyramid. Through the use of a spark chamber, placed in a known subterranean burial chamber beneath the pyramid, it is possible to detect the "muons" penetrating the pyramid's walls. Any voids within the pyramid will be manifested by a greater number of "muons" being recorded by the spark chamber. Through this method the scientists and archaeologists hope to discover and map out in a non-destructive manner the location of the hidden chamber they believe exists somewhere in the huge structure. This suggested project, I think, is an outstanding example of science in the service of the humanities, helping man to solve some of his past and present mysteries.

Of course the pyramid project is a specific, isolated example of the symbiosis I referred to earlier, but the integration of our Two Cultures is also under way in more general terms. One indication of this — oddly enough again involving high energy physics — relates to the search for a site for the AEC's proposed 200 BeV accelerator.

As you well know, the possibility of having a scientific laboratory of such importance and economic impact in their area has created keen competition among many communities across the nation. What is less well-known — but has even greater significance — is the unexpected *effect* of this competition on the various communities.

The process of assembling information brought together in each community an unusually broad group of local leaders, state officials, university presidents, professional men from a number of fields, and industrialists. Immediately there resulted a cross-pollination of ideas, an exchange of views, a frank discussion of problems — in short, these communities engaged in an unprecedented self-appraisal. In many cases, I think the people were led to see some of the hitherto unrecognized assets and shortcomings of their area.

As local resources were being mustered, it suddenly dawned on community leaders that the overall attractiveness of an area must be measured by factors which transcend economic considerations. Critics of intellectual and cultural activities were transformed into champions when it became evident that a strong academic base and a broad cultural environment were more highly regarded than amusement parks or dog racing. University presidents suddenly found that their pleas for understanding of long-range goals — pleas that had long been ignored — were blossoming into legislative appropriations. In some cases for the first time in history, legislatures specifically earmarked appropriations for their university's research programs.

This local self-appraisal brought a greater interest in racial harmony. It brought renewed support for education at every level. It brought a greater appreciation for the public library, the symphony, and even children's ballet classes.

The very process of assembling information opened important new channels of communications within the community. And if the people can maintain their momentum, not only will the community be a better place in which to live — it will be in a stronger position to compete for other scientific or technical installations.

Our site search has, I think, helped in another way which is important to me personally. It has helped change the public image of the scientist from a cold, detached individual to that of a very human person with a deep interest in his family and in his community.

All of this indicates, then, that the Two Cultures are, indeed, becoming integrated, and this will be mutually beneficial to each.

Finally, as part of the new blending of scientific, social and cultural interests which seems to be forthcoming today, and in turn influencing it, I think we may see in our country a new rise in the quality of leadership. This should come about as, in this process of better balancing human values with the growth of science and technology, we develop men and women who have both the broad social outlook and the knowledge to make the practical day-to-day decisions in keeping with that outlook.

Perhaps also in a society which fosters this combination of wisdom and knowledge, and which tries to keep its focus on individual human values within an evergrowing nation, we will be able to develop some immunity to the "Anti-Leadership Vaccine" which John. W. Gardner has described so effectively as one of today's new problems in education.

To me, all these aspects of our new awareness, and the new actions we are taking as a result of it, are signs of an important new period of maturity for our country, and I believe they will not only reflect in our building of a Great Society here but will have a profound influence in our relations throughout the world.

In October of 1963, the late President Kennedy flew to Amherst College in Massachusetts to participate in a ceremony honoring the poet Robert Frost. I think that the talk he gave at that ceremony best reflects what most Americans feel should be the goals of this country and our role in the world. And since it bears so profoundly on what I have been trying to say, I would like to conclude by quoting some of this most impressive statement.

"I look forward to a great future for America, a future in which our country will match its military strength with our moral restraint, its wealth with our wisdom, its power with our purpose.

"I look forward to an America which will not be afraid of grace and beauty, which will protect the beauty of our natural environment, which will preserve the great American houses and squares and parks of our national past, and which will build handsome and balanced cities for our future …

"I look forward to an America which will reward achievement in the arts as we reward achievement in business or statecraft …

"I look forward to an America which commands respect throughout the world not only for its strength but for its civilization as well. And I look forward to a world which will be safe not only for democracy and diversity but also for personal distinction."

The Many Faces of Change

Commencement address at the San Diego State College, San Diego, California, June 10, 1966.

To begin with, while constant change is a natural phenomenon, there is also a natural resistance to it. Man, who above all living things is most aware of change, is paradoxically the greatest innovator of change and at the same time a great resister to it. Perhaps it is when change foreshadows an encounter with the unknown that he fears it most. In his book *The Ordeal of Change* Eric Hoffer poignantly describes how, in his experience as a migrant field hand, even the change from the picking of a familiar crop to a new crop he had never picked before had in it an element of fear. He goes on, of course, to discuss many other aspects of change but particularly how, for those societies steeped in the traditions of ancient cultures and trying to enter the mainstream of twentieth century technological progress, facing the drastic change necessary is a traumatic experience. The drive for, and reaction to, such change is responsible for much of the unrest and upheaval in the world today and will be probably for some time to come.

Yet associated with change there is hope, and hope is an element essential to human survival and progress. Why this ambivalence toward change? Why is change so meaningful to man? The reason is that we, of all nature's creatures, are the only ones who play a major role in our own change, our own evolution. Man, the only form of life which intentionally accumulates knowledge and transmits it to its offspring, is responsible for a creative evolution which has so surpassed natural evolution, which has lifted him so far above other forms of life, that he now completely dominates that life and much of his environment.

Remarkable as our creative evolution has been to date, all indications are that our present state of accomplishment is only a prelude to what the future holds. On the basis of where we stand today — of what we know, what we are capable of learning and doing — man's future seems fraught with the most amazing possibilities and probabilities. The changes which can, and no doubt will, occur within your lifetime — perhaps within the next few decades — could be the most significant changes ever faced by man. And how we deal with them and direct them could decide the future of the entire human race for some time to come. This may seem like a far-reaching statement but there are, I believe, good reasons why it can be made with some confidence. The best one is that we now seem to realize that the scientific and technological progress we have made within just the past few decades places us in a position where we more or less control our own future on this planet, and there seems to be little doubt that we are now making a conscious effort to do just that.

I said "more or less" control our future because there is a significant caveat involved here. It is frankly and simply a matter of whether, for all our knowledge and techniques, we can prove ourselves rational enough to plan and work together for our ultimate survival and advancement. The gap which still exists between the rationality of our scientific and technological work and some of our social and political thinking must be narrowed. The global civilization we have created, the power now at our disposal and, in general, our rising dominance on the earth, make this a necessity. There are signs that we recognize this in much of our national planning and in the many worthwhile programs of international cooperation which are taking place at present, but it seems we still have a long way to go in solving some major problems, the most important of these being, no doubt, the resolving of international differences in a peaceful manner. Unfortunately, in the solving of many of our major problems today, time is running out.

Time is a most precious commodity in dealing with most areas of change today because the nature of our interdependent technological society is such that when changes take place in it, they take place rapidly and they tend to produce a chain reaction. And in this new society, as in a nuclear chain reaction, a period of criticality can be reached which, if uncontrolled, may have disastrous effects.

If I have been speaking rather generally up to this point, let me be more specific. Let's explore for the next few minutes some of the major developments and changes which seem to be in store for us, and in doing so let's view them in terms of the challenge and response mechanism which, to a great extent, has been the basis of our creative evolution to date and should continue to be to a great extent in the future.

Perhaps we should begin by taking a very objective look at ourselves and our relationship to our environment. The best way to do this might be to place ourselves in the position of an astronaut returning from a distant planet and looking down at Mother Earth. From this position we can develop some interesting perspective and perhaps, in the absence of radio communications (we have turned off our set for a while), do some quiet thinking. There is no need to dwell on the matter of the insignificance of Earth in the cosmic scheme of things. It is home and, therefore, most significant to us. But if we begin to think about what makes it home some remarkable and disturbing thoughts can be developed. The earth is approximately 8,000 miles in diameter and has a mass approaching 7 sextillion (10^{21}) tons, but of this large mass all life and life-supporting material rests on its surface in the form of a very thin skin. So thin, in fact, that if all this material were stripped from the surface of the earth, rolled into a ball and placed alongside the sphere which held it, it might appear as a tiny pea alongside a large apple. This is the way it has been for millions of years and this is essentially the way it will be for millions of years to come if we behave sensibly.

But switching from our role as astronaut to one somewhat akin to that of the Jolly Green Giant, let's examine that tiny pea of life and life-supporting material in some detail. (And at this point we ought to spread it back over the earth where it belongs.) While the amount of life-material has remained constant, remarkable changes have taken place on it. Various ages and forms of life have come and gone on it. The latest to arrive, of course, was man. And actually we have had very little effect on Earth until quite recently. Now (within recent decades) things are beginning to happen which will change all that. The changes involve basically two kinds of related growth.

We can view human development from two standpoints as it relates to the rest of the material. The first concerns the multiplication of individuals — a subject of popular discussion these days but, more importantly, one evoking much serious thought and action. We all know it has taken a million years for the human population to reach its present figure of about 3.5 billion but that in only the next thirty years this figure will double and that, if unchecked, it would continue at a geometric rate of increase until, in a relatively short period of time, the earth would be stacked with nothing but people. Of course we realize that such a situation could never arise because before it did all sorts of pressures would develop which would either destroy us or drive us into some action for survival. The most obvious one, naturally, would be a relative decrease in the food supply. Mass starvation is the major spectre which Malthus conjured up in his *Essay on Population* 168 years ago as a controlling factor in population growth. Certainly hunger haunts much of the globe today and enormous efforts are under way to change this situation. But food supply is not the only problem to be concerned about with the increase of population. I will be speaking about others shortly.

The current pressure of population and our projections of its growth have already served as a call to action to most nations to control their populations. I do not want to go further into detail on the population problem *per se*, because I think most of you are familiar with the subject, but suffice it to say that the population explosion is a primary human challenge and one to which we are responding — hopefully — in time. Even with the success of this response, we know that we must expect by the end of this century a world population at least double that of today.

Directly associated with the multiplication of human beings on this earth is another problem, or series of problems, which Malthus, and probably a good many people less than 40 or 50 years ago, could not have foreseen. For the world is not only populated with individuals, it is populated with organisms far more consuming than people — though they are composed of people. It is populated by societies — and particularly today by the scientific and technological society. While people have almost always banded together in societies, those societies have never seriously affected the total relationship of human life to its environment. The story is quite different today. Gathering and applying knowledge at an increasing rate, the technological society of the Twentieth Century is giving man the ability to extract from, use, and change the materials of his environment to an incredible degree. And as population increases (even though controlled), as more of mankind become members of the technological society, so will the relationship of man to his environment, to his life-supporting materials, change. We are just now beginning to recognize the challenges of that change and in the future they will be enormous. Let's explore some of them and see how we might respond.

If I use figures relating to the United States, it is not because they are typical — but for precisely the opposite reason — because they represent something of the goals for which a good part of humanity aims, and, therefore, will give us a good idea of what will have to take place in a world of both increasing population and rising standards of living.

First let's look at energy. For about 6,000 years of recorded history the prime source of man's energy was that which he could generate by his own muscle power. In many areas of the world — perhaps as much as two-thirds — a large part of the productive power is still this same muscle power. In our own technological society, however, human labor provides only a few ten thousandths of one percent of the work performed in all our industry. The rest is done by the more than 10 trillion mechanical–electrical horsepower hours used by industry each year. This figure, of course, rises every year. In addition to this mechanical energy used by industry, we use an even larger amount of energy for process and space heating and for transportation. In the United States, with only 6 percent of the world's populations, we use about 35 percent of the world's energy output. Our total energy requirements have been increasing at a rate of about 3.5 percent a year.

Where does all this energy come from? By far the greatest amount comes from heat — heat generated by expendable but not unlimited fuels. Our industrial revolution, of course, grew on the burning of fossil fuels, and our technological revolution of today is still based largely on this source of energy, primarily converted to electricity. It is estimated that between now and the year 2000 we will consume about 500 billion tons of coal and that the supplies to meet this demand are more than adequate. Estimates also indicate huge reserves of oil. But can we really be sure how far these supplies will go? In the past we have consistently underestimated our needs for energy. Perhaps we are still doing so. Can we really predict just what level of industrialization the world will reach by the year 2000?

With a world of six billion people striving for a standard of living similar to what most people in the U. S. and Western Europe enjoy today, and this will be rising too in the future, think of the energy demands which would have to be met. And future demands will arise out of problems we are only beginning to face today. A much larger percentage — perhaps as much as a third or more — of the world's land mass must be made arable and more productive. Among other things, this calls for the bringing of water to arid lands and the production of incredible amounts of chemical fertilizers — both calling for huge expenditures of energy. A far greater surface of the earth will have to be made habitable — and comfortable — if we are to have the "elbow room" necessary to preserve our survival, not to mention our sanity. This will, no doubt, call for the construction of cities in regions where few people would care to live now — cities which might have to be lit, heated or air-conditioned almost continuously.

In addition to these demands on energy, other demands for energy are arising to which we previously had not given much thought. One of these relates to the fact that our new dominance over our environment prohibits our unlimited expending of natural resources, and particularly the uncontrolled disposal of waste in the environment. We now realize that we can no longer merely take any material we want from the earth, change its form or composition, use what we want of it and then indiscriminately try to return to nature what is left. The day is almost at hand when we will no longer be able to burn, bury, stack, store or dump into our soil, atmosphere or waterways everything for which we have no immediate further use. There are two main reasons for this. One is that nature cannot absorb and reprocess our waste as rapidly as we need to dispose of it, and hence we will be overwhelmed by all types of environmental pollution if we do nothing to prevent it. Secondly, our natural resources, though vast, are not unlimited and some day in the future we are going to have to recycle and reprocess almost everything we use.

I mention all this in regard to energy because I think it brings to light the tremendous demands for energy resources which are inevitable in our future. In view of this it is most fortunate for man that within the past twenty-five years he was able to discover a new source of energy — the nucleus of the atom — and to develop it for economic use. Much of the change we will see in the future will depend on the use of cheap energy, primarily in the form of electricity, and much of this will depend on the progress made by nuclear energy in supplementing our fossil fuels and some day replacing them as the prime source of energy in the world. In reaching this position, nuclear power will help to solve some other problems we are facing. Because nuclear power plants release no combustion products into the atmosphere, their use in place of additional conventional plants will help to control air pollution. Electric-powered means of transportation — both private and public, automobiles and trains — receiving their electricity from cheap nuclear power will further help to reduce such pollution. In addition, the use of uranium and thorium as nuclear fuels will release fossil fuels, which are valuable sources of hydrocarbons for the chemical industries, and for which there are an evergrowing number of new uses.

A major change, then, which we can expect in the years ahead if we respond wisely to our needs, is one relating to our development and use of energy. I think we will see the most economic and efficient use made of all sources of energy — water power, fossil fuels, solar energy, and the atom. Nuclear energy will probably make its greatest strides with the development of breeder reactors which will be able to produce, from the world's large supply of uranium and thorium, more nuclear fuel than they consume while generating electricity. Eventually we will be able to develop controlled fusion and thus be able to harness, as an almost limitless source of energy, the heavy hydrogen atoms in the world's vast oceans.

Another major factor we must deal with in our future relations with our planet — in the days of great change ahead — is our use of resources other than those used for energy. One of the basic reasons for man's advancement has been that his creative evolution has given him the ability to manipulate and transform the materials of his environment to such a great degree and with such ingenuity. We have discovered all the natural chemical elements — and created quite a few synthetic ones — and we have done remarkable things with them. We are today synthesizing products from inorganic materials which must make nature jealous, and we have arrived at a point of remarkable understanding and use of organic materials and the processes of life.

This progress is bound to continue, and, even in view of our increasing numbers in the decades ahead, we can have an abundance of materials to work with provided — as I pointed out before — we learn to reuse almost everything. Some day in the future the word "waste" may apply to very little that man has finished using. Waste will be considered far more as it really is — a temporary state of material between uses in various forms. Man will, to a greater extent, take over nature's job of recycling materials because there will come a day when basic material resources will be relatively so scarce that it may be easier, quicker and cheaper to do this than to explore and dig deeper for more materials.

Another reason why we will have to exact such controls over the materials we use is that in our abundant use and change of form of these materials, we will have to maintain throughout our environment the balance of nature necessary to support life as it exists here on earth. Today, because of our growth and dominance on earth, we are concerned as we have never been before with the total ecology of our planet.

In line with this thinking, I would like to mention a few aspects of some problems I touched on before — those involving the two natural resources we most depend on, our water and our atmosphere.

To meet the needs of our future growth and accomplishments, we are going to have to greatly increase our control and use of the world's water supply. At the moment we do very poorly in this endeavor. It is ironic perhaps that on a planet about two-thirds of which is covered with water, less than one percent of that is fresh water available to us. More than 97 percent is in the oceans, of no immediate use for drinking or irrigation. Approximately another 2 percent lies frozen in glaciers and ice caps. What fraction of one percent that does arrive on the land, in the form of rain and snow, is poorly distributed as far as a great many of us are concerned, and we have not learned to control it very well nor use it very wisely. All this will have to be changed in the future because in the next 20 years the world's demand for fresh water will double.

The consumption of water varies tremendously in different areas of the world. In some tropical countries the average per capita consumption is less than 5 gallons a day, whereas in a certain area here in California — not too far from here — the per capita use of fresh water is the highest in the world — 500 gallons a day! What is most significant about the consumption of water is not our personal use of it but its relation to agriculture and industry — the two areas of accomplishment whose progress will most affect man's future. In our country, close to 50 percent of our water is used for irrigation, while 40 percent is used by industry. The latter fact is not surprising when one learns that it takes 770 gallons of water to refine one barrel of petroleum, 65,000 gallons to produce a ton of steel, and 250 tons of water to make one ton of paper. The generation of electricity is also a great consumer of water, since for each ton of coal burned in a steam-power plant 600 to 1,000 tons of water are required. The fact that we are largely a nation of meat-eaters has a significant effect on our agricultural use of water, as it takes 3,000 gallons of water to produce just one pound of beef.

A new awareness of man's water needs in relation to the world's current and future growth has already started a flurry of increased national and international thinking and activity concerning the development and use of water resources. This, of course, has been prompted further by serious droughts in many areas of the world and by a reaction to the disgraceful water pollution we have allowed to develop in our own country.

What changes can we expect to see regarding man's use of water in the years ahead? We will see, first of all, a vast new effort toward increasing man's knowledge of all water resources on the earth. Such an effort has already begun by UNESCO's sponsorship of an International Hydrological Decade. During this decade scientists of 70 nations will pool their efforts and facilities to map climatic conditions all over the world, to study precipitation, groundwater levels and stream ecology, and the measurement of water's capacity for self-purification. We are going to see extensive programs for the most efficient control and use of water that reaches the land. Huge new dams and flood control projects will rise in many parts of the world. Lakes, rivers, waterways, and watershed areas will have to be made or remade by man. We will have to undertake continental engineering projects of enormous proportions to accomplish this. We will have to discover and use more of the large sources of water under the land. We will have to develop large-scale, efficient, and economic desalting plants, probably nuclear-powered, to get fresh water out of the sea and possibly pump it great distances inland too. And perhaps of greatest importance we will have to learn how to use and reuse our water supplies over and over again before the water travels back to the sea. This will involve an all-out war on every type of water pollution and waste — a war which we can and must win quickly.

Just as we must change our ways in the management of water resources, so must we face up to a new relationship with our atmospheres. Air pollution has already become a serious problem in many urban and industrial areas. With the growth of fuel-burning industry and transportation, this problem could grow to disastrous proportions — in some places it is already approaching this — unless we quickly change our attitude toward it and our techniques in combating it. In the long run the problem is not only one of direct health hazards, for it has been predicted that, at the rate we are currently adding carbon dioxide to our atmosphere (six billion tons a year), within the next few decades the heat balance of that atmosphere could be altered enough to produce marked changes in the climate — changes which we might have no means of controlling even if by that time we have made great advances in our programs of weather modification. I, for one, would prefer to continue to travel toward the equator for my warmer weather rather than run the risk of melting the polar ice and having some of our coastal areas disappear beneath a rising ocean. If we are going to have undersea cities in the future, as some predict, let's have the kind we can build and inhabit rather than new versions of Atlantis resembling Boston, New York or San Diego.

Up to this point, I have devoted a lot of attention to our relationship with our environment. And I have done this because, as I indicated at the outset, our creative evolution, as well as our increase in number, will be seriously affecting this relationship. I also indicated that the reason for that creative evolution — for our ability to grow and effect so many changes on earth — was that we alone among all forms of life have this ability to gain, accumulate, and pass on new knowledge. While the growth and application of that knowledge have been taking place for thousands of years, the rate at which it has been taking place has been increasing phenomenally in recent years. I think that if it were possible to plot a graph on the growth of human knowledge and match it against those curves showing the growth of population and of man's consumption of resources and production of goods, we might see a remarkable relationship — a relationship in which knowledge was the prime mover.

I do not think it politic to devote much time to knowledge at a commencement ceremony — not that your interest in it or pursuit of it should end today — but in relating to my theme of change, I must say a word or two about the Information Explosion, certainly the greatest source of influence on future change. The growth of knowledge in our era makes lifetime learning a necessity. Educational obsolescence will be an ever-increasing factor in future individual advancement. It is already said that to keep up in his field a Ph. D. today must re-educate himself at ten-year intervals. All this, of course, forecasts vast changes in educational and learning methods and techniques.

Today we are creating waves of new knowledge which would all but engulf us and become useless if we were not also developing new methods of handling them. For those who insist on statistics, let me point out a few confined only to scientific knowledge. At present, each year, about 300,000 articles are published in some 6,000 scientific and technical journals in the United States alone. On a worldwide basis a total of about 35,000 journals publish some 2,000,000 articles a year written by some 750,000 scientists in about 50 languages. Looking ahead only 50 years, it is predicted that in the U. S. alone some 1,300,000 scientists will be publishing annually some 3,000,000 articles in about 60,000 journals, while around the world a total of about 8,000,000 scientists may well be writing over 20,000,000 articles in some 350,000 journals — if indeed "writing" is then still a major form of recording and communicating knowledge. And recording and communicating all this information will not be the only problem — interpreting it, correlating it, and putting it to use will be among the others. In view of all this, is it any wonder that such importance — almost reverence — is attached to the work and development of the computer today?

If any one invention is to be credited with guiding the future of man in the decades to come, it will probably be the computer. The day is not far off when almost all fields of human endeavor will be in some way influenced, or directly controlled, by the computer and assemblages of its solid-state cousins. Today the computer already plays a role in such varied fields as medicine, law, meteorology, humanistic studies, freight transport, air traffic controls, and a cross-section of research, industry and business too numerous and too obvious to mention here. What also may be obvious here, but is important to emphasize, is that with the use of the computer and associated electronic controls, man now has the artificial nervous system to match and coordinate the mighty mechanical muscles his use of the power machine and energy resources provides. The computer also gives him the ability to look into the future to learn where and in what way he will need to apply his intelligence and energy.

All this places man himself in an entirely new light, and, viewed in this new light, many seriously question what his future role will be — not only in relation to his environment and the products of his creation but to his fellowman. And this brings me to some final thoughts on change.

Probably the greatest challenges we face in the future relate to decisions we will have to make and which no computers will make for us. These are decisions related for the most part to human values and goals. Ultimately, they will determine how we live, whether our creative evolution is successful in the truest sense of the word. We know we will be able to create great urban complexes to house and care for our growing populations. We know we will be able to increase the speed and ease of our transportation so that no places will ever be more than hours or minutes apart. We know that there will come a day when almost anyone on earth will be able to communicate with almost anyone else — almost instantly. We know that at that time we will be able to see in our own homes events taking place anywhere in the world and perhaps far out in space. We know life will be much longer, much freer from suffering or discomfort, and that, for most, it will be mostly hours of leisure.

But, what will fill those hours of leisure? Will they be filled with some purposeful activity, creative pursuits, real pleasures and reasons for living or will they be spent in just existing, in a meaningless passing of time, in an idle, perhaps drugged, state? When we can see places anywhere on earth in our own homes, will many of us want to travel there? And what will our great new cities be like? Will they be centers of harmonious social relations, of learning, of art and cultural traditions, or places of unrest and violence with few sanctuaries for peace of mind and nourishment of soul? Will they be well-conceived, spacious complexes of housing, parks and market places, attractive and pleasant places in which to live and grow, or will they be ill-planned, unattractive sprawls of buildings, crowded, and confining to body and spirit? In short, will our new cities be our homes or our prisons, and will our new lives, filled as they may be with the wonders we have created, be filled with purpose and a joy in living?

All these are among the greatest questions that lie ahead for us and perhaps the greatest challenges to our creative evolution.

I have spoken to you today of change in terms of questions to be answered, problems to be solved and challenges to be met because I know that at this stage in your life you are up to and worthy of hearing the case stated this way. It would have been far easier for me to sketch for you only the outline of a "Gee Whiz World," to paint a bright picture of incredible progress, and to view the future only through rose-colored glasses. If I did so, you may have been amused and, to some extent perhaps, amazed, but I don't think you would have been convinced. You are already too wise for that. You know there is much work ahead. You know the world will be what you make it.

Today more than 2,000 of you graduate from this college. Across the land this year approximately 540,000 other students from some 1,700 other colleges and universities will receive their bachelor's or first professional degree. And during the next ten years another 7 or 8 million will follow, all joining you, all taking their place alongside you to help mold the future of this country and the world. If it is a formidable task which confronts your generation, surely it will be an exciting one and one proving worthy of your finest efforts. I say this because I know you will succeed. And as you encounter the many faces of change in the rapidly changing world you create, I think you will see on most of them something resembling a rewarding smile. It will be similar to the one worn by that astronaut returning to earth. Because if we do our job well here, we should always be able to feel in returning to this very special planet that there is still no place like home.

A New Era of International Science

Remarks made at the International Conference on Nuclear Physics, Gatlinburg, Tennessee, September 13, 1966.

The internationality of science is far from being a new phenomenon. I'm sure that all of you are familiar with the history of early exchanges of ideas and information among scientists of different nations. Within this history are even stories involving such exchanges between nations at war with one another; for example, during the 18th and early 19th centuries French and English men of science were allowed to safely cross each other's national borders though they were politically and militarily "enemy lines." It was also during this era that Sir Humphrey Davy received an award from the French government while his country was at war with France and that Benjamin Franklin, during the American Revolution, requested American naval ships and privateers to grant the British vessels of Captain Cook safe passage because of the scientific nature of Cook's expedition to the South Seas and the Antarctic.

But all this was true in part because of the nature of science in those days, because it had yet to be strongly allied with technology and thus had very little to do with people, political activity or the power and prestige of nations. All that, of course, has changed — but, relatively speaking, only very recently.

The change is related not only to the alliance of science and technology but also to the explosive growth of science in general. The history of physics offers some interesting examples of scientific progress particularly as it relates to the internationality of science.

Prior to the 20th century the body of knowledge in physics grew from theories and experiments of men of many countries — among them Galileo, Gilbert, Descartes, Huygens, Newton, Faraday, and Roentgen. The close of the 19th century found many scientists — physicists in particular — in much the same frame of mind as Alexander the Great when he sighed that he had no more worlds to conquer. They too believed that the last frontiers of their world were in sight. After all, what could come after the discovery of the electron, X-ray, and radioactivity?

Seldom has resignation been so quickly shattered. Einstein's theory of relativity and Planck's equations of quantum action revealed how much still remained unexplored in the scientific world and gave rise to important future discoveries. As a consequence, there were the discoveries of the neutron by Bothe, Joliot-Curie and Chadwick, deuterium by Urey, Brickwedde and Murphy, slow neutrons by Fermi, Segré and their co-workers, and fission by Hahn and Strassmann, all this continuing to prove the internationality of science and the advantage of a free exchange of information within the scientific community.

But, exciting as these advances were to the scientific community, they did little to capture the imagination of the public or change the rather isolated position that science held in society at that time. The events of World War II shattered that isolation and lifted the scientist to a new position among his fellowmen. With the development of nuclear fission, radar, the rocket, and the computer, science became an important instrument for national survival. By the end of the war governments were convinced of the importance of science, at first primarily for defense, but soon for its applications to health, welfare, economic growth and even national prestige. By 1950 our new scientific revolution had begun in earnest. What is perhaps most remarkable, and ironic about this revolution, which has had such impact through practical applications, is that it was initiated by what was at the time one of the most pure and theoretical of sciences — nuclear physics. It literally took the awesome explosion of a nuclear weapon to awaken the world to the power of scientific knowledge and the world has not been the same since — nor has science!

The general effect of this rude awakening was two-fold. An immediate result was that it tended to emphasize nationalism in some areas of science by making them a matter of national concern related heavily to defense and security. But at the same time it gave the scientists around the world the greater prestige and resources they needed to grow, and science springing from a universal heritage and being international in nature was bound to rise above the confines of nationalism. The result has been remarkably favorable to international cooperation in nearly all areas of science and I believe this may ultimately be mankind's greatest blessing.

I base this belief on two trends of thought. The first, and more obvious, is that international cooperation in science will accelerate those advances of mankind which if applied wisely and equally around the world will help to eliminate the causes of political and economic strife. Stated more simply, the more we can do through cooperation in science and technology to satisfy basic human wants throughout the world, the more human conflict we are going to eliminate.

My second trend of thought is that internationality in science extends the rational processes of science to other human activities in all countries, and that the ascendancy of scientists within their respective countries will influence national leaders and their people to deal with problems in a more rational, and hence more peaceful and productive way. I certainly do not think that science or scientists hold the answers to all human problems, but if we view science in its broadest terms, that is, as a highly organized and penetrating pursuit of knowledge and truth, some good is going to come by having the attitudes and approaches of science applied to other areas of human activity.

Bearing all this in mind, it is obvious that our international scientific conferences, which are growing each year in number and size, are becoming more than meeting grounds for the exchange of scientific ideas. But rather than dwell on this theme — which could be another talk in itself — let me stay with my main topic and review at this point some of the international scenes relating directly to scientific exchange. Let's see, as the expression goes, "where the action is" and why it's there.

While much emphasis is being placed these days on international cooperation in the applied sciences and technology, I think it is vital that we recognize, support, and expand the excellent international cooperation taking place in fundamental research. It is an obvious fact that our future world is going to be built more and more on a foundation of scientific knowledge, and basic research will continue to form the building blocks of that foundation. To advance science and take advantage of what it has to offer, all nations of the world must be able both to contribute to and borrow from our stockpile of human knowledge. Not all can contribute equally, but even those who cannot make a major contribution during any one era are obligated to their own people to keep up with and use this knowledge to its fullest.

As Alexander King, Director of Scientific Affairs for the Organization for Economic Cooperation and Development (OECD), put it: "On the one hand scientists of a particular country contribute pieces to the mosaic of knowledge; equally the technologists and entrepreneurs of the same country are free to exploit it through technology. In all but the largest countries in the world, only a small fraction of the new discoveries exploited originate in the country in question; yet unless there is considerable research effort and deep understanding of the problems and techniques of fundamental research, a country will not possess a breadth and depth of scientific understanding to enable it to select from the world's store of knowledge those elements which are relevant for its development."

What is not said in this statement, but which must be made clear, is that one never knows when those who have been borrowing from the "store of knowledge" may come forth with some major contribution to it. Provided we keep the channels of information open to all and nurture all scientific talent, that flash of insight which has led to so many great scientific discoveries may occur anywhere, at any time. No nation has a monopoly on scientific talent.

The case for international cooperation in *fundamental research* then lies in the fact that we need each other to advance science so as eventually to advance ourselves. Besides needing each other intellectually, a situation has arisen today in the physical sciences which demands that we cooperate economically in our pursuit of knowledge. I refer, of course, to the growth of what we call "big science," those disciplines which demand large facilities and expensive equipment, beyond the financial means of many individual scientific organizations and even many nations. All research today, by the way, continues to grow in cost. In the United States research costs for a given project rise at the rate of about 5 percent a year.

Science is meeting the need for cooperation in basic research in many ways. Most obvious, of course, are the exchanges of information and meetings of such groups as our own important organization, the International Union of Pure and Applied Physics. As you may know, 34 countries are members of the union and work is conducted through its commissions, each for the study of a different aspect of physics. The union has one affiliated commission — the International Commission for Optics — and 14 special commissions including those on Symbols, Units and Nomenclatures; Thermodynamics and Statistical Mechanics; Cosmic Rays; Very Low Temperatures; Physics; Low-Energy Nuclear Physics; Nuclidic Masses; and Physics Education. With its meetings and publications the union has played an important role in the dissemination of scientific information and in encouraging international cooperation in physics.

It is essential that we have an efficient body to review, consolidate and organize worldwide information in physics, as in all science. It is impossible to predict the rate of growth of scientific knowledge, but it has been pointed out that, if the output of scientific papers continued to follow the current curve, the mass of the total volumes of *The Physical Review* alone would exceed that of the earth itself in the mid-21st century. I do not mean to alarm those subscribing to, and keeping all back issues of, *The Physical Review*. Physics, of course, is one of those areas of science which today call for the economic cooperation I mentioned before. This is particularly true in the special discipline in which all of us here share an interest, nuclear physics.

Fundamental nuclear research is characterized by incredibly small phenomena and fantastically large and expensive machines. Fortunately, the international pooling of resources provides one solution to this problem. The Research Center of the European Organization for Nuclear Research (CERN) is an outstanding example of what can be achieved by international cooperation in the world of high-energy physics. The center and its impressive facilities are maintained by 13 European nations which also maintain active collaboration in fundamental nuclear research with similar centers throughout the world. CERN's 600 million electron volt proton accelerator stands as a monument to the international character of the organization. The magnet was made in France; the main generator in Germany; the coils in Belgium; the vacuum chamber in Sweden; the pumps in Germany; and the radio-frequency system in the Netherlands. The 28 GeV proton synchrotron is a most significant facility at CERN both in physical magnitude and scientific reputation. I am pleased to add that there is some family pride involved in this machine, the sister to our Alternating Gradient Synchrotron at Brookhaven. CERN is going forward with the construction of storage rings designed to allow experimentation with colliding proton beams which, under special conditions, would be equivalent to experimenting with a machine of 1.6 trillion electron volts (1,600 GeV). The success of the CERN cooperative endeavor points a way to the future for larger and broader programs involving more complex and more costly facilities which, in turn, will demand the pooling of both financial and scientific brain power resources.

Another example of international cooperation in fundamental research is the newest establishment of the International Atomic Energy Agency, the International Center for Theoretical Physics at Trieste. The center is the first scientific research institution to be established under the sponsorship of the United Nations and represents a new kind of international collaboration. The three main objectives of the center are: international cooperation, emphasis on the role of developing countries, and the advancement of theoretical physics. It was generally agreed when the idea of the Center was first raised that theoretical physics offered the most fruitful field since work of a high order has been done by scientists from a number of the developing countries. There is the added advantage in focusing upon theoretical physics that in many of the developing countries it is often one of the first subjects to be studied at an advanced level, with the practical advantage that progress might be made quickly without the need for a large outlay for laboratories.

All too often scientists from the developing countries, after completing research in one of the scientifically advanced countries, would find themselves confronted with the choice of leaving their own country permanently or living in virtual scientific isolation. Now such a scientist can come to the center for a stay as a fellow or associate and meet some of the outstanding minds in his own and closely related fields.

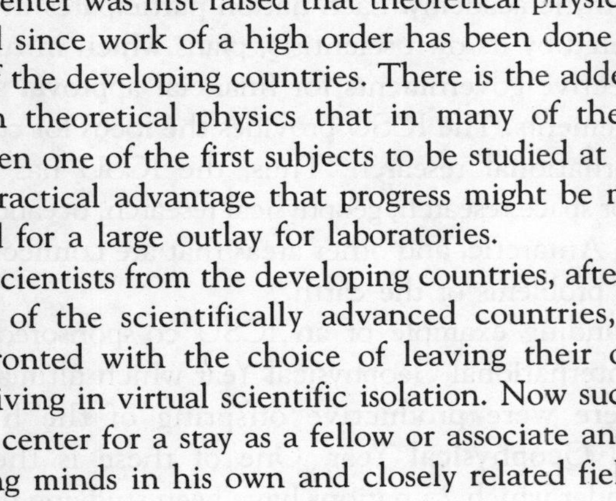

International Atomic Energy Agency symbol. Reproduced by courtesy of the International Atomic Energy Agency.

So far I have spoken mainly of physics, but international cooperation in science is growing in all basic disciplines. An organization responsible for a remarkable amount of worldwide cooperation in scientific investigation is the International Council of Scientific Unions (ICSU) which, of course is the parent organization of our International Union of Pure and Applied Physics.

The successes of the ICSU attest to the value of the cross-fertilization of scientific ideas. Each of its 54 member countries is represented by its principal scientific academy. Each nation participates in accordance with plans formulated by its own scientists, plans which in turn are submitted to their respective governments for financial approval and any required formal arrangements. The ICSU provides the focus for coordinating these areas of international research. Thus, the ICSU has created working committees for space research, geophysical research, oceanographic research, studies in the Antarctic, and other areas that are connected with regional or worldwide problems of the earth.

The outstanding example of an ICSU co-sponsored project was, of course, the International Geophysical Year which ultimately involved 66 nations. There were productive offspring of the highly successful International Geophysical Year. One of these is the Indian Ocean Expedition under which 25 nations have been studying that body of water which covers 14 percent of the surface of the earth. In the investigation of the solar terrestrial relationships, 60 countries cooperated in the International Year of the Quiet Sun. And the Upper Earth Mantle Project has been undertaken to study the earth's upper mantle and its influence on the earth's crust. Unique among the research programs carried out during the International Geophysical Year is the Antarctic Research Program whereby 12 nations — including the United States and the Soviet Union — agreed by treaty to reserve the Antarctic for 30 years solely for scientific research. During this time all territorial claims to any part of this vast continent will be held in abeyance.

Under the auspices of the ICSU, the World Meteorological Association, consisting of 126 member nations, is planning a world weather watch which in the near future may make it possible to receive reliable daily forecasts as much as two weeks in advance and further the goals of large-scale weather modification and control. Toward this end, two World

Meteorological Centers have already been established in Washington and Moscow, and the third is to be established in Melbourne, Australia.

These worldwide research programs attest to another fact of life concerning scientific investigation today and that is that in much of our research the entire earth, its land, oceans, atmosphere, and all its life, becomes our laboratory and the source of our studies. Therefore, national boundaries must be overlooked — as they are by the object of our investigation, nature itself.

The latest research program to be coordinated by the ICSU is the International Biological Year. Thus far, 43 nations are participating in a worldwide study of organic production on land, in fresh waters, and the sea, in order to estimate how man can most effectively manage his environment.

Up to this point I have spoken mostly about international cooperation in fundamental scientific research. But as I indicated at the outset, the impetus and growth of most of our science today spring not from the basic curiosity of man about his world but from man's needs and from the belief, only recently engendered in our history, that science can radically better our lives here on earth. To this end international efforts in applied science and technology are making tremendous strides, and these efforts are only a beginning to what can and will be done in the future.

Let me mention only briefly some of the major international organizations and the programs in which they are involved today.

As we all know, the production of sufficient food is a major problem in the world now and will be growing in proportion. The problem of hunger in many areas of the world requires our immediate attention and a concerted effort. Under the leadership of the United Nations Food and Agricultural Organization, a "Freedom from Hunger Campaign" has been launched which involves 80 countries working with UNESCO, UNICEF, the World Health Organization, the International Atomic Energy Agency, the International Labor Organization, and the World Meteorological Organization. This combination of international scientific and educational groups is exploring every natural, technological, economic and social means of providing basic food staples for human beings everywhere.

Closely related to the world's shortage of food is the scarcity of clean water, and one of UNESCO's most significant programs confronted with this problem is the International Hydrological Decade. During the period 1965–1975, it is planned to establish a global network of stations to measure and track water in its cycle from rain to the underground water table and back into the atmosphere. In addition, studies concerned with the expansion of fresh water supplies and how we can recover them are also being carried out.

As a further example of international scientific cooperation in meeting the world's water problems I should mention the First International Conference on Water for Peace scheduled to be held in Washington, D. C., next year. The purpose of the conference will be to report on the action taken and the progress made in meeting man's needs for pure water. It will represent a major expression of many nations to help each other scientifically, technically and economically, as well as a major effort to determine the current scientific and industrial capabilities of the world community in the development, conservation and use of the world's water resources.

Returning to the work of UNESCO, let me point out that since its establishment UNESCO has sought to promote science and technology for the benefit of the developing countries and has focused attention on two important questions in this regard: (1) the development of a coordinated national scientific community in each of its member states and (2) the initiation of research programs and the necessary planning and financing of work involving the natural resources and industry of each country.

The creation of the International Atomic Energy Agency in 1957 was a giant step toward the goal of productive joint efforts in the civil uses of nuclear science. Ninety-six nations have now joined together in this agency to coordinate nuclear policies and programs around the world. The agency is much more than an opportunity for dialogue and contact, and in the words of its statute, the objectives are "to seek to accelerate and enlarge the contribution of atomic energy to peace, health and prosperity throughout the world," and to insure "that assistance provided by it or at its request or under its supervision is not used in such a way as to further any military purpose." In addition to its all-important safeguard function the Agency has, through its technical assistance activities, made excellent progress in promoting the application of nuclear science and technology to the economic progress of the developing countries.

Some of the principal fields included in this assistance are radioisotope production and utilization, reactor operation and use, radiochemistry, radiation physics, solid state physics, food preservation, health physics, and radiation protection.

Last year, during a trip around the world, I had the pleasure of visiting many scientific centers, which were sponsored or guided by the IAEA, in both developed and developing countries, and I was highly impressed by what this international organization has helped to accomplish.

While I do not have time to go into detail on their many activities, no list of international organizations making worthy contributions to science and technology would be complete without the names of such organizations as the European Atomic Energy Community (EURATOM), the Organization for Economic Cooperation and Development (OECD), and the new European space organizations — the European Launcher Development Organization (ELDO) and the European Space Research Organization (ESRO).

Toward the beginning of my talk I stated that international cooperation in science may become man's greatest blessing. In concluding, let me reinforce this belief. As our U. S. Ambassador to the Federal Republic of Germany, George McGhee, has recently pointed out, "It is roughly estimated that 20,000 American scientists now join with some 80,000 foreign colleagues at approximately 2,000 international scientific meetings each year." All these meetings, all these contacts — intellectual, technical and personal — surely represent more than the mere exchange of information. They are exchanges which spark creative scientific thinking, they are forms of what we might broadly call "cultural exchange" and they are meetings of minds which more and more are influencing the courses of nations and the course of mankind. As the role of science grows, as scientists become more important in the affairs of men, not only in scientific and technical matters, but in those social and political areas which today's science greatly affects, the internationality of science may offer us a very special kind of "spin-off." It may offer us the kind of understanding which, coupled with the physical benefits provided through scientific progress, will lead the world to a lasting peace.

We who are privileged to meet here in this International Conference on Nuclear Physics should be grateful for the international heritage of our field of study. I hope we will all continue to work together in pursuing our common goals, in exploring the mysteries of nature, in seeking basic truths. In this way I believe we will best advance our own aims, the position of science and the cause of all mankind.

Goals in Public Understanding of Science

Remarks made at the Arches of Science Award Banquet, Pacific Science Center, Seattle, Washington, October 19, 1966.

Why should it be so important that our people understand something about science? Why should people know what a scientist does, what motivates him, and the implications of scientific results? What are our goals?

In my opinion, there are two major, interrelated goals. The first is philosophical, involving the quality and the dignity of human life and the fullest use by man of his capacities; in sum, the increase in man's significance. The evolution of cultures from the primitive to the sophisticated, from intellectual poverty to intellectual riches, from needless fears and superstitions to the relative security of relative understanding; these forward movements have been paced by the painstaking ascent of the mountain of knowledge. Our lives are better for knowing the sources of lightning and earthquakes, and for understanding that the earth, far from being the center of the universe, is but a tiny speck in a pageant of immense grandeur. We are enriched by knowing why a tree is green and how it captures and stores the energy of the sun, and thereby grows. To know that we inherit our individual characteristics from a lovely, orderly strand of molecules and to comprehend the arrangement of those molecules seems to me to be a triumph for the human spirit, entirely apart from the future practical nature of this knowledge.

Reprinted with permission from *Science News*, 90, 354 (1966),

All of this knowledge, it seems to me, increases and in no way diminishes man. For me, the beauty of a starry night or a forest or a rainbow is enhanced by an understanding of them. Nor does scientific knowledge, in my opinion, reduce appreciation of a poet's sonnet, a musician's theme, or a painter's canvas.

A second major goal, not unrelated to the first, is concerned with the continuation of freedom and of the most effective functioning of democratic government in a period of revolutionary social change. This revolution is well-named "The Scientific Revolution," since the engines that drive it are science and technology. In the last three decades, science, once a peripheral preoccupation of a few intellectuals, has emerged as a central force in domestic and world affairs.

The outline of this revolution has been drawn by many thoughtful people. Knowledge, exploited by engineering and invention, is the capital of this revolution. We support science perhaps as no society has ever supported it at any time in history, and the main reason unquestionably is the hope of practical applications. Science has responded with an unprecedented outpouring of knowledge, sufficient to be satisfying to scientists and, one would think, to those who require a return on their investment. Today, flourishing industries, employing thousands of people, are based on knowledge that did not exist a decade ago. We use science and technology as instruments for improving the material quality of human life, both in our own country and abroad.

Moreover, we are dependent upon science and an aggressively exploitative technology for the continuation of the kind of society in which we now live. And we must redouble these efforts if we are to realize present, and perhaps better, living conditions for immensely larger populations in the world of the future.

All of this means, of course, that some of the most important decisions affecting our lives and those of millions of people around the world revolve around science and technology. Our form of government is flexible enough so that we have, I believe, an excellent cros-section of representation in making these decisions in the executive and legislative branches of the government. This representation is similar in kind to that employed in technical decisions in finance, foreign affairs, agriculture, and other fields.

Yet, the heart of freedom and its agency, democracy, is widespread, informed participation in the processes of government. I fear that the central forces of change in today's world — science and technology — are but dimly understood, or not understood at all, by the majority of the electorate. We cannot hope that now, or perhaps ever, the majority of people will be able to explain the quantum theory or the third law of thermodynamics. Nor is this necessary. But I believe there are minimum levels of understanding that can be conveyed to all citizens and there is more sophisticated knowledge that can be conveyed to a large percentage of our people. Such understanding can, I believe, help our society to react more intelligently and quickly, and in a democratic way, to challenge and opportunity as we travel farther along the road of the Scientific Revolution. We may learn, for example, to better adapt to swift social changes like those forced by automation, more readily accept necessary technologies such as automation and nuclear power, and perhaps even support science for the right reasons.

In reaching our goals of enriching human life and strengthening the democratic process in a period of revolutionary change, what can we hope to achieve in public understanding of science?

There are certain fundamental ideas that, I believe, virtually every citizen can grasp and that represent minimum attainable levels in public understanding. The minimum level that can be achieved almost universally, I believe, is an understanding of what science is, what technology is, and the difference between the two.

As we all know, science is a search for an understanding of new knowledge about nature. Generally, it does not have an immediately practical goal, although sometimes applications may be easy to foresee. The essence of basic research is freedom of the scientist to pursue his curiosity where it leads him. New knowledge is hard to acquire; nevertheless, it is necessary in order to synthesize the bits and pieces into general laws representing major progress.

On the other hand, technology, or more correctly, engineering development, involves the transformation of the knowledge we gain from basic science into useful things. Our radios, television sets, automobiles, synthetic fibres and plastics, nuclear reactors, and moon rockets are all engineering developments rising from towering edifices of basic knowledge erected by chemists and physicists.

Now, these distinctions between science and technology would seem to be very elementary. Certainly, a general understanding of these elementary principles is important to the national interest. Yet it seems clear that we have failed to impart these minimum facts. A farmer who knows that he must set aside a portion of his crop as seed for the following year may not understand that science is the seed of the technological future. Businessmen who fully realize that an investment in capital equipment today may not bring a return for years may be hard put to see an analogy in the relationship between science and technology. The connection between an investment in a basic research project and an ultimate gadget is seldom specifically apparent. Many citizens who can appreciate the connection between the scientific research of the past and the way they live today have difficulty projecting from a theoretical study today to an unimaginable product tomorrow. There is no value in criticism of this difficulty. By and large, the intangible future has always been hard for man to grasp. However, the future depends upon faith that, in the progress of science, history will repeat itself; that nature is only beginning to yield up its most fundamental principles; and that the new knowledge we gain will enable us to realize our dreams for the world of tomorrow. Teaching widely this lesson — that today's intangibles in science are tomorrow's realities in technology — is one of the important tasks we face.

Confusion over what science is and is not can do considerable mischief. Not uncommonly, men are hostile to what they do not understand. Many people, among them some otherwise well-informed individuals, are critical of government support of science. Not infrequently this criticism is based on two mistaken notions: first, that the $15 billion annual federal budget for research and development is all for "science"; and second, that "science" may not be "producing" enough "gadgets" to justify the outlay. Actually, only about $1.5 billion, or about 10 percent of this sum, is allocated for basic science, with the remainder being allocated for applications of knowledge. It would seem to be hazardous to provide less for the search for new knowledge, the "seed" of our technological future. Surely, the level of support of scientific research by the government is fair game for searching exploration, but it is in the interest of us all to conduct such examination from an informed position.

Lack of understanding causes still other mischief. Science receives much credit for its accomplishments, of course, but it also is the whipping boy for those who would like to find a simple explanation for man's destructive weapons. It is true, for example, that nuclear weapons were an outgrowth of new scientific knowledge. But so is the nuclear reactor, which promises to perpetuate a technological civilization that is dependent upon the production of huge quantities of energy. In other words, knowledge is born without moral properties. It is man who applies knowledge, and he applies it according to his acquired patterns of behavior. The point is that misunderstanding on this can mislead us about causality. Man, not knowledge, is the cause of violence. We are not likely to bring the nations of the world to agree to an embargo on man's curiosity or its pursuit through science, even if such a course were desirable or had any chance of succeeding in its purpose. There is more hope in directing our energies at the right target — the causes and cures of man's misapplication of knowledge.

While I have placed considerable emphasis on the most elementary needs in public understanding of science, I do not mean to underrate other aspects of the cultivation of science by the public. A large portion of the public is interested in and capable of absorbing considerable scientific knowledge. Cultivation of science at many levels by increasing numbers of people is surely important in achieving our major goals of enriching human life and strengthening democratic institutions in an age of science.

What about the means of conveying an understanding of science? Unfortunately, as we all know, there is no *deus ex machina* that we can invoke to equip everyone with knowledge of science. We would abolish all our difficulties, of course, if we had a device such as that described by Jonathan Swift in his satire on a project of the Grand Academy of Lagado in "Gulliver's Travels." You may remember that the head of the project explained to Gulliver that "Everyone knew how laborious the usual method is of attaining to arts and sciences; whereas by his contrivance, the most ignorant person, at a reasonable charge, and with little bodily labor, might write books in philosophy, poetry, politics, laws, mathematics, and theology, without the least assistance from genius or study." Gulliver then relates how the Lilliputian scientist led him to a frame, about 20 feet square, about which his students were clustered. "The superficies were composed of several bits of wood, about the bigness of a die, but some larger than others," Gulliver relates. "They were all linked together by slender wires. These bits of wood were covered, on every square, with paper pasted on them; and on these papers were written all the words of their language, in their several moods, tenses, and declensions, but without any order."

Some of the students by turning cranks on this machine changed the pattern of the words while others transcribed the bits and pieces of sentences that appeared. Gulliver was shown several volumes in a large folio, already collected, of broken sentences, which were eventually to give the world a complete body of all arts and sciences.

Unhappily, I fear that further research and the accumulation of considerably more knowledge will be required before technology can give us the solution to our problem in the form of a practical device resembling Swift's prophetic contrivance or its successor, the modern teaching machine. Meanwhile, we will have to rely on more prosaic methods requiring hard work by all concerned. Our effort to achieve widespread understanding of science must be pluralistic, to use a fashionable and highly descriptive term. We cannot neglect any avenue or medium at any level. The need for minimum understanding is with us now. We must attend to it now, and at the same time build toward rising levels of understanding in the future.

A great deal of dedicated work has been done for many years, and is being expanded. A commendable contribution is being made by newspapers, magazines, television, radio, and through educational films to give the public a sense of the significance and scope of scientific discovery. Considerable credit belongs to the National Association of Science Writers and the Council for the Advancement of Science Writing for improving and increasing the nation's corps of professional writers on science for the mass media. Private foundations and the government have supported these and other efforts at popularization. Science Service, of which I have the honor to be president, has done yeoman's service for over 40 years in popularizing science through its news services and *Science News Letter*, now *Science News*. The scientific societies have also established helpful programs, among them the American Association for the Advancement of Science, the American Institute of Physics, and the American Chemical Society. Government agencies, including the National Science Foundation and the Atomic Energy Commission, have made significant efforts to explain science to the public.

Occupying the middle ground is the unique *Scientific American*, serving highly motivated, intelligent laymen as well as scientists, engineers, and other technical people. I had the privilege of being among the company of those who encouraged Gerard Piel and Dennis Flanagan in launching this valuable publication in 1948.

I wonder if there are not some gaps in the popularization of science in the mass media for communication that might be filled to the profit of media management and the public. For example, some surveys indicate that the public is interested in reading more science than they are getting in the newspapers. Is there a place, for example, for more regular column or feature material that would make good reading and that, being free of the restriction of the news lead, would provide a larger opportunity for revealing more about the rather interesting processes and personalities of science? Is there a place for a national science newspaper? Might it not be profitable to publish a high-quality popular science magazine that would interest a mass audience? In connection with these thoughts, I believe it is germane to remember that the technically oriented community is growing larger every day, and will inevitably grow larger in the future.

While the mass media are the primary instruments for informing the present generation on science, I have always believed that there should be intensive concentration on developing a future electorate well versed in the principles and processes of science. I believe we can all take heart in the significant programs that have been developing in recent years to make science a real part of the education of the young. These include experimental programs to develop materials and techniques for teaching science to elementary school children. The curricula of high school science have been immensely improved by the work in physics of the Physical Sciences Study Committee and Educational Services, Inc.; by the work in chemistry of the Chemical Educational Material (CHEM) Study and the Chemical Bond Approach Project; and by the work in biology of the Biological Sciences Study Committee. Science Service's extensive program with its local science fairs and its International Science Fair, and with its more than 25,000 youth science clubs having over three-quarters of a million members, is making an important impression on a national and international basis.

New centers, with dynamic concepts for imparting knowledge of science, are beginning to rise. The Pacific Science Center is an outstanding example. The Lawrence Hall of Science, in Berkeley, holds great promise, being in part a national research institute on the teaching of science at all levels. The new Hall of Science of the City of New York is also incorporating programs to assist the schools in the improvement of science teaching. All of these fine centers also provide science learning opportunities for the public at all age levels.

While this work is starting to pay off, we must not neglect other opportunities. Apparently thousands of young people still emerge from colleges and universities with little capacity to discriminate between science and invention. No doubt a part of the reason is that teachers in the lower schools had, perhaps through no fault of their own, little or no instruction in science themselves. Perhaps this can be remedied in part by programs now under way to assist teachers in science instruction.

But surely our future teachers and leading citizens should gain in college some feeling for the processes and values of science. Many institutions of higher learning have taken steps to provide interesting science instruction for the generalist. But I fear that in many places the only science available is designed for the student who intends to become a scientist. As a former university administrator, I recognize the difficulties. But it seems clear to me that in our time, especially, institutions of higher learning have an obligation to fill what is sometimes a serious gap in general education.

I should like to touch on one other subject that is pertinent to the matter of public understanding of science, and it involves the scientists themselves. Science long since has departed the ivory tower, as any researcher can testify who has been called to a congressional hearing. I believe there is wide appreciation among scientists of the value of public understanding of what the scientist is doing. Scientists have been among those who have taken the lead in many efforts to improve public understanding. The ethical code and the culture of science, however, often inhibit the efforts of scientists in this field. I would hope that senior scientists might help to break down needless inhibitions among their students and junior colleagues. I would hope that senior scientists would encourage those among their juniors who have a talent for popularization to cultivate that talent in acceptable ways.

In conclusion, I offer my congratulations to this community for contributing, through the Arches of Science Award and the Pacific Science Center Foundation, to one of the most important educational problems of our time. I believe that we are only at the beginning of the Scientific Revolution. As this revolution accelerates, so will our dependence on the processes of science and technology. We can expect change to be swifter and more profound. The decisions we must make will increasingly involve science and applications of new knowledge. Wisdom in self-government requires that we as a people know more and still more about those processes of the Scientific Revolution that deeply affect our lives. Let our learning, however, not be only a duty. For in the learning, I am sure, larger numbers of us will come to enjoy the rich experience of seeing the wonders of our environment through the revealing eyes of science.

Youth — A Vote of Confidence

Commencement address at the University of Miami, Coral Gables, Florida, January 27, 1967.

Commencement address by Seaborg at the University of Miami, Coral Gables, Florida, 27 January 1967. Audience photo. Reproduced by courtesy of UM Publications.

As some of you may have seen in a cover story done by a major magazine last summer (1966), my generation — that is, everyone older than Jack Benny — has been named the "Command Generation." Now that's a very impressive title, and one would imagine that the cover picture which accompanied that story would have been a picture of the President or perhaps one of our military leaders or captains of industry. But as I remember, it was a picture of Lauren Bacall. The point of the article, however, was not to prove that women are taking over, but only that those of us in the 40 through 60 age group — women included — were still the country's leaders in most fields of endeavor.

I bring up this article, which I'm sure was excellent therapy for many of us here, because what it did not say, as well as what it said, bears an important message for all of us here. What it did say, among other things, was that middle age was a time of success and accomplishment, that the middle-aged were the leaders, the decision-makers, the high-earners of the country, and so on. It named astronauts and artists, senators and salesmen, engineers and executives, and even some Nobel Prize winners — all at their prime between 40 and 60. What it did not say, and what I would like to emphasize today, is how these people achieved their success. They arrived at their current status not merely by aging but because they were exceptional in their special fields — and particularly so in their youth. It was the attributes of youth — the energy, the questioning and open mind, the drive and determination — all combined with their specific talents which brought these people to their positions of power and prestige in middle age. In many cases their greatest accomplishments took place in their younger years and launched them on the road to an easier success later.

As a sort of footnote here let me add that the magazine which devoted its cover story to the Command Generation just recently did a cover story on the generation under twenty-five. Referring to you as the "Now Generation," they collectively named you, girls included, "Man of the Year." The article, incidentally, did more than give equal time to, or placate, the publication's young subscribers; it was strongly flattering of today's youth. But let me return to my original thought about the evolution of the "Now Generation" into the "Command Generation."

Youth in History

I know from my own experience that the young years are extremely productive years, and I am inclined to agree with Benjamin Disraeli's statement that "Almost everything that is great has been done by youth." Let me back this up with a few examples in history: Alexander the Great ruled Macedonia at 20 and the civilized world at 27. By the age of 26 Michelangelo had sculpted the famous Pietá. Pascal had written a book on geometry by 16 and invented an adding machine at 19. When he was 27, Charles Darwin began the observations which led to his theory of evolution. At age 21, Michael Faraday started the electromagnetic work which led to his invention of the electric motor. James Watt patented the first practical steam engine at the ripe old age of 33. Einstein began work on his theory of relativity at 26. Alexander Graham Bell patented the first telephone at 29. The Wright Brothers initiated their aircraft experiments in their 20's. And if you're still not convinced — Mozart completed his first symphony at the age of 8. The full list is, of course, much longer and more impressive. It also includes accomplishments of the youth of today.

The Young Shall Inherit the Earth

For a moment let me present some statistics on the position of youth in the world today. I think they are enlightening. Numerically, youth dominates the scene. While one-fifth of the population is between 40 and 60 years of age, over one-half is under 25. There are about 25 million teenagers in the country today. By the mid-1970's more than two-thirds, and perhaps three-fourths, of the U. S. population will be under 35. And on a worldwide basis it is forecast that, by 1986, 35 percent of all the people alive will be less than 15 years old. (It is also interesting to note — and certainly thought-provoking — that China today has more children under 10 than the entire population of the Soviet Union.)

Now all these statistics on youth bear more significance than just promising a great market for transistor radios, surfboards and miniskirts. I think they point out, among other things, that "the *young* shall inherit the earth." And with it they are going to inherit a world of both enormous problems and incredible possibilities. Furthermore, I believe that it is today's "Pre-Command Generation" — those of you now in the 20 to 40 year group — who will, before the end of this century, determine the course of human history, perhaps for centuries to come. You may argue that every new generation determines the course of history as it comes of age, but this younger generation is in a very special position. You have arrived on the scene at a time when the Scientific Revolution of the past few decades — forged under the current Command Generation — is effecting the rapid evolution of a truly global civilization. And because of this you are inheriting the Earth morally as well as physically. For as Marshall McLuhan has often pointed out, in this electronic age we are developing a global consciousness. And he questions, "Can we possibly have a global consciousness without also developing a global conscience?" I think it is clear today that we are well on our way to having both.

The Now Generation

Perhaps it is this fact which heightens so much of today's differences between the younger and the older generations — differences which basically have always arisen between youth and their elders but which are so sharply in focus today. While those of us in the Command Generation feel we are struggling to narrow the gap between our ideals and the harsh realities of the world we experience every day, younger people want only the best of all possible worlds — and they want it now. Seeing more clearly than ever the world as it is, and the world as it should be and could be, rubs raw the impatience of youth. And your reactions disturb, anger, puzzle, and exasperate many older people.

But if one has the imagination to project himself into your position, it is not difficult to imagine the feelings that many of you young people share. You are being asked to enter this world of developing global consciousness and conscience, of immense new physical possibilities confined by barriers of social inertia, of affluence existing adjacent to poverty, of knowledge and concern amid ignorance and indifference, of political turmoil and philosophical upheaval, of all this tossing about on a sea of change. You feel the urgent need for coping with this, for working to bring order from apparent chaos, to control and direct the tide of change. The whole thing cries out for action — meaningful action.

Yet there seems to be little opportunity for it — especially for you. Change seems to be made through the Establishment, the Power Structure, the Command Generation — and there are less flattering names — but they do not seem to listen to you. They only ask that you learn all there is to know (and today that's far more than they ever learned in school), join the rank and file and work toward goals you have not set and perpetuate values you wish changed. Unfortunately, if you believe all this, it is only a few steps further to being convinced that no meaningful participation is possible and that a psychedelic experience has more meaning than those experiences in the world of reality. Fortunately, most of you do not go this far, but recognize that the real world still holds some excitement and purpose worth pursuing.

On the other hand, many of you feel the necessity to go still another route. Feeling powerless as individuals, but not about to retreat, you turn to group action. I will not go into details on the pros and cons of this. Naturally, much good can be accomplished in a democratic society by concerted action. And such action carried out in a responsible and constructive fashion should be encouraged. But I, for one, believe that most accomplishments of youth throughout the ages have come from the attributes of youth being applied through the individual. Perhaps in a free society the qualities which characterize youth become more productive when expressed through individual efforts. Let me spend a minute or two examining some of these qualities, because they are your greatest assets and they should be recognized and encouraged to the fullest.

The first one that comes to mind is sheer energy. I have occasion to speak about the usefulness of nuclear and other energy resources almost every day. And in this country we are all inclined to think of mechanical power as doing most of our work — which physically it does. But in a broader sense, human energy is still the most productive force in the world and youth has most of it. Nothing is needed more today than the stamina and determination to see a difficult job through, to generate new ideas and find acceptance for them, to rebuild, renew, and revitalize places, organizations, and people, and to battle ignorance and injustice. All this is an endlessly wearing and enervating task demanding the vitality of youth.

In addition to physical energy, youth has the emotional energy — the ardor — to keep things moving. If I may use a somewhat mixed metaphor, of which I'm sure Marshall McLuhan would disapprove, youth is a hot item in a cool world. The cool world resists change — particularly social and cultural change. It tends to support the status quo, to perpetuate old institutions, to resist new ideas, and it takes a "show me" attitude, if it is willing to listen at all. But youth is restless and easily bored. It looks for and thrives on change. And change — rapid change — is the order of the day.

I often speak of this change and our new awareness of it. But the other day I was reminded that to some extent the forces of change have been recognized throughout the ages. In fact, some years ago Nicholas Murray Butler (the former president of Columbia University) stated that, at some time in the Garden of Eden, Adam must have paused and said, "Eve, we are living in a period of transition." But they were a young couple so I'm sure they looked forward to this change — even though it involved some temporary hardships.

But change today is different. It is rapid to the point of being "explosive." And it produces crisis after crisis because its effects are disruptive to old habits and traditional, highly-ingrained thinking. Today's rapid change may be painful to many of us, but it is necessary. Most importantly, it is necessary to deal with it effectively — to recognize it, to control and direct it and, above all, to use it to shape the future toward human goals. We have an unprecedented opportunity to do this today, but it is essential that we give great thought to and reach some agreement on those goals, and that we have determination, the staying power, the "guts" to follow through on our beliefs. There is a vast gap between expressing dissatisfaction and impatience with the way things are going and taking the positive, well-planned action toward constructive change.

In dealing with change youth has a decided advantage. It produces and accepts change more readily because of two things. The first is it has less fear of failure and therefore is more willing to take risk. The world is filled with people who say "It can't be done," and they will give you many reasons, even from their own experience, why it can't. But young people, not having the benefit of that experience, are willing to try for themselves. And if they have the knowledge and skill to go with their courage — perhaps foolhardiness — and stamina, they may succeed where others have failed.

At this point I must stress the knowledge and skill aspect because these days they make a bigger difference in the margin between success and failure. Last September, in speaking at Rutgers University, Vice President Humphrey made what I feel is a very cogent statement when he said: "The world is filled with the noise and clash of opinion. It is woefully shy of dependable knowledge." So I would not play up courage at the expense of knowledge. It takes both to do the job.

Another factor which gives youth an advantage in producing and working with change is the ability to innovate, to use imagination and a fresh outlook. Oscar Wilde is credited with having said, " In America, the young are always ready to give those who are older than themselves the full benefit of their inexperience." While there may be some truth in this, I think it is for the very reason that the young in America are not afraid to speak up that we have progressed so far. For, aside from the lesser fear of failure which I mentioned before, there is a freshness of outlook which

Commencement address by Seaborg at the University of Miami, Coral Gables, Florida, 27 January 1967. Seaborg at podium. Reproduced by courtesy of UM Publications.

is still present in the inexperienced. And sometimes it is this very quality which allows the imagination and intuition to produce a breakthrough.

In listening recently to an old recording of a lecture given in the early 1930's by Lord Rutherford, one of the founders of nuclear physics, I was struck by a particular phrase he used. In discussing details of the nucleus of the atom, of which so little was known at that time, he said, "... but we can look at it with the eye of imagination." "*The eye of imagination*" — this is something which youth is more apt to have and which it should try to retain and make the most of. The world can always use more people who think and act creatively and who can keep that creativity alive as they grow in knowledge and experience. These are the people who contribute the most to society and, I believe, also have the most rewarding lives. They are the ones who stay alive, in the fullest sense, as long as they live.

There is one final element in which most young people today are not lacking and which is so important. In the most old-fashioned and perhaps "squarest" sense, it would be called "brotherly love." Today it is more often referred to simply as "human concern." Whatever we call it, it is the most vital element in human nature and especially today, when man is becoming so powerful through his science and technology. If science and technology amplify the power of man, then certainly concern should focus that power on the most human and humane goals. I think most of today's youth have that concern and are determined to keep it alive and growing.

Admittedly, these are trying times for anyone who wants to maintain a high ethical standard. As John W. Gardner has written in his book, *Self-Renewal*: "One difficulty is that man's moral aspirations rise more rapidly than his performance. Given relatively modest advances in the justice and decency of human life, he is soon dreaming that he can fashion a perfectly just and decent world. And this is a conviction that is certain to bring some measure of disillusionment."

I hope that, on this important day for you, you will rise above any disillusionment or doubts you may have about the state of the world and the state of man, and turn every fiber of your being toward helping shape them toward your ideals. If ever there was a time to act positively and constructively toward those ideals, to which so many give lip service but fail to act, it is now — and you are the "Now Generation."

Today's "Command Generation" feels the hot breath of youth on the back of its neck. But it is not altogether an uncomfortable feeling. In a way it is good to know you're right behind us. And when the time comes — and it is almost at hand — I have a feeling you will do as good a job as you think you can — perhaps even better.

Women and the Year 2000

Remarks made to the Woman's National Democratic Club, Washington, D. C., February 6, 1967.

When Mrs. Horace Busby invited me to speak on this occasion, she wrote that I should talk "about the year 2000 and the changes that will have been *wrought* by then in the lives of women ..." As soon as I saw the word "wrought" in her letter of invitation, I realized that in preparing my talk for today I would have to take greater cognizance of the feminine point of view. For, as you know, the word "wrought" — to a woman — means fashioned, as in the case of jewelry, or elaborately ornamented, as in the case of tapestry. But to a man and a chemist like myself "wrought" means hammered or beaten into shape by tools, or "wrought iron," which contains less than three percent carbon and has one or two percent of slag mixed with it.

There is, however, one reason why a male chemist like myself would be willing to forego an opportunity to speak about metallurgical chemistry in favor of talking to an audience more interested in jewelry and tapestry. For as Dr. Samuel Johnson said about two centuries ago, "One of the last things we men are willing to give up ... is the supposition that we've something to say of interest to the opposite sex." My wife and two daughters know that I am deeply committed to this principle.

I don't think that Dr. Johnson claimed to have any special knowledge of the women of the eighteenth century and I certainly don't consider myself an expert on the woman of today. I haven't read any of the writing by women on the modern woman, such as the books by Betty Friedan and Simone de Beauvoir. Neither have I read volume two of the Kinsey Report. And my librarian told me that I am so far down on the waiting list for the recent best seller by Masters and Johnson that all of her copies of this graphic study on sex will probably be worn out before I get a chance to read it.

As to the future and women's lives in the year 2000, I think that one of the most interesting things in life is how unsure we are of the future and how often the future fails to bear us out. It is not easy to think tomorrow's thoughts today, for the future is hidden even from those who make it. I should also say that many of the predictions that I shall make this afternoon will be well on their way toward implementation much before the year 2000.

But we must not merely look toward the future; we must act toward it. Since the future is the past come home to roost, we must build a better future by deepening our understanding of the past and by taking a firm grip on the present. The men and women who can build a better future are those who know that greater things are possible and that they themselves can help bring them about.

The idea that I want to stress most of all in remarks this afternoon is my conviction that a better future for our country depends to a very significant extent on making a more effective utilization of the talents and abilities of American women. On this objective I am optimistic about the prospects for improvement. In the year 2000, I think that our society will be much more enlightened than it is today with respect to the role and position of women. By the turn of the century, we will be doing much more to help our women appreciate the diversity of roles they can play and the growing values of high-level skills and education. We will be doing much more to encourage women of all ages to prepare themselves to cope more effectively with the challenges of the present and future. At the same time, I think that women themselves will have greatly enlarged their ambitions as individuals, will be aspiring to a broader role and higher position in our society, and will be acquiring much better training toward meeting these expanded responsibilities.

A More Professional Woman

But this brighter future for the women of 2000 will not simply evolve out of forces already set in motion. We must create the conditions that will make it a reality, and I believe that we will do what is required.

One area that bears heavily on the future of women and one that must be improved is education. Since I have spent most of my career in college and university education, I am alarmed especially by the declining percentage of women in proportion to our total population who are college graduates and holders of advanced degrees. Among people who are twenty-five or older, only three-fifths as many women as men are college graduates. In 1930 women earned 40 percent of all masters and other second level degrees, but only 32 percent in 1965. Among the recipients of doctoral and equivalent degrees during this same period, the percentages of women fell from 15 to 11 percent. Perhaps these trends are partly responsible for the fact that during the last 25 years the number of college and university faculty positions held by women has declined from 28 to 20 percent. These figures reflect what I consider to be a deplorable and needless underdevelopment of women's talents. They indicate a direction that our society must begin now to reverse or for which we must pay a high price in future years.

By the year 2000 I hope that we will have developed different patterns in the education of our more talented female youth. The prospects for broadening educational opportunities (especially if our national wealth continues to grow as expected) and for strengthening our society's appreciation of the value of college and postgraduate education for women should result in the preparation of a larger percentage of women for professional careers. If this occurs the role of women in the professions could be considerably more prominent in the year 2000 than it is today.

During the past 15 years there has been a significant and steady decline in the percentage of women among all workers in professional, technical, and kindred occupations. Currently women comprise only three percent of our nation's lawyers and only six percent of our physicians. One development that would help to alleviate this situation would be a reorientation of the thoughts of many people about the kind of curriculum that girls should pursue. Many parents — especially many fathers — hold the assumption that girls cannot excel in mathematics, that it is not quite feminine to major in the sciences, or that an engineering or a medical curriculum should be strictly a male endeavor. But experience has clearly shown that girls can excel in these subjects and that women can succeed in scientific, technical, and medical careers. By the year 2000, I think that this fact will be generally accepted, and that women will be taking much better advantage of educational and professional opportunities in these fields.

I also think that in the year 2000 we will attach more importance to education for married women — both as preparation for community service and for reentry into the labor force when their children reach an age where mothers are able to work outside the home. In 1964 and 1965, I had the pleasure of working with an outstanding lady who has very successfully combined family life (including four children) with public service and career, Dr. Mary I. Bunting, who took a one year leave of absence from her position as President of Radcliffe College to serve as a Member of the Atomic Energy Commission. Among "Polly" Bunting's many contributions to the AEC was the strong support and encouragement that she gave to our programs for equal employment opportunities for women and for the part-time employment of professional women.

By the turn of the century, housewives in general should have more time for study due to automation, more money for educational expenses due to the expected rise in family incomes, and more opportunities for continuing their education due to the increased availability of local colleges and universities. There will probably be many day care centers to enable student mothers to undertake part-time study, and local colleges may even have supervised nurseries and playgrounds to keep young children occupied while mothers attend class. For mature women wishing to enter the labor force but without marketable skills, there will probably be scholarships and educational guidance counseling of both an academic and vocational nature.

This broadening of the educational opportunities for women will enable them to make an even stronger contribution to civic affairs in the year 2000 than they are making today. I believe that women make a very special contribution to public life because their sense of the responsibilities of child-rearing and their desire to cultivate excellence in family life are preeminent qualities that cause women in public life to foster a broad concern for human welfare. It is natural that those who bring life into the world should attach the greatest importance to making a better life for all people. Perhaps James Stephens, the Irish poet, articulated another reason why women have made such an important contribution to the development of social and humanitarian policy when he said, "Women are wiser than men because they know less and understand more."

A More Political Woman

One reason I believe that the active and constructive participation of women in public life will be extended and strengthened by the turn of the century is that prejudice against women in politics — although still very apparent — is diminishing. I also believe that the impact of other factors affecting political participation by women will be reduced. Today the relatively low proportion of women in public office reflects, to a considerable extent, the low proportion of women who are prominent in those private occupations that normally lead to political activity and advancement. But women's prominence in the professions is likely to increase, and more married women will plan ahead for a career after the age of their children permits it.

Moreover, during their early married lives women's political participation is limited by the need to work near their homes. But the very high proportion of political work that is done by women at the precinct level demonstrates their interest and skill in political activity. During their earlier married years more women will apply themselves to a grassroots political apprenticeship of service as members of school boards, town councils, state legislatures, and so forth.

As a result of these changes that I think will occur in women's lives — a steady erosion of the prejudice against women in politics, a stronger career orientation on the part of women, and increased participation by women in local affairs — the opportunities for women to contribute to public life will be substantially broadened.

During the twenty-first century, I think that, compared to our times, many more women will be elected to serve in the United States Senate and House of Representatives and in state legislatures. I also think that the range of government positions to which women are appointed will be wider. Especially in some state governments, convention has thus far tended to exclude women from offices except those dealing with juveniles, school affairs, health, welfare, and libraries. But this should change, and by the turn of the century there will be many more women in federal and state legislative offices, in the upper levels of both federal and state executive branches, in federal, state, and local judicial offices, and in the higher diplomatic positions. If this prediction comes true, I think our society will benefit substantially as a result of making a greater utilization of the abilities of women in government and public affairs.

Women's role in politics will not be the only aspect of their lives that will change. There will also be important advances in other areas of importance to women. One effect of these changes will be a reduction in the amount of time that women will have to spend on traditional household tasks. This will permit women to devote more attention to activities outside the home, such as those I have been discussing.

More Free Time via Robots and Trained Apes

By the year 2000, the widening impact of the computer will undoubtedly be extended into the home. For example, some people foresee small household computer consoles connected to a large complex of central station computers that would help with family budgeting, tax calculations, school work, purchasing, and menu planning by providing such assistance as banking and credit, library and reference sources, and mail order and shopping services. It is also possible that homes of the future will be equipped with something like a teleprinter, so that the family's daily newspapers, magazines, and books will be brought to the home not by old-fashioned delivery but through facsimile reproduction or some other form of high-speed printing.

Many homes in the year 2000 will probably have automated kitchens. The housewife may make out her menu for the coming week, put the necessary foods into storage spaces, and give instructions to her kitchen computer. At the prescribed time mechanical arms would get out the preselected food, cook it, and serve it.

As for grocery shopping, housewives who prefer not to go to the store may be able to shop by videophone, which will enable them to check the prices and quality of groceries at the local supermarket and to place an order without leaving the house.

By the year 2000 or not too many years thereafter, housewives may be relieved of many of the ordinary household chores of today. Many wives will probably have a robot "maid" that is trained to the requirements of her particular home and programmed to carry out several standard household operations when switched on by the housewife. Professor M. W. Thring of the University of Sheffield thinks that the commercial availability of robot maids is a real possibility as soon as the next twenty years, provided that the necessary applied science and financial support are made available.

These household robots would not look at all like a human being; instead they would be shaped like a box and have one large eye on the top, several arms and hands, and long narrow pads on each side for moving about. Those of you who are now thinking that you would be terrified to have such a mechanical being moving about the house (even opening doors and going up and down stairs) should remember that it can be switched off, unplugged, or stopped without difficulty at any time, and even directed to put itself away in a cupboard. One can imagine an advanced, multi-arm version of the robot housekeeper that would be able to do a variety of tasks simultaneously — such as washing windows, sweeping floors, vacuuming rugs, dusting furniture, cleaning woodwork, brewing coffee, and picking up your husband's clothing. By the year 2000 you may have even trained your husband to pick up his own clothing. But I don't think that anyone will ever devise a robot that could fit two- and three-year-old children into snowsuits and overshoes.

For housewives of the twenty-first century who prefer animate rather than mechanical domestic servants, there may be a choice other than the robot. About two years ago, the RAND Corporation came out with a *Report on a Long Range Forecasting Study* (by T. J. Gordon and Olaf Helmer) which forecasts future developments in a number of important areas. The RAND panel mentioned that by the year 2020 it may be possible to breed intelligent species of animals, such as apes, that will be capable of performing manual labor. During the twenty-first century, those houses that don't have a robot in the broom closet could have a live-in ape to do the cleaning and gardening chores. Also, the use of well-trained apes as family chauffeurs might decrease the number of automobile accidents.

By the turn of the century consumer goods will have become even more abundant in a society of continuously rising affluence. And like today, the preferences and tastes of American women will determine the form of many of those products. The latest improvements will come into service as a result of frequent model changes and the introduction of new products. Advertising will continue to create new demand, and consumers will continue to enjoy the ads even more than they enjoy the products.

Fashions and Foods

As to women's fashions in the year 2000, I am sure that there will be even more opportunity for ingenuity and one-upmanship by fashion conscious women than there is today. Low cost, disposable dresses made from new synthetic fabrics could enable women never to be seen more than once in the same outfit. Wedding dresses made of these new materials would no longer be made to be worn by granddaughters, but fathers of the bride will welcome the decline in initial costs of dresses their daughters would wear only once. There is one thing I am certain about with respect to women's clothes in the year 2000: those designs that are the most fantastic to us men will continue to become universally accepted among women.

The women of the twenty-first century will also probably purchase and serve food produced by a very highly mechanized agricultural industry and by radically different techniques, such as large-scale ocean farming and the fabrication of synthetic protein. Radiation will be used to preserve the quality of many food products that housewives will buy at automated supermarkets. And doubtlessly, even the preparation time for minute rice and instant coffee will be shortened.

"Spare Parts" — Medicine and Personality Control Pills

As for some of the advances women can look forward to in medicine, by the turn of the century there will be considerable progress in some very important areas. And it is even possible that there will be breakthroughs resulting in a major impact on health and life expectancy. In the coming century many more people will live to the ripe ages of, say, 75 or 80 or beyond, and older people will live out their lifespans in greater comfort than they do today. But unless there are major breakthroughs in our understanding and control of aging, the larger number of people who will then celebrate their ninetieth or one-hundredth birthdays will still be likely to owe more to their parent's genes than to medical science.

By the year 2000 we should have better control over infectious diseases. While total eradication of infectious diseases is not likely, I do look for significant advances in immunization, and it is possible that general immunization against bacterial and viral diseases will be available.

I also believe that we will make remarkable progress against cancer. It will be possible to lower the incidence of cancer by reducing or eliminating many deleterious factors in our environment. And there are also possibilities for immunizing people against viruses that may cause some kinds of cancer and for developing new and more highly effective chemical and biological approaches to cancer therapy.

In the case of children with defects due to heredity and metabolic anomalies, our knowledge of genetics, biochemistry, and molecular biology should increase to the point where most of these personal tragedies can be prevented and many of those that do occur could be corrected if treated at an early age.

There will also be a rather sophisticated medical technology concerned with both natural and artificial replacements of damaged body components — what is sometimes called "spare parts medicine." Organ transplantations will be common practice. A variety of prosthetic devices will have become feasible, such as artificial hearts powered with radioisotopes, plastic heart valves, artificial kidneys and lungs, and artificial limbs. Some artificial limbs should even be controlled by the brain — that is, they may be operated by electric impulses generated in appropriately connected muscles by the person thinking of the movements he desires to make.

The use of drugs for personality management will be more effective and more widespread than it is today, at least for people suffering from depression and other emotional disturbances. For those people who are mentally healthy but chronically irritable, there has been discussion about the possibility of an "anti-grouch" pill. Some people have already argued that a variety of personality control drugs should be made available to the general public as a way of ensuring that husbands are ambitious, wives understanding, and children well-behaved. Of course, we would be wise to allow a few disagreeable characters to carry on as such, if only to provide material for TV and the movies. A world in which there is no one to laugh at or criticize might prove intolerable.

One of the best medicines will continue to be relaxation, and by the year 2000 many women in the United States will find it commonplace to be spending an occasional weekend in Europe or Asia. Future developments in aviation technology are likely to have a profound effect on foreign travel. By the turn of the century the hypersonic transport aircraft, successor to the supersonic jet, will undoubtedly be in commercial service. These hypersonic planes would travel at speeds of 4,000 to 6,000 miles per hour and would reduce the flight time from New York to London, for example, from the present six hours to about 30 minutes — which is less time than it takes me to commute from my home to my downtown office. Passengers on these hypersonic aircraft will feel about the same as they do today on present jets; that is, there probably won't be any sensation of the increased speed.

The Challenges of the Future

The twenty-first century is closer than we think — closer than the first inauguration of President Franklin D. Roosevelt. One of the reasons I am grateful for your invitation to speak to you today about the year 2000 is that I share your feeling that the time has already arrived when we must think more about the kind of world we want in the future and what we should do now to make it a reality.

Some of the problems that will confront the women of the year 2000 will be how to make good use of the increased leisure time that will be available due to automation, how to keep physically fit in a society with so many labor-saving devices, how to effectively combine home and family with careers and contributions to public affairs, how to keep education and skills up to date, and how best to transmit the values of our society to the succeeding generation.

In future years we are going to be confronted more and more with the dilemma that modern science and technology favor discoveries and inventions with potentially radical implications, while many other forces in society lean in conservative directions. In the years to come, many of the most significant issues and developments will involve a clash of these two opposing tendencies.

Women of the twenty-first century will be involved with these problems more deeply than ever before, and the contribution that women make to their solution will be greater than ever before.

Time, Leisure and the Computer: The Crisis of Modern Technology

This article is based on a speech delivered at a conference on "The University in a Changing Society," celebrating the Centennial of Howard University, March 1, 1967.

Almost every time I pick up a newspaper, magazine or book these days I see some statement about "the wonderful world of computers" — computers that perform the most complex calculations in a few seconds, computers that make up corporation payrolls and review a nation's tax returns, computers that diagnose diseases, computers that help design, produce and market new products, computers that control air and auto traffic and operate bakeries, computers that hire and fire, read and write, learn and teach, and even play cupid — though fortunately not yet to other computers, just among people.

What I have yet to come across in all my reading about computers — and what is most sorely needed — is an ultimate computer that will tell us where all other computers are leading us. Such a computer may someday be built and programmed. In the meantime it is still left to us to try to answer this incredibly difficult question.

Why is it incredibly difficult? Because the ultimate potential of the computer puts us to the test as human beings. It brings up questions we have lived with for centuries but never been asked to answer fully or act upon if we believed we knew the answers. It gives us new freedom and yet tremendous responsibilities which, if not acted upon, could result in a loss of almost all freedom. It presents us with choices and decisions of enormous consequences. It offers man a remarkable new chance to shape his own destiny, but asks him to be God-like enough to select that destiny without much margin for error.

I will not attempt to discuss all the potentials and consequences of the computer and the cybernated society. What I would like to do is dwell on the larger, more total aspect of its potential uses and tie in its consequences with the role of the university.

To begin with, let me state that cybernation — the complete adaptation of computer-like equipment to industrial, economic, and social activity — will represent a quantum jump in the extension of man. The Industrial Revolution amplified (and to a large extent replaced) man's muscle as a productive force. Now springing from our Scientific Revolution of recent decades is what is being called our Cybernetic Revolution. With the fullest development of cybernation controlling the enormous energy at our command today we could be faced with prospects which challenge our very relationships to such basic concepts as freedom and the nature of work and leisure.

With the coming Cybernetic Revolution we will probably be forced to a more rational approach in setting up national goals and promoting the welfare of the individual within the total framework of society. Part of this will come about because of what many people now fear — a large increase in productivity by fewer and fewer workers.

The time may someday come when the computer will enable the production of enough consumer products and the performance of enough individual services so quickly and efficiently that it may require little if any sacrifice to perform the additional work necessary to achieve most of our desired social goals. Cybernation will, of course, also be effective in this area, and enhance its productivity as well.

Eventually, in the era of abundance we are seeking, many new economic and social phenomena will take place. No one will be deprived of the means to an adequate income and, above this, there will be many incentives for creativity and productivity on numerous scales. Probably the highest status and rewards will not come from money or material possessions. Their value will some day become almost meaningless. In this regard, it is interesting to recall what the great economist, John Maynard Keynes, wrote in his 1932 *Essays on Persuasion*. Keynes said this: "When the accumulation of wealth is no longer of high social importance there will be great changes in the code of morals. We shall be able to rid ourselves of many of the pseudo-moral principles which have hag-ridden us for two hundred years, by which we have exalted some of the most distasteful of human qualities into the position of highest values."

During the transition to a social millenium, which I do not claim to be just around the corner, a radical change in man's relationship to work will take place and the growth of leisure time will pose new problems to be solved.

Work has always been central to mankind's existence. It is so much a part of most cultures that its virtue has been written into the religious scriptures, extolled in the folklore and become a psychological necessity as well as a physical one. Cybernation is bound to alter this — perhaps slowly at first but eventually drastically.

Our ideas on leisure will also change. Most people today do not recognize the true value of leisure. Some of them are so conditioned to their daily work routine that they refer to their free time as "time to kill." But leisure can and should be a treasured commodity, and throughout history there are many examples of it as a highly creative force. Many of our greatest inventions and scientific concepts were arrived at during moments of leisure, some springing actually from playthings. A little leisure has always been treasured and there have been societies in which certain men and women lived in almost complete leisure, though at the expense of others' labor. But the idea of almost an entire civilization living in even relative leisure is beyond the comprehension of many of us, and still frowned upon by most others.

"Idle hands are the devil's workshop" we are told by those who see only disaster in mass leisure. That is because they see leisure only as idleness, unemployment, vegetative loafing or a prelude to restlessness leading in turn to a venting of energies in aggression and destruction. I do not doubt that if today we were to have sudden massive leisure — let's still call it "unemployment" — that all the horrors these people envision would come true. But it is a failure of imagination to believe that a transition to the Cybernetic Age cannot be made in which leisure can become central to man's existence and his greatest blessing. And we are going to have to prepare eventually, and perhaps sooner than we think, to handle in a meaningful, creative way, the growing leisure that is bound to evolve.

Consider for a moment the growth and scope of cybernation, the work it will be able to do. Consider that in 1960 there were about 300 areas of application for computers, in 1961 over 500 areas and today over 1,000 areas. Consider the huge influx of young people looking for gainful employment. Consider the reduction in work hours and the lowering of retirement age with a longer, more active, lifespan. Consider that computers are already being designed to program other computers. And consider Norbert Weiner's statement that the computer is a slave and those who would compete with slaves accept the conditions of slavery. Then ask yourselves whether the time is not coming, or at hand, when we must reconsider our attitudes toward leisure and develop bold new concepts in preparing to handle it on a massive scale.

A civilization equipped and educated to live in an era of relative leisure can be a new Golden Age — one without a slave base, other than those mechanical and cybernetic slaves produced by the ingenuity of a higher level of man. Such an age does not have to be, as a few predict, a civilization of drugged, purposeless people controlled by a small elite. But it could tragically become that if we did nothing but let ourselves be swept along by some of the forces in motion today.

I realize that I am charting a somewhat utopian course. Now, let me be quite realistic as I turn to the one force which I believe can do most to help us understand ourselves and our society and help to create and fulfill the highest goals which a Cybernetic Revolution might offer. I believe that force is education and that the university should play a leading role in shaping the Cybernetic Revolution, even more than it has in the shaping of the Scientific Revolution of recent decades.

Jerome B. Wiesner has stated that "Society is a Learning Machine," and that we receive feedback from the economy in the form of business success or failure, unemployment, crime, social problems, etc. He sees the computer as an instrument, which can help us to forecast, and give us lead time to handle, trends which could otherwise lead to catastrophe. But even without the help of the computer there are some trends today which we, as a nation, can recognize as we read our daily newspapers, our magazines and books, listen to our radios and watch television. Among these, I believe, are an increasing uneasiness about the state of our personal and community lives in a highly materialistic society, a concern over the individual's role in the growing complexity and impersonalization of that society, a groping for "national purpose," and a feeling that the unity of man, referred to by poets and philosophers throughout the ages, is becoming a reality with immense psychological and physical implications.

To me these feelings forecast the need for a huge reevaluation of our goals and values, and it will be in our universities where such a reevaluation will take place. Perhaps its seeds have already been sown in the current unrest and activities on the campuses of many of our universities. From this reevaluation, from the debates and soul-searching which take place, will evolve both a new understanding and reinforcement of those old ideals which are still valid, and new ideals and goals. Together they may provide us with something like a comprehensive philosophy of life to match the physical unity of mankind rapidly being fostered by today's science and technology.

As Sir Julian Huxley points out in his recent article "The Crisis in Man's Destiny," some of the self-limiting and even self-destructive trends of today are moving mankind into a new stage of human advance. I believe this is true to the extent that we may be about to witness, possibly within the next decade or two, the equivalent of a new "human breakthrough" — an advance to a new stage of social development, and hence human development, brought about by our reaction to today's trends. This might be the beginning of a dramatic evolutionary change in which Konrad Lorenz's "intraspecific aggression" among man almost disappears and he is united to find a new harmony within himself and with the nature of his planet.

In such a development the university, the greatest depository and dispenser of man's knowledge, should play a major role. In fact, I can see no other institution more logically equipped to be the central force in this evolutionary process, to develop, refine and pass on to the new generations a new heritage of a higher-level of mankind.

But to carry out such a monumental task, many changes will probably have to take place in the universities and our educational system in general. Let me briefly project a few thoughts in this area.

As the result of the explosion of scientific and technical knowledge of the past few decades, we are reaching new heights of specialization. There is no doubt that this trend is going to continue for some time, although computers are going to be extremely useful in reducing duplication of work and making specialization more manageable. But we are also recognizing more and more today a tremendous need for interdisciplinary thinking — not only in science and technology, but in all areas of our economic, social, and human development. Specialization has been giving us knowledge, but the world cries out today for more of something beyond knowledge — for *wisdom*. And this can better be achieved by the more interdisciplinary approach, the "systems approach," to dealing with our world and our lives. A shift in emphasis in our education will, therefore, be necessary toward bigger, broader thinking and action on a similar scale.

Most of today's schools and universities are involved to a great degree in serving the requirements of an industrial age, in fulfilling the needs of a society which has only been partly and indirectly of their making. In the future I think this role will shift to where the nature of society is determined more by the thinking of the university, and in which the industrial community will tend to serve goals created by that thinking. This is not to say that in the future the nation and the world will be under the leadership of a handful of college professors. I think even the academic community would view this prospect with horror. What I look for from the universities is the development of an education which turns out individuals of the highest intellect and broadest outlook, able to understand man and machine and live creatively with both. Such an education could not be expected in a four-year curriculum or even a six- or eight-year one. It would start as early as the beginning of school or sooner and involve continuing education of one type or another throughout a person's lifetime. And as Robert Theobald indicates, education in the age of the Cybernetic Revolution would not be directed toward "earning a living" but toward "total living."

This is a big order involving imagination, energy, and bold leadership from the academic world. But the time is certainly at hand for such leadership.

The coming Cybernetic Revolution which calls forth these new goals for education will also give education valuable new tools and technologies for pursuing them. The computer will make knowledge more accessible. It will perform miracles in compiling, organizing, and analyzing information. It should link the knowledge of the world's libraries and depositories of information into networks responding like a giant brain. And it should put at the fingertips of anyone who wishes to be a modern-day Faust all the knowledge he desires without selling his soul to the Devil.

I mentioned previously that we are not educated today to handle the growing amount of free time which the coming age will make available to us. The new goals of education should help prepare us to make the most of this new leisure.

At some point in time education and our new experience with cybernation will have entirely changed our current thinking about leisure and work. The distinction between them will almost have vanished. Work — if we still wish to call it that — will involve physical, intellectual, and artistic accomplishment, but mainly for pleasure.

With the growing use of computers to "shrink" time, we will eliminate what might be called "the whip of time" and thus considerably reduce ulcers and the need for tranquilizers. Some will say that as all this happens human incentives will diminish and we will completely stagnate. I don't believe this will happen at all. New incentives will arise as man moves up to higher levels of needs. The quest for new knowledge will always grow. The domain of science is practically boundless. We are only beginning our adventures in space and we still have a long way to go in understanding many things about this planet and the life on it.

Much has been said about the impersonalization caused by the growth of machines, but as a result of this growth I can see a new and better relationship arising among men. If in the past we have spent most of our time working with machines, serving and being served by them, naturally we feel a sense of isolation and alienation among them. But when machines have truly freed us from the necessity of work, perhaps we can better accept them for what they are and have the time to see and relate to other people in a different light. When we have more time to be with other people — not accidentally, on crowded buses, in elevators, in markets, and offices — but in places of our own choosing at our own leisure, we may feel differently toward one another. When we are less likely to be in competition with one another, much of the hypocrisy of society will vanish and more honest relationships will be formed. (When we do not have to worry about "keeping up with the Joneses" we might discover that the Joneses possess something far more valuable than their house, cars and wall-to-wall carpet.) And finally, when we can walk down the street — anywhere in the world — in a community free from want, where every human being has a sense of dignity not gained at the expense of others, we might not only walk free from fear but with a great feeling of exaltation.

If we can make the transition of living with and using the complex machines of the future in a *human-oriented* society, the rewards will be worth any effort we can make. As everyone knows, such a transition will not be easy, as it involves so much of what Eric Hoffer calls "The Ordeal of Change." But I think we will have to make such a transition eventually. We may have already begun to do so.

It has been said that man is now "inventing the future." If this is true, let us make the universities of the world the workshops of human ingenuity. And let us see that the tools we fashion are those which will serve the highest purpose of man. The time to do all this is not when the crisis of modern technology begins to overwhelm us. The time is now.

How to Become a Scientist

Address delivered at the International Science Fair, San Francisco, California, May 9, 1967.

The title of this talk is simply "How to Become a Scientist." I considered using a catchier title, "How to Succeed in Science Without Really Trying," because I thought it might cheer up any of you who might be worrying about final examinations or anxiously waiting to hear from the colleges that you are hoping will grant you scholarships. But when I was thinking about using the title "How to Succeed in Science Without Really Trying" I found that I just couldn't construct a talk around this theme, for I don't know of any successful scientists who got where they are today by means such as those used by Robert Morse in "How to Succeed in Business Without Really Trying" — that is, by washing windows in executive office suites, by ingratiating themselves with the secretaries of big-shot officials, or by standing in front of mirrors and singing "I Believe in You."

The methods I am going to advocate tonight are not as clever as the "success formulae" that appeal to the writers and composers of musical comedies. Becoming a scientist requires hard work and sound preparation, but these are methods for success that can lead to a career of great personal satisfaction and significant benefits for mankind.

Brainpower — A Natural Resource

It was only a century or two ago that modern science became an organized study of natural phenomena deduced from experimental evidence. Since that time, applied science and invention have influenced the life of the individual, the development of industry, the evolution of political societies and the course of history. In the past few decades, we have seen a great increase in the impact of science on our society owing to a new factor which was not previously present. This new factor is the systematic and intensive accumulation of new scientific knowledge — the result of basic or fundamental scientific research.

We now know that the search for new knowledge, if not restricted to subjects of foreseeable and immediate practical importance, results in an unexpected increase in our understanding of physical or biological phenomena. These increases, in turn, give rise to far-reaching practical applications which could not have been anticipated from the original basic research. Our scientific knowledge and technology are advancing at an explosive rate. The time lag between the discovery of a fundamentally new scientific principle and its application in engineering or medicine is now very short, and these rapid developments are changing the lives of all of us in many ways which we only dimly perceive.

Because of our inescapable dependence on modern science and technology, we must regard trained brainpower as a precious natural resource. The extent to which we discover exceptional intellectual talent, encourage and develop it, and provide conditions for its effective flowering will be a measure of our success in meeting the truly challenging problems which technology and population growth are posing for us in the remainder of this troubled twentieth century. I refer broadly to two types of trained brainpower. One is represented by the professional scientist or engineer, the other by the educated person in other fields who has mastered enough of the meaning and content of modern science to make valid judgments on the many questions raised by the influence of science on his field, whether it be law, medicine, politics, military affairs, industrial management or some other.

It is fitting, therefore, that considerable attention be paid to the early identification of intellectual talent. We are gathered to participate in one attempt at the identification of those who have exceptional aptitude for a creative and productive career in science. We are here to honor young men and women who have demonstrated by their conception and execution of some science project that they have a strong motivation and exceptional promise for a scientific education. I think it is quite proper that we pause on regular occasions to acknowledge the intellectual, esthetic and idealistic aspirations of our young people and encourage them by recognizing their academic excellence. Not unnaturally, a young person is influenced to seek goals which are recognized and respected. If praise is reserved only for athletic prowess or monetary success, who can blame him if he seeks these even if it means sacrificing a great potential in some other field.

The Scientist — His Training and His Work

I would like now to mention briefly a few important steps in the education of a scientist, and a few attractive features of the life of a research scientist.

The work of a research scientist is of great interest and importance. Often it is difficult to relate his work to matters of dollars and cents, but in the value system of the scientist the subject he has under study — whether this be the origin of the solar system, the biochemical basis of heredity, or the nature of a meson — is of great significance. I think that personal success in making money is often overrated in our standard of values. I believe that most people have a deep psychological need to feel that what they are doing is of some importance. The scientist feels the satisfaction of this need, and this gives drive and zest to his efforts. This is particularly true if, from time to time, his efforts are rewarded by the thrill of discovery. In his search, he knows that in the final analysis his success as a scientist is measured against the criteria of nature rather than the judgments of persons. Science is self-correcting in that spurious results will sooner or later be unmasked by new experiments or the attempted verification of previous conclusions. Hence it can more readily establish truth and confound error than other more abstract fields of study. A scientist who is correct can prove he is correct, and by a proper marshalling of experimental evidence can convince his colleagues regardless of their superior reputation, seniority, or rank.

Scientific research is a stimulating activity requiring constant assimilation of new facts, theories and techniques. This feature appeals to many gifted persons who would be stifled by the repetitive, dulling routine of many other occupations. A career in scientific research has many deep intellectual satisfactions which appeal strongly to the person of superior natural abilities.

But many of our most successful scientists owe their great success not so much to superior brains as to their capacity for sustained work. Many people of quite superior promise never have that promise realized unless they are fortunate enough to be in an environment where they are continually prodded into activity. In science, the intellectual satisfactions, the thrill of discovery, and the sense of worthwhile effort are a rich reward and a strong stimulus to continued work. Scientists and engineers are definitely not clock-watchers. The large majority of my personal acquaintances work in establishments where the doors of the laboratory are never locked and the lights frequently burn late into the night.

Mastery of Subject Knowledge Comes First

Let us turn now to a consideration of how an interested young person may make a career in this exciting field.

Science is an organized body of knowledge and a method of extending or revising that body of knowledge by observation, hypothesis formation, and experiment. The training of a scientist is a two-step process — the mastery of a body of knowledge developed by previous workers and the mastery of a technique for extending that knowledge. I sometimes liken the role of the scientist to that of a mountain climber who with great care and exertion achieves some great prominence from which he is able to perceive immediately and clearly new vistas which are hidden from the sight of those down in the valley below, even though many of those in the valley may have better eyesight. New science builds on the work of the past. The scientist of today stands on the shoulders of the giants of yesterday. Therefore, the first task of any serious aspirant toward a career in science is to master the recorded history of the past. The task of the effective teacher is to organize and present that history concisely and effectively. Our libraries are so filled with an enormous accumulation of facts, hypotheses and theories that the complete mastery of any science or even one major branch of a single science is a quite impossible task. Hence there must be a judicious selection of material — a judgment concerning the relative importance of various fields of study at various stages of the education ladder, and a continuing judgment concerning the relative importance of facts, laws and correlations.

We are greatly helped in this effort by the great unity of much of science because of the fundamental laws or generalizations underlying all natural phenomena. In physics and chemistry, for example, we are enormously aided by the laws of thermodynamics, and the laws of conservation of momentum and energy. We are also aided by the circumstance that many decades of past work in science may become logical and clear once some satisfactory organizing law or theory is evolved. By intensive study of the organizing law, the new scientist may understand immediately many hundreds of individual facts which were quite mysterious to the past generation of scientists. New mathematical techniques may also make it possible to explain or quickly derive numerous experimental facts which could only be understood at the expense of great labor by previous students. Therefore, it is the purpose of high school and undergraduate education in science to teach with the utmost economy of effort the organizing principles which may unlock for the student the important heritage of the past. Because the relative importance of facts is subject to rapid obsolescence, the goal is not to teach facts alone, but a system of understanding facts so that the new knowledge can later be absorbed.

Transforming Trained Intelligence Into Creative Intelligence

We turn now to the next step in the making of a scientist, namely, his training in the methods of carrying on scientific investigation. The principal centers for this type of instruction are the graduate schools of our great universities. The graduate holding a bachelor's degree in science has a good knowledge of his field, is able to perform a variety of laboratory tasks and may, with on-the-job training, develop into a creative research scientist. But as a general rule we rely on the rigorous training of our graduate schools to convert the trained intelligence of the holder of a bachelor's degree into the creative intelligence of the research scientist. The Ph. D. degree is the symbol of this creative intelligence.

The chief instructional technique of the graduate school is to put the student in the laboratory under the supervision of a master scientist to do a piece of original scientific investigation of a problem of considerable importance and difficulty. Here he learns a variety of experimental techniques. He learns the importance of asking big questions. He learns how to set up a meaningful experiment and how to extract correct answers from the data he collects. He learns the importance of letting the unexpected result lead him to new conclusions or at least to new experiments. Graduate research is a rigorous, demanding experience which makes an enormous change in the scientific effectiveness of the candidate. At the better institutions, the research interests of the professor are very advanced and are likely to be in a frontier area far beyond the material currently appearing in undergraduate textbooks. The professor is also greatly stimulated by the student. The student is usually brilliant, comes to the problem with a rather different educational background from that of his professor, and is eager to work hard to find out what the experiments will reveal. With this fresh outlook it frequently happens that he contributes greatly to the success of the research and may transform it into an advance far greater than might reasonably have been expected at the initial stages of the work.

The American people can regard with great pride the graduate schools of our great universities. They meet all the qualifications of excellence. They train virtually all of our great creative scientists. The scientific research done under the student–professor symbiotic working arrangement adds a major fraction to our truly important new knowledge. I could cite many, many examples of this. The selection of any scientific problem involves a judgment concerning what is worth investigation. In the first rank graduate schools it is the big problems which are attacked. It is free research. The fact that about 50 living Americans hold the Nobel Prize in chemistry, physics or medicine attests to the excellence of our graduate schools of science. Unfortunately, there is not a sufficient number of such schools. A study made by the President's Science Advisory Committee included recommendations that it be our national science policy to strengthen these schools, to increase their number and to work out ways to extend the mutual creative stimulation of graduate education and research into more of the laboratories in which our fundamental research is done.

A potential young scientist listening to these words may be hesitant to embark on such a forbidding and rigorous course of training. A few words of comfort may therefore be in order. Your decision whether or not to consider graduate training in science can be deferred to the middle or later undergraduate college years. By then your decision will be based on a clearer knowledge of your abilities, interests and intellectual performance. Should you then decide that your talents and interests lie in other directions, there is a strong likelihood that the science courses which you studied will nevertheless form a significant part of your total education. While graduate training usually takes from three to five years, the extra expense attached to this period can be greatly lessened by the many teaching and research appointments and fellowships which are available. In many fields the student can expect to pay all his living and school expenses from such appointments.

Science Careers and Women

I believe the young women in this audience can take a great deal of satisfaction in the fact that there is now greater opportunity and encouragement for women in today's world of science. Lise Meitner's significant contributions to our nuclear age — in particular her role in the identification of nuclear fission — should dispel for all time the old chestnut that only boys are capable of scientific work and girls should always aspire to become English teachers. Four years ago we in America had our first woman Nobel Prize winner in physics — Maria Goeppert Mayer — who shared the physics prize with Professor J. Hans D. Jensen of Heidelberg for work revealing the onion-like structuring of atomic nuclei.

Even closer to home for those of us who work in the Atomic Energy Commission is the fact that a woman microbiologist and college president, Dr. Mary Ingraham Bunting, has served as a Commissioner of the Atomic Energy Commission.

Qualities of a Scientist

A further comment that I should make is that there is plenty of room in scientific research for those who are not in the genius category. All Ph. D.'s are better than average in intelligence, but few of them indeed approach the manifestly superior intelligence of Enrico Fermi or John Von Neumann. There is a large volume of work which needs to be done and which can be done very well by the trained scientist of more modest endowments. Furthermore, much of the potential of a Fermi or a Von Neumann would be lost were there not many other scientists to try out their suggestions or to turn up new phenomena and new data for them to study and consider.

Creative research calls for a combination of qualities, only one of which is superior intelligence. It is difficult to specify the combination of characteristics which may be of crucial importance for success in solving a specified scientific problem. Perhaps an intuition or a "feel" derived from lengthy experience with certain types of phenomena, perhaps special knowledge of a new instrumental technique, perhaps a natural manual dexterity in some important type of laboratory manipulation, perhaps an unreasonable stubbornness in seeking a better explanation of some phenomena which others have passed by as "explained" may be the key to a fruitful series of developments. I should find it difficult to explain the secret of the effectiveness of the many able scientists I know or the relative ineffectiveness of others who seemed to hold out great promise. Maybe it's just plain hard work, because, without downright hard work, there can be no success in a scientific career. All the productive scientists have been "smart," it is true, but the ways in which they have been "smart" have differed greatly. And, as I have said, they are all dedicated hard workers.

Those of you who decide to become engineers will need a strong foundation in the physical sciences. Today, new discoveries in the nature of matter, the structure and the electrical and magnetic properties of materials, etc. are rapidly reduced to engineering application. To carry out this reduction rapidly and surely the engineer must have a sound and sophisticated knowledge of science and an adaptability which in general will be achieved only through a curriculum that includes some graduate training in the sciences.

A General Education First; Then Specialization

In talking about the requirements of a scientific career and the education of scientists, I want to stress a point that I think teachers of science often fail to appreciate and students frequently overlook. This point is that specialization — although indispensable for the graduate student and the research scientist — is often followed to detrimental extremes in high school and college. Over-specialization makes it more difficult for students to become broadly educated and well rounded individuals. It also impairs the ability of scientists to live up to their special obligations as citizens. Over-specialization can even make it more difficult for a student to become a scientist and to choose eventually the field of scientific specialization that is best suited to his particular abilities and interests.

You may now have your mind made up to be a nuclear power engineer, an expert on snakes, a brain surgeon, or a designer of missiles. Fortunately for you, when you get to college you will not be able to sign up for a course in Snakes 1A, Missiles 1A, but will be required to pursue a regimen of courses of a more basic and general nature. You will get your chance for specialization — even extreme specialization — later on, but that lies years ahead of you. It is far better that your training be much broader and more fundamental in the beginning, because you are at present ill equipped to make a legitimate selection of a specialty. You are but dimly aware of many of the major wonders of mathematics and science; and the expert who knows these well finds it hard to convey to you his feelings about them in words which will at this time have meaning for you.

If you become deeply committed to a particular discipline early in your student career, and if that discipline turns out to be less attractive than you anticipated, then you may find it difficult to change. For instance, you may have to make up courses that you should have taken originally; and this often turns out to be difficult. Sometimes it seems to the student to be so difficult or time consuming that he takes another alternative — by getting out of science entirely.

Another consideration worth keeping in mind is that science is rapidly changing in ways that make specialized fields, such as biology and physics, related to each other as they never were before. Today people educated primarily in physics are going into biology, and increasingly biologists are perceiving the applications of mathematics and physics to their problems. Since it is becoming more and more difficult to predict what specialized training will be most useful, I hope that you will try to acquire as broad a background as possible. So take courses in mathematics and in science subjects that are outside the field you now think you want to specialize in.

The Humanities Are Important Too

But science and mathematics are not the only subjects that the aspiring scientist needs to know. Students should not specialize in science to the extent that the humanities are neglected. The significance of science and technology and their role in society cannot be understood apart from the human values and social institutions that are affected — often drastically — by the dynamic forces of scientific discovery and technological change. In our rapidly changing world, the scientist has a special responsibility to think about the impact of science on people and to communicate to the public about science. Scientists must be concerned about the world in which science is used, and they must understand the values, the sensitivities, and wants, and the needs of people. Only the humanities and the social sciences can provide this essential, broad framework of understanding. The insights about the past and the present that are expressed in the humanities are essential for making sound judgments about the future — such as decisions involving the control of our environment and the way we use the material resources that science and technology have given us.

If those of you who become scientists also recognize the humanistic aspects of science, you will be able to look beyond the immediate results of scientific endeavors to an appreciation of their consequences for man and society. You will then be better prepared to fulfill your civic responsibility of helping to clarify to the public the broader implications of scientific and technological developments. Your ability to do this will not only make you a more valuable citizen; it will also raise the stature of science and scientists in our society.

There is one subject in the humanities that is essential to any scientist, whether he is a public spirited citizen or the type who lives in his own private world. That subject is English. Unfortunately, many advanced students of science are deficient in the fundamentals of English expression. This is commonly acknowledged by professors of science who read papers written by their graduate students. It is remarkable that some scientists and students of science who can readily appreciate the logic and order of mathematics as a mode of expressing a concept or a conclusion fail to understand that clear and grammatical English is equally indispensable for giving structure to thought and to communicating with other scientists and intelligent laymen.

English has an importance that is similar to mathematics in the following respect: it has applications over a wide range of subjects in many fields of specialization. So I would urge that one of the best ways for broadening — and thereby acquiring a stronger basis for eventual specialization — is to develop the facility to express your thoughts in both oral and written English.

The importance of effective and accurate English expression was recently illustrated by an experience that a certain professor had when he gave a lecture to a student audience at one of our colleges. During the course of his remarks, the professor pointed out that while some of today's students are members of what is variously called the Beat, Hip, or Mod Generation, others may belong to an Inarticulate Generation. After he had finished his talk and stepped down from the platform, a girl student, wearing long straight hair, a mini-skirt, and sandals, elbowed her way through the crowd to where he was standing and told him earnestly:

> You know, Professor, I really agreed with what you said about us being inarticulate. I've often felt something like that myself, only I've never been able to put it into words.

Science for Non-Scientists

While scientists need a deeper awareness of the outside world, it is equally important that non-scientists learn more about science. In the education of the non-scientist, there is a need to include courses that will insure a better understanding of science as part of the general high school and college or university undergraduate curriculum.

We live in an age in which, for better or worse, the influence of science is pervasive and revolutionary. It is a part of our culture which is shaping nearly every aspect of our lives and our institutions. We can no more ignore it than the man of the middle ages could ignore the Christian church or the feudal system. Properly nurtured and employed, science can provide us with marvelous tools for the solution of many of the weighty problems of our physical and social world. The promise of the future lies in the hands of the dedicated and the educated. A most significant part of that promise lies in the field of science. I would invite the earnest young men and women who are prepared to study and work hard to join in the exciting and rewarding profession of science and the scientist.

The Promise of the International Atomic Energy Agency

This article is adapted from an address delivered before the International Theoretical Physics Conference on Particles and Fields at the University of Rochester on 31 August 1967.

Earlier than most men, scientists recognized that the successful occupancy of this tiny planet by man could best — or perhaps only — be accomplished through international cooperation. Indeed, I suspect that many scientists are inclined to believe that the very existence of the term "international cooperation" implies an artificial and unfortunate separation into arbitrary components of what is obviously a single, interdependent physical and biological system. Yet that separation exists, and it is to the credit of science and scientists that they have found ways to minimize its impact in their own affairs. Important as international scientific cooperation is in its own right, perhaps its most profound current significance is that it is showing the way increasingly to cooperation between nations in matters which can affect the well-being and even the life and death of millions of human beings. By the example of institutions such as the International Theoretical Physics Conference, by the active participation of scientists in the affairs of government, and in many other ways, scientific internationalism is transforming relationships between countries. It is doing this far more quickly and more profoundly than most of us, even those close to the scene, realize. We can only hope that it is working this transformation quickly enough.

Reprinted with permission from *Science*,
Vol. 158, No. 3798, pp. 226–230, October 13, 1967.

Seaborg delivering the speech at the Third International Conference on the Peaceful Uses of Atomic Energy, Geneva, Switzerland, 31 August 1964. This is an occasion that Seaborg is talking about in his speech to the International Theoretical Physics Conference on Particles and Fields, Rochester University, Rochester, New York on 31 August 1967. Reproduced by courtesy of the International Atomic Energy Agency.

Formation of IAEA

What I have in mind is perhaps nowhere better illustrated than by the International Atomic Energy Agency. I believe its accomplishments and its prospects for the future hold deep significance for progress in international institutions, both scientific and otherwise. The IAEA was born out of a realization that the promise and the problems of the peaceful uses of nuclear energy recognize no national boundaries. The United States' proposal to create this agency in 1953 marked the turning point in the postwar evolution of nuclear energy from its preoccupation with military uses to an emphasis on peaceful uses open to all. In little more than three years, not an extended time for the formation of an international organization, the agency was established. It was established at a time when achieving international agreement in even the most minor matters was far more difficult than it is today. I believe it is possible to say that the creation of the agency was one of the early and significant demonstrations that a widely favored international objective can be achieved even in a climate of controversy between the major powers.

Today, the IAEA has achieved the status of a healthy youngster in the family of international organizations. It is 10 years old, with 97 members, and still growing. With its professional staff of 300, it is participating — often taking the initiative — in a broad spectrum of international cooperation in peaceful uses of nuclear energy. In addition to these statistics, I would like to review its programs and some of the possibilities that I see for the agency in the future.

Like the Atomic Energy Commission of the United States and of a number of other of its members, the IAEA has two broad lines of activity. On the one hand, it actively promotes the peaceful uses of nuclear energy among its members. At the same time, it does all in its power to see that these applications are carried out without risk to world peace or to health. In all of its activities, the agency makes a special effort to meet the needs of its lesser developed members in the peaceful uses of nuclear energy, but it also provides services of great importance to even its most advanced members.

IAEA Programs

If we examine each of the major peaceful applications of nuclear energy, we find the agency actively at work helping its members to learn, evaluate, and take advantage of the benefits to be gained. Perhaps this is a good point for saying a word about the role of nuclear science and nuclear research centers in lesser developed nations. There are some who say that such nations are not ready for advanced scientific work or that they could use the required funds in a more productive way. On the basis of my observations in many developing countries, I cannot agree with this. I have found that nuclear research facilities are often the focal point of an entire country's best scientific activities. They provide a powerful influence for good, not only by advancing the scientific status of the country, but also by building a broad scientific base and by acting as a powerful force for the retention of scientists in their native lands.

In the field of nuclear power, the agency has for a number of years been the principal sponsor of the exchange of information on an international basis. Since 1959, 20 conferences and symposia on various fields of reactor science and technology have been held under agency auspices, not including the 1964 Third Geneva Conference on the Peaceful Uses of Atomic Energy, for which the IAEA had responsibility for technical organization. This year, the agency's conferences on reactor topics will have included containment and siting of nuclear power plants, neutron thermalization and reactor spectra, physics and related safety problems of fast reactors, international extrapolation and comparison of nuclear power costs, heavy-water reactors, and the use of plutonium as a reactor fuel. It is hardly necessary to point out that this list goes to the heart of many of the most active and pressing problems of reactor technology today. International scientists know that no price tag can be placed on the benefits gained from the timely exchange of scientific results and the stimulation of direct contact with fellow researchers which conferences such as these afford. The Secretary General of the United Nations has asked his Scientific Advisory Committee to advise him on the question of a possible Fourth Geneva Conference in 1969 or 1970. I personally feel that there would be considerable merit in holding such a conference with the IAEA once again providing the technical leadership.

The agency's technical meetings are not limited to conferences and symposia held primarily for the exchange of information. The IAEA has made effective use of a tool which might be more extensively used by other international scientific organizations — panel meetings at which small groups of experts meet to discuss and prepare recommendations on how specific technical problems might be attacked. Recent or scheduled panel meetings in the reactor field cover topics such as lattice constants of mixed systems, long-term reactivity and burn-up predictions, and long-term reactivity control. The recommendations of these panels provide useful guidance to the agency and its members in scheduling future research work, and often avoid costly duplication of effort.

Desalting of Seawater

There is one area related to nuclear power in which the agency's efforts to encourage international exchange and understanding have been of paramount importance. This is the field of large-scale desalting of seawater. I know of no more exciting prospect for the future — a future which I believe is not distant — than that of nuclear reactors producing unprecedented amounts of energy for the production of electric power and the desalting of water in some of the earth's most arid regions. This is not a futuristic dream. A decision has already been made to proceed with the first such plant, one which will produce 1.8 million kilowatts of electric power and ultimately 150 million gallons of fresh water per day for the Los Angeles area.

Since 1963, the IAEA has served as the principal forum for the international exchange of information on the technology and requirements for large-scale nuclear desalting. Through eight meetings of its Panel on Nuclear Desalting, it has brought together and established invaluable continuing contacts between both the technical leaders and the policy officials from all parts of the world who are concerned with the life-and-death subject of water. But its activities in this field go far beyond this exchange. Right now, a tripartite study under IAEA sponsorship and chairmanship is being made of the technical and economic feasibility of a large-scale dual-purpose nuclear power and desalting plant to serve the arid regions of Mexico and the United States surrounding the Gulf of California. This plant, which might produce as much as 1 billion gallons of water per day, could be the largest international project in history and could point the way to closer international cooperation in many fields.

We may find nuclear desalting playing a vital role in promoting international cooperation and the solution of international disputes earlier than many had thought likely. On 19 June, President Johnson, speaking on the Middle East, said: "We here will do our share, and do more, to see that the peaceful promise of nuclear energy is applied to the critical problem of desalting water and helping to make the deserts bloom." Other U.S. leaders have expressed their support for this solution of some of the obstinate problems of the Middle East. If this concept is to be implemented, the IAEA will undoubtedly play an important role.

The participation of the agency in the current study of nuclear desalting for the Gulf of California, as well as in earlier desalting surveys or studies in Israel, Chile, the United Arab Republic, and Peru, illustrates a vital role of the IAEA in the field of nuclear power. Many nations considering the possible application of nuclear power in their economies find it invaluable to have the objective advice and assistance of an international organization of which they are themselves members and which has nothing to sell. The agency has undertaken nuclear power studies or surveys in more than ten countries including Peru, Korea, Turkey, Thailand, the Philippines, Finland, Tunisia, Chile, Pakistan and El Salvador. Last year, in cooperation with the United Nations Special Fund, it completed a detailed preinvestment feasibility study of nuclear power for Luzon in the Philippines. The agency's studies have been characterized by an informed and objective approach, which ensures that they will be of genuine value not only to the country concerned but also to international financial institutions and other interested countries.

These agency studies of nuclear power feasibility are parallelled by other services to countries undertaking reactor construction, such as assistance in evaluation of both proposals and the safety problems of proposed reactors. I believe it is hard to overestimate the importance of having services of this type available for nations whose size or current state of development makes it impractical for them to maintain their own capability in these fields, or who feel the need for objective and expert independent advice.

In the field of reactors the agency's activities are not confined to nuclear power and desalting. In small but significant ways, the agency is helping to advance reactor technology through the direct support of research on a regional or international basis. For example, it is making a valuable contribution to the collection of reactor-physics constants through its sponsorship of the NORA project, a cooperative undertaking in Norway to develop the fundamental parameters of light- and heavy-water reactor lattices.

Radioisotopes and Radiation

In another field that has already achieved greater practical impact, the applications of radioisotopes and radiation, the agency also has a strong program. These versatile tools are used extensively in research, industry, agriculture, and medicine for a multitude of very valuable purposes. Many of these agency's programs involving radioisotopes and radiation are devoted directly or indirectly to increasing man's food supply, and no more critical need than this exists in the world today. Such projects as determining the most efficient use of fertilizers and irrigation water, eradicating insects such as the Mediterranean fruit fly by radiation sterilization, and breeding improved plant varieties through radiation are already yielding practical results, and rapid progress is being made toward much greater and wider application. Much of the agency's work in the field of agricultural applications of radioisotopes is conducted through an office operated jointly with the Food and Agriculture Organization, another member of the United Nations' family. This cooperative endeavor between the two international organizations is in itself a valuable demonstration of effective international administration.

In a more indirect way, the IAEA is contributing to the alleviation of hunger through its hydrological studies, for, in much of the world, water is often the missing vital ingredient for agricultural productivity. Measurement of tritium and other isotopes in naturally occurring water constitutes one of the most powerful tools ever devised for the understanding and measurement of the water cycle. For more than 6 years the agency has supported, jointly with the World Meteorological Association, a worldwide program of isotope-concentration measurements in rainfall, which provides basic input data for the extensive studies of water movement now going on as part of the International Hydrological Decade.

The agency has supported studies of groundwater movement, employing isotope concentrates in river basins in Austria and Czechoslovakia and in springs and lakes in Turkey and Kenya. It has also supported studies of snow hydrology in Czechoslovakia. Much of the experimental and analytical support for these projects is undertaken in the agency's own laboratory at Seibersdorf, near Vienna, an institution of growing importance and service to the agency's membership.

Medical Research and Services

The agency's program in the medical uses of radioisotopes reflects the increasingly vital contribution which these substances are making to modern medical research and practices. In the past 6 years, as part of its technical assistance program, the IAEA has provided experts and consultants to 37 countries concerned with the medical applications of nuclear energy, conducted 12 training courses in the applications of radioisotopes, and provided $300,000 worth of equipment for use in this field. Its work in the field of standardization has been significant in such projects as the international calibration of measurements of uptake of thyroid radioiodine. Also it has been active in the field of the sensitivity of biological material to radiation, one of the most important areas of radiobiological research. This field is directly related to fascinating developments in use of radiation to facilitate the transplantation of organs between genetically dissimilar people by suppression of the immune reaction. Radiation is already being used to suppress the immune reaction in kidney transplants, and this application may be extended to the transplantation of other organs. Various aspects of this subject will be treated in the agency's panel and conference program this year and next.

One of the most important of the IAEA services to its members in the field of medical applications is that of providing adequate numbers of trained personnel. A training program, which includes fellowships, regional training courses, and advanced research assignments, is responsible for a major proportion of the qualified personnel working in the field of nuclear medicine in the developing countries.

As I have observed firsthand in my own visits to a number of these countries, medicine and related biological sciences are normally the first scientific pursuits undertaken in a developing country. Thus, aside from its obvious humanitarian importance, medicine plays a major role in the formation of a country's scientific structure. The large role being assumed by nuclear medicine in developing nations, with the assistance of the IAEA, is helping to ensure that medicine constitutes a sound foundation for future scientific progress in these countries.

Basic Research Contributions

Understandably, much of the IAEA's scientific program is directed toward applied work, but the agency recognizes that basic and applied science are indivisible and that the sound development of its members requires effort in both areas. There is no better example of the IAEA's contribution to basic research than its International Center for Theoretical Physics at Trieste, Italy. Inspired and directed by Professor Salam, Trieste has quickly assumed a place among the foremost theoretical physics institutes in being today. More importantly, it has filled a gap in theoretical physics that perhaps could be filled only by an institution operated under the aegis of an international organization such as the IAEA. For the first time, it affords a place where the growing number of theoretical physicists from the less advanced countries can work as a matter of right, keeping abreast of developments in the mainstream of physics and profiting from the guidance and stimulation of their colleagues from more established centers.

The agency has many other activities in basic research, covering the whole range of its technical programs. It has actively assisted in ensuring the effective use of the research reactors located in several developing countries by conducting regional seminars in research reactor utilization, by providing expert advisers under its technical assistance program, and by research contracts. Its active programs include symposia, research contracts, and fellowships in such varied fields as the thermodynamics of nuclear materials, the effects of ionizing radiation on DNA, neutron diffraction studies, low-energy physics, and plasma physics, to name only a few.

Health and Safety Measures

In one field in particular, the agency's services are as directly valuable to its most advanced members as to its least developed ones. I refer to the critical problem of ensuring that nuclear energy activities do not endanger health or safety. By their very nature, these problems are international in scope. The consequences of a reactor incident or other potential release of radioactivity might well not be confined within the boundaries of the country in which it takes place. Efficient development of the peaceful uses of nuclear energy involves large-scale shipments of radioactive materials across national boundaries. The disposal or management of millions of curies of radioactive waste, some having a half-life of many years, is a problem of worldwide concern, not only to us but to future generations as well.

In most nations, the regulation of nuclear energy activities to ensure adequate health and safety precautions is a major governmental responsibility. The IAEA is in many respects fulfilling this function in relation to the international health and safety aspects of nuclear activities. Even though the agency lacks direct regulatory authority except in relation to its own projects, its advisory standards on radiation protection and transportation of radioactive materials have come into wide use and have helped to rationalize many international transfers of radioactive substances. The United States has adopted regulations very similar to the IAEA standards for the international shipment of several categories of radioactive materials, and is working to extend these regulations to additional categories within the next year or so.

Peaceful Uses of Nuclear Energy

The IAEA role in helping to ensure that peaceful nuclear activities do not endanger health and safety in the usual sense has a counterpart in another novel and indispensable program. This is its responsibility for ensuring that nuclear materials devoted to the peaceful uses of nuclear energy are not diverted to military purposes. No greater risk to public health and safety could occur, in the broad sense, than the unauthorized use of these materials for the development or production of nuclear explosives or other military purposes. The world's leaders are in agreement that one of the most urgent tasks facing us today is the conclusion of agreements — specifically a nonproliferation treaty — to avoid the acquisition of nuclear weapons by additional countries. The IAEA is already playing an important role in the accomplishment of this objective, and its responsibilities under a nonproliferation treaty could well be vastly increased. The agency's part in avoiding the further proliferation of nuclear weapons consists of the development and application of a system of safeguards, measures designed to detect the diversion of nuclear material to unauthorized uses. The development of these safeguards and their application by an international organization constitutes in itself an unprecedented step forward in international relations, a step which I believe has received far too little notice.

The agency's safeguards system includes actual on-site inspections by international civil servants, with broad powers of access and investigation, to detect any military use of nuclear materials committed to peaceful uses. Surely this constitutes a meaningful step toward the eventual world rule of law, which may someday ensure that international disputes are settled without recourse to hostilities between the parties.

Today IAEA safeguards are in effect in a growing number of activities for which a nation, voluntarily or in agreement with a supplier of nuclear assistance, has requested the agency to apply its safeguards. Agency safeguards are also applied whenever the agency itself has supplied nuclear material or other substantial assistance. Agency agreements involving the application of safeguards in 27 nations, covering more than 60 reactors, have been approved by the Board of Governors.

This is an impressive record, but a fundamental problem still persists. Even those nations which have requested the application of the IAEA safeguards to some or all of their current nuclear programs remain free to undertake the development of nuclear explosives in the future if they desire.

Fortunately, most nations have recognized that their own self-interests are much better served by refraining from a costly program of development of nuclear weapons which could only serve to stimulate similar activities in other nations — with a consequent loss, rather than gain, in security for all. However, as long as there is no restraint on individual national action in this dangerous field, the decision of a few, or even of one more nation, to follow the path of independent capability in nuclear weapons could precipitate a widespread breakdown in the current consensus against proliferation of these instruments of mass destruction. I believe the ideal solution to this problem would be the widespread adoption of a treaty of nonproliferation, providing for effective international safeguards such as those I have described, which would be applied or verified by the IAEA. In this way, every nation can be assured that its neighbors, like itself, are not constructing nuclear explosives and that the peaceful nuclear programs of all can go forward without the risk that they will be turned to destructive ends.

This development, which I believe lies within our grasp, would surely place the IAEA in the forefront of institutions responsible for world peace and would constitute an impressive demonstration of the contribution which international scientific cooperation can make toward the maintenance of peace.

International Meeting Ground

A major accomplishment of the agency, which was foreseen by its founders as one of its principal purposes, is the function of serving as a constructive and cordial meeting ground of scientists and political leaders of East and West. On the basis of my own experience, the agency has been exceptionally successful and profoundly important in fulfilling this purpose. Its success cannot be measured simply in statistical terms by the numbers of meetings, conference, and symposia attended by scientists from both East and West. Rather, its success must be viewed in terms of the quality of the contacts which have taken place and the relationships which have been established. I know of no other international organization in which there has been as much constructive cooperation between scientists of East and West, for such an extended period of time and on such a wide variety of topics, as in the IAEA.

I have personally valued the opportunities which my annual attendance at the General Conference of the IAEA has given me to meet and hold cordial discussions with my counterparts from many nations — both East and West. This experience is duplicated many times over in practically every conference and activity of the agency. Of equal if not greater importance are the lasting relationships which come out of these contacts. I am convinced that the strong scientific interchange between East and West in the peaceful uses of nuclear energy, which is so much in evidence today, would be much less intense were it not for the opportunity to meet and know each other which the agency has afforded to myself and my counterparts, our predecessors, and scientists in many different fields.

The IAEA represents, in an institutional sense, one of the most significant and promising developments in many years in international relations — the emergence of science and scientists as an important factor in the determination and execution of national policy. Scientists are by their very nature internationalists. They deal on a daily basis with laws and phenomena which are universal in their occurrence and application. They attack problems with a presumption that solutions are possible and that solutions will not be arbitrarily limited by political boundaries. Finally, they bring to the solution of problems the long history of scientific objectivity and rationalism and has served us so well in advancing the frontiers of knowledge. In short, there is an international language of science which, as a minimum, makes it possible for scientists to discuss with each other problems that often defy communications between people of other disciplines.

I do not suggest that science holds easy solutions to all or most of the profound problems which beset us. But I do suggest that through its capabilities for bettering the tragically inadequate conditions under which so much of the world's population lives today — through its methodology of problem-solving which has application to issues far beyond the normal realm of science and, perhaps above all, through the ability of scientists themselves to communicate with their fellow scientists of every nationality — science is one of the powerful forces that we can put to work to build a better world. The International Atomic Energy Agency is perhaps a test tube in which this hypothesis is being proved.

Nuclear Energy in Latin America

Remarks made at a Symposium on Nuclear Energy and Latin American Development on the tenth anniversary of the Puerto Rico Nuclear Center, San Juan, Puerto Rico, October 30, 1967.

It is both an honor and pleasure for me to be with you this evening and to share in the celebration of the tenth anniversary of the Puerto Rico Nuclear Center (PRNC). I believe it is fitting that you have chosen to celebrate this event by holding a symposium on nuclear energy and Latin American development. This is a subject of growing importance to this hemisphere — and therefore to the entire world.

I was told that my address here tonight was to be the "keynote address" of the symposium. If this is so, I would suggest that the keynote to be struck at this splendid three-day meeting of so many eminent scientists from so many countries should be the keynote of cooperation. It was this spirit of cooperation that in ten short years has helped make the Puerto Rico Nuclear Center a recognized center of excellence available to all Latin America. It was this spirit of cooperation that was responsible for President Kennedy's call for an Alliance for Progress between the Americas — an Alliance to which President Johnson has repeatedly reaffirmed my country's steadfast support. Most recently, at the Punta del Este conference in Uruguay last April, President Johnson, together with your esteemed Presidents, gave new emphasis to the Alliance and particularly to the importance of regional cooperation in the pursuit of scientific and technical progress.

At Punta del Este our Presidents proposed a Regional Scientific and Technological Development Program to implement such cooperation. And, as you know, work toward this end in the nuclear field is being carried on by the Inter-American Nuclear Energy Commission (IANEC). Last summer, to follow up, on behalf of my country, this call for cooperation, I had the pleasure of taking a trip to six nations of South America — a trip to view firsthand some of the scientific and technological progress that was being made in Latin America. It was truly a rewarding trip, leaving me with many favorable impressions that I wish to reflect in my thoughts tonight.

During my visits to these nations I had the opportunity to discuss with the heads of the various atomic commissions, agencies and laboratories the importance of developing nuclear energy in Latin America on a cooperative and regional basis in implementation of the conclusions reached at Punta del Este. I also had the opportunity to see many of the fine nuclear facilities in Latin America. These facilities, their dedicated staffs, the quality and scope of work being carried on by them, and their plans for the future projected to me a promising picture of the atom's role in the development of this hemisphere.

After personally reviewing the facilities, the accomplishments and the continuing programs of many of the nuclear centers of Latin America, one can appreciate the rationale behind the recent IANEC recommendations of having several of these laboratories, which are truly centers of excellence,

From L to R: Puerto Rican Governor Roberto Sanchez Vilella, Seaborg, Dr. Jaime Benitez, Ambassador Sol Linowitz, and Dr. Henry Gomberg, San Juan, Puerto Rico, October 30, 1967. Reproduced by courtesy of the U. S. Department of Energy (formerly the U. S. Atomic Energy Commission).

serve as regional centers for multinational cooperative programs. The Puerto Rico Nuclear Center, whose tenth anniversary we are celebrating with this symposium is, I believe, an outstanding example of one of the laboratories that could be considered one of these regional centers. I will be mentioning others later.

Although I did not have the opportunity to visit Mexico or the countries of Central America on my recent trip, my fellow commissioners and other AEC officials have visited these countries on many occasions. I might mention the many visits of Commissioner James T. Ramey and other Commissioners to Mexico and our most cordial and productive relationship with the late Nabor Carillo Flores, former Commissioner of the Mexican Nuclear Energy Commission. In addition, Commissioner Gerald F. Tape has visited Costa Rica and Panama to participate in the opening of our Atoms at Work Exhibit in those countries, and one of our former Commissioners, Dr. Samuel M. Nabrit, visited Nicaragua to inaugurate the Atoms at Work Exhibit there.

I might digress for a moment here to say a word or two concerning these exhibits, which I consider to be unique and an example of a special kind of cooperation between our countries. These exhibits that the USAEC has conducted in cooperation with many Latin American countries are unique in that they perform an educational function at many levels. They include some training for young scientists, lectures and classroom work for high school students, and exhibits, demonstrations and films to inform the general public about the benefits of the peaceful atom. In the seven years that "Atoms at Work" has toured Latin America, it has appeared in Buenos Aires, Argentina; Rio de Janeiro, Brazil; Lima, Peru; Mexico City, Mexico; Santiago, Chile; Bogota, Colombia; Montevideo, Uruguay; San Salvador, El Salvador; Guatamala City, Guatamala; San Jose, Costa Rica; Managua, Nicaragua; and Panama City, Panama. It will open on November 6 in Quito, Ecuador, and plans have already been made for it to appear within the next year or so in Caracas, Venezuela; Cordoba, Argentina and São Paulo, Brazil. I believe that this exhibit, operated in large measure by personnel from its host country, contributes significantly toward an understanding of nuclear energy and hence toward support of nuclear programs in Latin America.

Perhaps this anniversary is a fitting occasion to recall and review a few facts about the past and present of the atom in the Americas. In addition to this being the tenth anniversary of the Puerto Rico Nuclear Center, we recognize another significant anniversary of nuclear history this year.

It is just a quarter of a century ago that the principle of nuclear fission was first applied to obtain a practical nuclear chain reaction. As you all know, this was accomplished by a team working in the United States under the direction of the great Italian scientist, Enrico Fermi.

Under the stands of Stagg Field at the University of Chicago, Fermi and his colleagues constructed and operated the world's first nuclear reactor. On December 2, 1942, a control rod was slowly drawn from a crude pile of uranium and graphite and the scientists present verified that a self-sustaining nuclear reaction could be carried out under controlled conditions, and thus the nucleus of the atom be made to give off energy which could be harnessed for useful purposes.

This was an event of overriding importance in the world of science and one which we are planning to observe with appropriate ceremonies in the United States in this twenty-fifth anniversary year. The focal point of the observance will be at Chicago on December 2.

The U.S. Atomic Energy Commission (USAEC) was established just four years after Fermi's historic achievement to assume responsibility for the development, use and control of nuclear energy in the United States. And just four years after the USAEC was established, the first atomic energy commission in Latin America was organized — the National Commission of Atomic Energy of Argentina (CNEA), of which my good friend and esteemed colleague, Admiral Oscar A. Quihillalt, is Chairman.

The CNEA was established in 1950 and the significance of this event to me is that Latin America so quickly joined the vanguard of world communities that recognized the importance of the atom and took steps to do something about it.

Today, national atomic energy commissions or governing agencies of one kind or another exist in fourteen Latin American nations and have responsibility for programs of varying degrees of advancement. The Inter-American Nuclear Energy Commission, organized within the framework of the Organization of American States, was established in 1959 to promote the benefits of the atom among its member states, and most Latin American nations are also members of the International Atomic Energy Agency (IAEA). Fruitful cooperation involving a variety of peaceful applications of the atom has taken place between the U.S. Atomic Energy Commission and virtually every nation of Latin America.

A moment ago I mentioned the early recognition that Latin America gave to nuclear energy as a means of promoting progress in this hemisphere. Let me take a few minutes to review briefly some of Latin America's achievements and continuing work toward that end.

I know that my fellow member of the United States Atomic Energy Commission, Commissioner Ramey, earlier today described in some detail the outlook for nuclear power in Latin America, particularly in connection with the energy center and dual-purpose desalting plant concepts. I should like to add my own expression of enthusiasm for these exciting prospects.

As you know, several Latin American nations are actively interested in the use of nuclear energy for power needs in the immediate future. Acting under the authorization of President Ongania and under the direction of the CNEA, Argentina has established an intra-governmental committee to consider proposals which have been submitted for the erection of a 300 to 500 megawatt station to serve the Gran Buenos Aires–Litoral area, the highest energy consumption area in that country.

Steadily increasing demands for power, I am told, point to the need for some 1,300 more megawatts of generating capacity in the Buenos Aires–Litoral area by 1972, with indications that much of this new demand must be met by thermal plants. Argentina's first nuclear power station is expected to come into service no later than 1972 to share the burden of meeting this load.

Brazil, too, is considering the use of nuclear energy for future power generation. As far back as 1962, a comprehensive effort was begun to determine the feasibility of nuclear power in that country. The following year, Brazil's National Commission of Nuclear Energy (CNEN) began contemplating three possible projects: a 300 MWe nuclear generating plant for South Central Brazil; a 60 MWe nuclear station for the southern region near Rio Grande do Sul; and a 50 MWe reactor for the northern region, possibly at Teresina.

In January 1965, president Castello Branco directed that a nuclear plant be established in the South Central region where approximately 75 percent of Brazil's industrial production capacity is located. Two feasibility study groups were established to consider the type of reactor needed, the best possible location for the plant and other technical details. One facet of these studies was, quite naturally, consideration of reactors which might, in time, make use of Brazil's considerable thorium deposits.

I am happy that our Oak Ridge National Laboratory has been cooperating closely with Brazil in these studies, which also include plans to consider all types of reactors for this project, as well as the possible thorium reactors of the future.

This matter of direct cooperation with a USAEC laboratory calls to mind several discussions I had while in Latin America concerning the value of "sister laboratory" agreements between U.S. and Latin American nuclear laboratories. The "sister laboratory" arrangement is a very effective relationship which establishes a special bond between a USAEC laboratory and a nuclear laboratory in another country. There already exists a "sister laboratory" agreement between the fine new laboratories of the Institute of Nuclear Affairs (IAN) in Colombia and the Puerto Rico Nuclear Center, an agreement which results in effective use of the IAN research reactor and makes results of studies there available at PRNC. More "sister laboratory" agreements between nuclear centers in Latin America and PRNC, and between Latin American laboratories and AEC laboratories in the United States, should prove mutually beneficial in the future.

As was discussed earlier today by Commissioner Ramey, Mexico and the United States are examining, under a joint study sponsored by the IAEA, the feasibility of a nuclear desalting program that would furnish both power and fresh water to the Mexican states of Sonora and Baja California and our own states of California and Arizona. We look forward to the success of these studies and, hopefully, to further cooperative efforts with Mexico that may set a pattern for progress in international cooperation in this important field of nuclear desalting.

I submit that these nuclear power beginnings are most encouraging. Taking into account the fact that in most Latin American countries fossil fuels are either scarce, difficult to recover, or in extremely remote areas, it is my belief that nuclear energy will play a major role in furnishing electricity and process heat in Latin America and will do so far sooner than had been expected only a few years ago.

While power reactors are being planned, more than half a dozen research reactors already are in existence in Latin America and are being used for a variety of important purposes including training, cross section measurements, activation analysis and the preparation of radioisotopes for use in medicine, agriculture and industry. During my trip I was fortunate to visit a number of these research reactors and learn of the work being done with them.

Through the courtesy of Chairman Uriel da Costa Ribeiro of the Brazilian National Nuclear Energy Commission I visited the 10 kilowatt Argonaut reactor at the Institute of Nuclear Engineering on the new campus of the Federal University in Brazil, and saw the 5 megawatt swimming pool reactor at the Institute of Atomic Energy (IAE) in São Paulo, Brazil. I also saw fine research reactors in Venezuela and Colombia, at institutions I will be discussing in a moment.

In Argentina I visited the 5 megawatt RA-3 reactor at the Ezeiza Laboratory and had the pleasure of touring, with Admiral Quihillalt, the CNEA Laboratory where work on reactor fuels, metallurgy and technical assistance to industry is taking place.

After visit to Ezeiza Laboratory near Buenos Aires, Argentina, 6 July 1967. From L to R: Admiral Oscar A. Quihillalt (Chairman, Argentine Nuclear Energy Commission [CNEA]), Mario E. Bancora (Commissioner, CNEA), Seaborg, Herman Pollack (Acting Director of International Scientific and Technological Affairs, U.S. Department of State), Myron B. Kratzer (Assistant General Manager for International Activities, AEC), Robert Hollingsworth (General Manager, AEC), Dan J. Beninson (Chief, Health and Safety, CNEA), and Juan E. Bazet (Chief of Protocol, CNEA). Reproduced by courtesy of Arnold R. Fritsch.

I believe that the nuclear laboratories in Argentina and Brazil offer excellent prospects for use as regional centers for multinational cooperative programs.

A moment ago I mentioned the use of radioisotopes. We find that widespread use of these important tools of the nuclear age is being made in Latin America today. Radioisotopes are, in particular, playing an increasingly important part in the diagnosis of disease and the relief of human suffering. Hospitals in virtually every Latin American country are using these invaluable aids in the diagnosis and treatment of ailments. A number of teletherapy units also are available for the treatment of cancer and for other uses, and courses in medical research and radioisotope training are widely available at leading hospitals and university laboratories.

The University of São Paulo in Brazil, for example, has an active nuclear medical program in which advanced techniques and applications of radioisotopes are employed. It also offers postgraduate courses in radioisotope tracers; radioisotope diagnosis; radioisotope applications in biological research; and radiation hygiene. I might add here that in visiting the Institute of Atomic Energy in São Paulo I was also pleased to see, as I saw in other nuclear facilities in Latin America, a USAEC depository library in active use.

In Colombia, where Dr. Tulio A. Marulanda, Executive Director of the Colombian Institute of Nuclear Affairs and his colleagues were our hosts, I had an opportunity to visit the Institute of Nuclear Affairs (IAN), which has new laboratories near Bogota to provide training in radioisotopes and radiochemistry. I understand that staff members of this fine laboratory, in addition to their work at the laboratory, contribute their services to colleges and secondary schools in Colombia. The IAN also has an extensive medical and biological program which is being carried out at Colombia's National Cancer Institute. A cobalt-60 teletherapy unit is available there for diagnosis and therapy, and courses in research and training in medical radioisotopes are also given. I cite these as but two outstanding examples of the work in radiomedicine being done in this part of the globe today.

In speaking of the progress of the peaceful atom, we should note some of the steps being taken in Latin America today to develop the use of radiation for food preservation.

Beginning this year, Argentina will make use of a portable food irradiator in studies of the sterilization of bovine blood as a possible protein source for human nutrition and in studies of the irradiation of such foods as fish, strawberries, and potatoes to prolong shelflife. The irradiator, now undergoing tests at our Brookhaven National Laboratory, is expected to be delivered to Argentina in November. I look forward with interest to the results of this cooperative program.

We also look forward to cooperation with Chile in its program of research on the preservation of foods, including potatoes and fish, and for a broad program in agriculture to be undertaken by the Chilean Commission of Nuclear Energy and Chile's National Institute of Agricultural Research. We expect that this program will involve a cooperative effort with Chile, and we believe that such cooperation would be beneficial to both countries.

Colombia likewise is interested in the use of radiation for the preservation of foods, notably potatoes, and similar interest is being shown in Venezuela in the radiation preservation of tropical fruits and vegetables. Other Latin American programs include several studies in the use of radioisotopes to trace plant uptake of fertilizers for the improvement of agricultural production.

Now I must mention another phase of radioisotope agricultural research and development of increasing importance in Latin America. This is the use of radiation to control insect pests by sterilization of males of the species.

A significant step in this direction was taken this past July when, at the invitation of Peru, representatives of Chile and Bolivia met in Lima, along with representatives from the United States, to discuss a joint three-way program by Peru, Chile and Bolivia to eradicate the fruit fly from their countries. Preliminary sterilization work with an irradiator using radioactive cesium-137 is already under way at Peru's Agricultural Experiment Station in Lima, and the release of the first batch of sterilized male files on an experimental basis is expected to take place later this year. Success in eradicating the fruit fly in these three nations would, it is estimated, save millions of dollars a year in damages to citrus, mango, apple and pear crops in Peru, Chile and Bolivia. I am informed, for example, that in the case of mangoes only one mango of every ten to twelve on the tree comes to market because of the depredations of this devastating pest.

By way of further illustration of significant activities in the use of nuclear energy in Latin America, I should like to cite a cooperative program that has been under way for some time at Turrialba in Costa Rica.

The program includes a number of individual studies of radiation and radioisotopes in attacking agricultural problems in a tropical situation. It has, moreover, the added value of training Latin American students in these techniques by having them engage in the actual research under the supervision of senior scientists.

There are four general areas of activity:

(1) Radiation biology and genetics designed to determine the radiation sensitivity of tropical plants.
(2) Biochemical studies on the pathogenicity of certain fungi tropical crops.
(3) Radiation sterilization of insect pests.
(4) Soil chemistry work on the problems of phosphorus in tropical soils.

I am happy to hear that the work is steadily generating increased interest and enthusiasm among several participating Latin American states.

Finally, as many of you are aware, a symposium to which representatives of all Latin American nations and the United States have been invited is to be held in Santiago, Chile, in January on nuclear radiation and agricultural productivity. Topics to be discussed will include studies on water, soil and fertilizers; animal research; entomology; and food irradiation.

From all of these developments, it seems clear to me that radiation and radioisotopes are beginning to assume an increasing importance in vital aspects of the life and economy of the Latin American nations. I am glad to note that a number of papers on the topics which I have touched upon so briefly here are on the agenda for tomorrow. I am sure that these discussions by experts intimately associated with the programs involved will be among the highlights of this symposium.

Again, let me touch on the theme of cooperation, and particularly on the sincere desire for it I found evidenced during my recent trip. In each country I visited I was extremely pleased by the frank discussions I had with the administrators and scientific staff of nuclear facilities.

These included my talks in Lima with Javier Correa Miller, Chairman of the Board of Control of Atomic Energy of Peru, and with Dr. Tulio Marulanda of Colombia, Admiral Quihillalt of Argentina and Chairman Ribeiro of Brazil, whose broad programs in the fields of nuclear energy I mentioned earlier.

While in Chile, where Benjamin Viel, President of the National Commission of Nuclear Energy (CNEN) and Efrain Friedmann, the Commission's Executive Director, were my kind hosts, I enjoyed a visit to the new Macul Campus of the University of Chile to see, among other things, the 22-inch cyclotron that had recently been installed there. This cyclotron, furnished through a cooperative arrangement between the University of California and the University of Chile, was one that had served as a prototype for the rebuilt 60-inch cyclotron moved from the University of California's Berkeley campus to its Davis campus. I believe that the 22-inch cyclotron will prove to be a valuable asset to Chile in its nuclear science work. I was especially gratified to learn of the fine cooperation between the University of California and the University of Chile, which includes exchanges of professors and other personnel.

In Venezuela, where I visited the excellent facilities of the Venezuelan Institute of Scientific Research (IVIC), I also had an opportunity to discuss several aspects of inter-American cooperation on nuclear matters with Dr. Marcel Roche, Director of IVIC. In view of the concept of regional cooperation I discussed earlier, I was encouraged to hear Dr. Roche's ideas on inter-regional cooperation involving the possibility of Venezuela, Colombia, Ecuador, Bolivia and Peru all using the IVIC as an Isotopes Training Center.

Now what is the place of the Puerto Rico Nuclear Center among the many cooperative efforts taking place in nuclear energy and nuclear science in the Americas?

As this symposium reminds us, it was ten years ago this month that the AEC and the University of Puerto Rico signed a contract under which the University agreed to operate the Center for the Commission.

Dr. Benitez, President of the University, gave us a splendid summary of PRNC activities and aims this morning in his opening symposium address on "The Mission of the Puerto Rico Nuclear Center."

The growth in size of the Center alone has been impressive, for during the past ten years the Center has been attended by more than 1500 students, including students from almost every country of Latin America, and scientists from several of those countries have served on its staff. And, as Dr. Benitez pointed out, its programs range from a study of the effects of radiation on a typical forest environment to studies, through the use of radioisotopes and other nuclear means, of the disease Schistosomiasis

Masoni, which afflicts an estimated 100 million persons in the world and for which no control presently is known.

Today, as we celebrate the tenth anniversary of the Puerto Rico Nuclear Center, we look upon PRNC as one of several regional centers of excellence whose programs have been particularly well-tailored to the nuclear needs of Latin America. And we look forward to the day when several of these centers, all cooperating, will cause a chain reaction of nuclear — and hence human — progress throughout the hemisphere.

As has been stated throughout this symposium, nuclear energy has a great contribution to make to Latin America. This contribution can be made through the use of radiation and radioisotopes, by the development of nuclear power for a variety of purposes including large central station power complexes, desalting and, hopefully someday, the development of small, economical plants for remote sites where an adequate supply of electricity would do much to raise the hopes and standards of living of a vast number of people.

Large amounts of capital and far-seeing management will be needed for the successful development of many of these projects, but equally important will be the need for large numbers of well-trained engineers and scientists.

The Puerto Rico Nuclear Center, with programs which now attract a growing influx of students from Latin America, including many young scientists from university faculties, stands ready to cooperate with other leading nuclear centers of the hemisphere to provide this training.

I think we can look forward to the next ten years of the life of PRNC as an era of great progress in nuclear energy in Latin America — one in which this fine facility will continue to make a noteworthy contribution. And at this time I offer my congratulations and best wishes for the future to Dr. Henry J. Gomberg, Director of the Puerto Rico Nuclear Center, his associates, and his excellent staff. I know that my fellow Commissioners and the staff of the AEC also extend their good wishes on this happy occasion. We also wish to express our appreciation for the fine leadership provided by the previous directors of the Puerto Rico Nuclear Center, John C. Bugher and Carlos F. Bonilla.

It has been a privilege to be here this evening to discuss broadly some of the nuclear activities in Latin America; to review some of the important developments which have taken place since the formation of the first nuclear energy control agency on the South American continent nearly two decades ago, and to look forward with you to the bright promises of the future.

As we continue to advance in knowledge and application of the peaceful power of the atom, it is my earnest hope that we shall do so in large part by means of the cooperative efforts that have marked so much of our progress in the past. Certainly, we have in the Puerto Rico Nuclear Center and in the other key nuclear installations scattered throughout the hemisphere the tools to do so.

All that is needed is the desire to work together, as our American Presidents have suggested, in regional developments that will help to ensure a better life for all. I believe we have this desire and willingness in abundance, as is evidenced by the participation of representatives from so many nations in this anniversary symposium.

Let us hope that future occasions such as this will see us drawn even closer together in working toward our common goals.

Twenty-five years ago, when the energy released from the nucleus of the atom was first controlled, men speculated on what this great new force could mean to mankind. Each year has added both new fears and new hopes to our outlook for the atom. But today I believe the hopes can be made to overshadow the fears. What nuclear energy can now offer mankind in the way of a positive, constructive force for peace and human betterment makes me believe that the peaceful atom will prevail. Let us strive toward that end — and let us do it together, so that we may live not only in peace and prosperity but with the dignity and self-respect of men who know they have helped to shape their own destiny.

The Positive Power of Science

Remarks made at the International Joint Conference of the American Geographical Society and the American Division of the World Academy of Art and Science, New York, May 1, 1970.

To speak to such a distinguished gathering, and one that over the past few days has immersed itself in matters of such importance, is indeed a challenge. This has been a conference with a difficult and most urgent task — that of considering "the environmental and social consequences of science and technology" and, in the light of these considerations, attempting to "formulate policy goals and strategies for the needs of man." Of course, the kind of thinking that has been responsible for initiating this conference and organizing its direction and content has been long overdue. Most of us, I am sure, realize that, whatever has been accomplished during these deliberations and discussions, we cannot be too self-congratulatory. There is a great segment of the public that would tell us we are "too late with too little." And a significant part of that group will go even further and say that we of the scientific community and our work are at the root of much of man's ills today.

Nevertheless, I am not here tonight to add to the despair and guilt of our times. I am not one who happens to believe that we can degrade the environment less by degrading man more. We are not going to save the

Reprinted with permission from the *Annals of the New York Academy of Sciences*, Vol. 184, pp. 682–687, 1972.

Also published in *Representative American Speeches 1969–1973*, ed. Lester Thonssen, Vol. 42, No. 4, p. 108 (N. W. Wilson, New York).

Earth merely by despising ourselves as its inhabitants any more than we are going to build a better society by belittling ourselves as individuals.

I think it is unfortunate — almost calamitous — that in the often sincere effort to establish a better perspective on our past errors and a sense of proportion about our present powers there are many who feel they have to go to the extremes of demeaning man, of denying reason and downgrading science. Tonight I plan to defend all three. I particularly want to speak up in behalf of science and urge that it rise to the crisis that faces it and man. In a sense it will be the success or failure of science — science epitomizing man's extended perception and power — that may well determine his survival as a species. But it is not science as we have used it in the past that will do the job. It must be science as part of a new human philosophy, as part of a new age of enlightenment that we are entering. And it is both the past shortcomings and successes of science that are moving us into that age. Let me explain.

What we are seeing today in all our social upheavals, in all our alarm and anguish over an environmental feedback and, in general, in the apparent piling of crisis upon crisis to an almost intolerable degree, is not a forecast of doom. It is the birthpangs of a new world. It is the period of struggle in which we are making the physical transition from man to mankind — a mankind that will be an organic as well as a spiritual whole on this earth. I see this transition as a natural evolutionary process, a continuation of the growth and growing complexity of life on this planet. I do not believe that this growth is malignant in nature. It will not destroy itself by devouring its host or poisoning itself in its own waste. Neither will it self-destruct after delivering its message. Rather it will self-adjust through listening to and responding to that message — one that for all the static surrounding it is coming through quite clear.

I would like to touch on several ideas related to my admittedly broad and optimistic outlook for the future. In each case I hope to bring in the relationship of science and technology to the more general idea.

First, let me dwell on the subject of growth. This is something that until recently was a major value, but of which we now seem to be developing a deadly fear. Both that growth and today's reactions to it, I believe, are natural and necessary. There seem to be many people today who, with no mean amount of eloquent hindsight, deplore the fact we did not, long before now, predict our population growth and our growing productivity with its accompanying waste, and somehow forecast our current environmental dilemma. Unfortunately, most great minds of the past foresaw only small segments of the evolving problem. And the values of the past were centered about unlimited growth because this seemed to be to the advantage of the individual and his society. In a world of seemingly endless physical frontiers, where the exploration and exploitation of these frontiers by new human creativity generated and fulfilled new human needs and values, few if any could think in terms of limits, balance and stability.

Many of us who today talk so glibly about our use of the atmosphere and oceans as "sinks," and who have embraced ecology almost as fervently as we have become embarrassed with economics, tend to overlook or forget the facts and feelings of the most recent past. It does little good to make scapegoats of our ancestors or of one or another segment of our society for the crises we face today. Our environmental crisis in particular could not have been theorized or accepted in the abstract before. It was an experiment that had to be lived in conjunction with the other problems of human growth that have evolved and which we must now move on to resolve. I think it was inevitable that man had to grow to this point. He is now entering what must be a period of tremendous maturity. And just as there is written within the genetic code some incipient biological mechanism that stops physical growth at a certain point in life, so I believe there is within our evolving mankind a well-coded message that is now being released.

Before I go on to discuss some of what that message means, let me backtrack to discuss the role of science and technology in the growth to which I have referred.

It is obvious that for some people science and technology are among the best scapegoats of the time. They are said to be the cause of most of our ills today. By conquering disease and extending life they have been responsible for an explosion of population. By increasing productivity and raising living standards they have been responsible for depleting resources and polluting nature. By expanding knowledge and emphasizing efficiency, they have been responsible for deflating myths and diminishing man. And by placing enormous power in the hands of man they have brought him to the brink of his own destruction. The list of accusations is endless and it is not fashionable today to attempt to answer the charges or put the matter in perspective. It is more fashionable to dwell on man's relationship to the natural world, to lament that he is not more like the animal life from which he descended and to wish for his return to a simpler, perhaps more primitive, existence.

This approach may be more fashionable, it may even contain a certain amount of wisdom we should heed, but it is not the whole message we should be hearing. The major part of that message, I believe, should tell us not to deny our dependence on nature but neither to deny our differences from what we left behind in our evolution. It is this recognition of what we are — with all its potential as well as its shortcomings — and the emphasis on what we must and can become that is important. The fact is that in this transitional period from tribal man to a truly organic mankind, and to a world in which we can live in harmony with each other and in balance with our global environment, we need a new level of excellence in our science and technology and a new degree of integration between them.

There is, no doubt, a great deal of pain and shock involved in this transitional period for we are breaking one set of long-established natural bonds and forming new ones. The whole process of change produces shock, reaction and readjustment. There is always a tendency to revert back, to flee from the new and challenging, before new understanding and confidence allow us to move ahead. We are living through such times. We are experiencing what Alvin Toffler refers to as "future shock," and it is often difficult to sort out our movement and its direction. For example, take note of the action and reaction of our youth as it resists some change with an anti-rational thrust — and often a flight from reality — while at the same time demanding realization of a new level of idealism that can only be achieved through change — change employing the highest form of rationality.

A similar dichotomy and flux exist in our confrontation with environmental problems. Many feel the need for simplicity, limits and balance. Yet we know that what we must do to accomplish the goals we express in these terms involves to some degree the mastery of a greater complexity, new growth and a dynamic rather than static type of balance. We really do not want to freeze the world as it is now or go back to "the good old days." What we want, if I read correctly the ideals that so many of you — and so many deeply thinking and concerned people around the world — feel today, is a world far different than we have ever known it. Today — at this conference and at similar meetings that are being held and organized — we are struggling, fiercely but fruitfully I think, to clarify our conception of that world and work toward its realization. We must not weaken or lose heart in this struggle.

What will be the outcome when we begin to succeed? What will the evolution of this new mankind mean? And what will be some of the manifestations that it is taking place and succeeding?

Perhaps most important, we will see the elimination of war as an attempt to resolve human differences. It will not be only that war becomes untenable as a form of such resolution. Neither will it be only that through the wisest and fullest application of science and technology we will eliminate most of the physical insecurity and want at the root of war. What may be most significant as both a cause and effect in establishing world peace will be a sublimation of man's territorial instincts, and the aggressiveness that is tied to it, to a new feeling, one of the communality of man in the possession of the entire earth. Men are already in some measure sharing the earth through international travel, communication and exchange of resources. As this sharing is enhanced by a parallel releasing of the age-old bond of fear of scarcity — and adjustments in the economic system we have built to institutionalize that bond — we will begin to see the true meaning of the brotherhood of man materalize. And as this happens the tribal loyalty that Arther Koestler has seen as the root of much of man's conflict will be broken and shift to a new global loyalty — a loyalty of man to all his fellowmen.

Concurrently with this establishment of world peace — and again as both a cause and effect of it — will be the closing of the chasms between the peoples of the world. Aurelio Peccei, Barbara Ward and many others have warned us that we cannot live in a world growing apart in the rate of development of its peoples. What I think we must, and will, see is a new concerted effort to raise the standard of living and productivity of the underdeveloped areas of the world while readjusting the growth of what many feel are becoming "overdeveloped" areas, harming themselves and others by some unwise management of their power and affluence. Anyone fully attuned to the problems of the world today can feel the need for this effort and readjustment. I think it is vitally important that we in scientific and technological fields do everything we can to encourage and work with those social and political forces that recognize this need and are trying to fulfill it. I will have more to say about such cooperative efforts in a moment.

Another manifestation of our evolving mankind will be the reduction, and eventual elimination, of environmental pollution. The organic mankind of which I have spoken could exist but momentarily on this earth if it were to act as a parasite or cancer. It must learn to exist as an integral and contributing part of the earth that up to now has supported it unquestioningly. This can only be achieved by the formulation and application of a whole new scientific outlook and new ecological–technological relationship. This relationship must be based on a non-exploitive, closed-cycle way of life that is difficult to conceive of in terms of the way we live today. We will have to achieve what Rene Dubos has referred to as a "steady-state world." We will have to think and operate in terms of tremendous efficiencies. We will have to work with natural resources, energy and the dynamics of the biosphere as a single system, nurturing and replenishing nature as she supports and sustains us. Such a system can be operated at various levels. A steady-state world does not have to be one in which mankind merely subsists and waits for natural evolution to take place. In fact, a steady-state world would be a challenge and stimulus to man's creative evolution — which I believe we should not deny is a natural process and which may be the highest form of natural evolution. Perhaps the organic global mankind which I have portrayed will be the acme of physical evolution of this planet. And I will not try to speculate beyond this point.

I have no doubt that many people envision the concept of such a complex, efficient and organic mankind as a nightmare — an anthill civilization in which individuals are mere automatons or mindless cells in an emotionless body. I do not agree with such thinking. What I see evolving is quite the opposite of this. It is a world in which the sphere of freedom of action and choice, individual creativity and sensitivity is enlarged by the growth and application of knowledge and by greater efficiency and organization. These elements buy time and provide freedom. It is ignorance, confusion and waste that enslave and eventually destroy.

Of course, new values and greater education must accompany the transition of this type of civilization. That is why I believe that the age of enlightenment we must enter must be one that combines scientific understanding with a new humanistic philosophy. We need both now to survive and grow.

What will be the role and direction of science in achieving this new age? I believe this conference offers us a good indication of both. But let me briefly sketch what I feel has been the movement of science and where it is going today.

Science, we might say, has become a victim of its own success. Or to put it more precisely, it has become a victim of its own single-minded success. I think this has happened in several ways. First, in going from the broad and general philosophy from which it originated into a growing number of more precise disciplines — each becoming more productive the narrower its focus became — science traded off wisdom for knowledge and, to some extent, knowledge for information. In recent years this process has been reversed and we are now seeing the growth of interdisciplinary sciences and a striving for a more all-encompassing grasp of the physical world and even broader relationships such as you have been exploring in this conference. This type of growth is essential if science is to be the guiding force — as it must be — behind our evolving mankind. Science must grow stronger by continuing to nourish and improve its individual disciplines. We need the specific knowledge they offer. At the same time it must grow wiser through its correlation of knowledge. And it must be able to transmit its wisdom in the most effective way to society.

We have a tremendous task before us in humanizing the focus and feeling of science while at the same time organizing and rationalizing the forces of humanity. In recent years we have not been too successful in either of these directions. That is the reason why we have been faced, and are faced today, with a decline in the prestige of science, an anti-rational reaction on the part of many of our disillusioned youth, talk of the "eroding integrity of science" and even a feeling of guilt and despair in much of the scientific community. We must move away from all this. We must work toward a unification of the scientific spirit and a restoration of our self-confidence as well as a new degree of respect for science on the part of those who have lost faith and hope in it.

Let me offer some specific proposals as to ways we might accomplish this.

I think we should establish more international interdisciplinary conferences — such as the outstanding one we have been attending here — and more organizations that integrate our various disciplines, within and outside of the sciences. I believe these conferences and organizations should bring together for positive, constructive exploration, discussion and action participants of varied interests, opinions and talents — visionaries and realists, environmentalists and technologists, ecologists and economists, theorists and activists. But the purpose of these meetings should not be that of many of the "confrontations" we are witnessing today. We should seek not the degraded power of polarization but the more beneficial strength of unity — that achieved through recognizing and working toward common goals.

In this regard I would like to see those scientists who in recent years have done a great service to man by calling attention to his environmental problems now contribute an even greater service by joining their colleagues in a concerted effort to solve those problems.

I think that out of such conferences as we have held here should come concrete programs or ideas that can be acted upon, as well as the broader policy-setting type of recommendations. We must give our activists something constructive to act upon and encourage the idea that many small positive measures can add up to a significant force. They can develop an important momentum and a spirit that in time can become overwhelming in this total effect.

I think that the information and programs generated at these conferences must be brought to the attention of the public and the world's political leaders more successfully. We have been very unsuccessful in communicating with the public, in bridging the gap between C. P. Snow's Two Cultures. Now we must not only communicate, we must involve. We should particularly encourage the participation of youth in scientific and technical activities. Merely to decry their alienation, to speak of their immaturity or their unrealistic, "non-negotiable" approach to achieving their ideals, is pointless — more than that, it is disastrous. We must at all levels engage them in the realities of life, not to blunt their ideals or enthusiasm, but for the purpose of capturing what is good and constructive in them, of harnessing their energy and creativity, of growing with them.

If some sparks must fly between the gap of our generations, let us not use them to ignite conflagrations but rather to fire an engine of human progress. We in the scientific community in particular need our young people working with us, and it is one of the tragedies of our time that so many of them have become cynical about the accomplishments and prospects of science. I believe we can win many of them back, especially by showing them how effective we can be in working toward the solution of our environmental ills. We must prove to them that science and technology are among man's most creative and constructive forces — when they are used by creative and constructive men.

Finally, I think that in bringing together the many forces I have referred to tonight, and in emphasizing the importance of their working together, we must establish the leadership and goals to direct and sustain their efforts. Never before has the world had such a desperate need for greatness, for inspiration, for a vision. The cynics today will tell us that any vision we would have now would be a delusion. But I cannot agree. I feel as it says in Proverbs: "Where there is no dream — the people perish." Let us create that dream then and work to achieve it — not only that man shall live but that mankind shall be born.

A Scientific Safari to Africa

Article prepared for Science,
published 7 August 1970.

As the technicians removed the sample containers from the target holder, the laboratory director explained how the radiation-attenuated organisms of the parasitic disease leishmaniasis were being tested to determine their immunological potential. Success would offer a means to control the vicious disease leishmaniasis, which disfigures and blinds millions in the tropics.

The scene I was witnessing and the explanation I was receiving were occurring not in a highly advanced U.S. medical research laboratory but on the outskirts of Addis Ababa, in a modest building that houses the Institute of Pathobiology of the Haile Selassie I University. The explanation of the application of one of the most advanced and progressive techniques for the control of parasitic diseases was being given by Aklilu Lemma, dean of the Faculty of Sciences of the university and director of the laboratory. I was visiting Ethiopia — and five other African countries — with several other U.S. scientists and scientific administrators[1] to become better acquainted with African science, to establish contacts between African and U.S. scientists, and to identify ways in which closer and more effective cooperation with science in Africa might be achieved.

In almost 9 years as chairman of the Atomic Energy Commission, I had already visited over 50 countries, ranging from the technologically and scientifically advanced nations of Europe to the developing states of Asia and Latin America. Despite my long-standing interest in Africa, I had

Reprinted with permission from *Science*,
Vol. 169, pp. 554–561, August 7, 1970.

avoided a visit there because I doubted whether the state of scientific — and especially of nuclear energy — development would justify the trip. Finally, however, conversations with other government officials led to the idea that a visit to Africa by a senior U.S. government science official would be beneficial. Secretary of State William P. Rogers gave the plan his enthusiastic support.

I was to visit Africa in my capacity as a scientist, rather than specifically as chairman of the Atomic Energy Commission. My purpose was to make a broad-ranging evaluation of science on that continent and of the prospects for beneficial scientific cooperation between the United States and the countries I would visit. An itinerary was drawn up which included six countries: Morocco, Tunisia, Ethiopia, Kenya, the Congo (Kinshasa), and Ghana. Both northern and sub-Sahara Africa and both English- and French-speaking areas were represented. A small group of outstanding scientists in the fields of medicine, biology, and agriculture was selected and was supplemented by appropriate government officials who could follow up any leads for increased cooperation.

Our group departed on 3 January 1970 and returned 13 days and 21,720 miles later. My predominant reaction on this trip was surprise — surprise that Africa had come as far as it has in science, in self-government, and in eagerness to move ahead. Although some of the countries we visited had only a handful of college graduates just 10 or so years ago, I returned with the conviction that, in many respects, Africa is moving faster (although from a more elementary base) and has more unclouded prospects for technological and economic progress than many developing nations in other parts of the world.

If the progress in Africa is surprisingly impressive, the needs are even more apparent. The efforts to develop a radiation-attenuated vaccine for the leishmaniasis organism is an example. In a well-equipped laboratory, the irradiation would be conducted with a small cobalt-60 gamma source, rather than with the electron accelerator with maximum energy of 300,000 electron volts that Dr. Lemma was using, in order to ensure uniformity of dosage throughout the entire sample. Dr. Lemma was well aware of this deficiency in his procedure and was anxious to correct it.

If we want to seize the opportunities that exist to help Africa in a number of modest but highly effective ways, much of the initiative must come (in light of current budgetary restrictions) from nongovernmental individuals, institutions, and resources in the United States. This article is one of several steps that I am taking to inform the American scientific community of the opportunities for exciting and productive cooperation with emerging Africa. If it generates even a handful of new contacts between American scientists and their African counterparts, its preparation will have been worthwhile. Throughout the article I have identified people and institutions in Africa in the hope that interested American scientists will contact them directly.

What We Saw

Throughout Africa, the scientific emphasis is on agriculture and medicine and, most importantly of all, on science education. Since science ministries generally do not exist, responsibility for scientific research is spread among the "using" ministries and the universities. The universities in particular are the focal points of scientific achievement and planning for the future. The heads of the university science faculties are often, therefore, the "de facto" senior science officials of the countries we visited. The ministers of agriculture and health are also influential in scientific affairs. I was gratified to discover that the heads of governments themselves have a lively interest in science and a sharp recognition of its importance to their national development. Emperor Haile Selassie, for example, who contributed one of his palaces with its grounds to provide the core of Ethiopia's Haile Selassie I University, told me that he believes science and technology will play a crucial role in Ethiopia's, and indeed in all Africa's, development.

Morocco

Morocco was a more traditional land than we had expected, despite its many years of close contact with European technology through association with France.

Its economy remains primarily agricultural, and King Hassan himself has directed that emphasis in the development of Morocco be placed on agriculture. Its modest scientific program is thus pointed in this direction, but public health and mineral exploitation are also important.

The Minister of Agriculture, Mohamed Benhima, explained to us that the most critical problem in conducting agricultural research programs is the shortage of personnel, including both professional scientists and skilled technicians. This refrain was heard over and over again during our African visit. Because of this shortage, there is heavy dependence on foreign scientific personnel. Local officials and foreign personnel are cooperating in strenuous efforts to train local scientists and to prepare them to assume leadership roles as expeditiously as possible.

Agricultural research work in Morocco is conducted in the National Institute of Agricultural Research (INRA), founded during the French rule, and in the biology department of the Faculty of Sciences of the Mohammed V University, both in Rabat.

Albert Sasson, chairman of the biology department of the Faculty of Sciences and a native of Morocco, is typical of the competent, highly qualified, and enthusiastic young scientists who are now making their way upward in African educational and scientific institutions. In addition to providing inspiring leadership to this department, Professor Sasson is personally engaged in research on the physiology of olive and citrus trees, a subject of immediate relevance to Moroccan agriculture. Other areas in which the department is active include the effect of calcium on nitrogen metabolism, nitrogen fixation by legumes, and the mycology and microbiology of arid soils.

Morocco has important mineral resources, principally phosphates, and reasonably good prospects for future discoveries. This fact has given rise to an interesting research project involving radioactive dating of Precambrian rocks by the argon–potassium and rubidium–strontium methods. This work is performed at the laboratory of the Ministry of Commerce, Industry, Mines, and Merchant Marine, under Minister Mohamed Jaidi.

Minister Jaidi is also responsible for an interministerial committee to explore and coordinate Moroccan programs in the peaceful uses of nuclear energy. The committee has a small full-time staff headed by Ahmed Majid. From these officials we learned that the maintenance of foreign equipment is a common problem due both to inadequately trained manpower and to frequent shortages of spare parts. On later laboratory visits, we sometimes saw equipment that was inoperable because parts as simple as transistors or vacuum tubes were unavailable.

Over the last several years, Morocco has received significant assistance in the peaceful uses of nuclear energy from the International Atomic Energy Agency (IAEA) in Vienna. This was our first direct encounter with the agency's technical assistance program in Africa. Throughout our visit, we learned that the agency's modest program of providing equipment grants, fellowships, and experts was the principal link of African scientists to the possibilities of beneficial uses of nuclear energy.

The medical uses of radiation and radioisotopes are the most intensive current use of nuclear energy in Morocco. A Soviet-built cobalt-60 teletherapy unit supplied by the IAEA is in use in Casablanca at the Radiological and Cancer Control Center of Morocco, and a small medical radioisotope clinic is functioning at the Avicenne Hospital in Rabat. Research efforts at the Rabat clinic are interesting and advanced; the work there on the use of technetium-99m to identify tuberculous disease of the spinal cord brought us into direct contact with a problem that is commonplace in Africa. The project is now under the direction of a talented Frenchman, Robert Granier, who is scheduled to leave soon with no well-qualified Moroccan available as a replacement.

The Faculty of Sciences of Mohammed V University has an enrollment in excess of 1000, of whom about half are candidates for science degrees; the remainder are enrolled in science courses required by other faculties. Later in the tour, Dean Ben Abdeljlil pointed out that the principal function of the African university in science today is to produce science teachers. Lacking scientific and technological tradition, these countries desperately need an exponential growth in the number of students with elementary scientific knowledge, not only to provide a nucleus of future university science candidates but, perhaps even more importantly, to develop a public capable of accepting and absorbing scientific and technological innovations. This gap is one that we who come from a society oriented for centuries to the application of technology can only dimly understand.

Tunisia

Our stop in Tunisia brought us into contact with the most scientifically oriented country of our entire visit.

Tunisia has a seething fervor for education, and the greatest enthusiast and leader is President Bourguiba himself, who has assigned it the highest of national priorities and nearly a third of the national budget. Despite Tunisia's population of only 4.5 million, the University of Tunis has an enrollment of 9000, of whom 2600 are studying sciences.

The university is beginning to move from its location in downtown Tunis to a strikingly handsome new campus on the outskirts of the city, built with help from several nations including the United States, which financed construction of the building that houses the law school, and the Soviet Union, which financed the engineering building. The Faculty of Sciences, headed by Dean Adnan Zmerli, is awaiting completion of a magnificent building at the new campus, built with aid from Kuwait. As in Morocco, however, most of the foreign faculty members are French, many of them under French government assistance programs.

Tunisia's foreign minister, Habib Bourguiba, Jr., son of the founder and president of the nation, represents the younger generation of highly educated and informed men who are now in positions of governmental responsibility in Tunisia. Minister Bourguiba spoke to us eloquently of the role of education and science in Tunisia's effort to transform itself into a modern society.

Like Morocco, Tunisia operates a National Institute for Agricultural Research, which dates from the days of the French protectorate. There we reviewed a project, under M. B. Cheikh and F. Soria, aimed at controlling the Mediterranean fruit fly by sterilization of the male. These flies annually wreak some $3 million in direct damage to Tunisia's citrus crop and cause much higher indirect losses by inhibiting citrus culture in the areas of heaviest infestation. The technique involves release of a large population of radiation-sterilized males in a breeding area; they compete with the normal males to breed with the females and thus reduce the population of each succeeding generation.

This method has successfully eliminated the screwworm fly from large areas of the United States. However, careful and sophisticated work is required to extend the technique (if it is applicable at all) to new species.

Medical research and educational facilities in Tunis were well equipped but understaffed; obviously, outside cooperation in the form of visiting experts could be helpful. The U.S. medical research ship *S. S. Hope* was in port at Tunis at the time of our visit and was providing assistance and training to Tunisian physicians.

One well-established science in Tunisia is archaeology. In modern Carthage in the suburbs of Tunis, the Bardo, an outstanding museum of

Carthaginian and Roman antiquities, houses perhaps the world's largest and finest collection of Roman mosaics. Its director, Mohammed Yacoub, is a fascinating source of knowledge of ancient Carthage and its art.

Ethiopia

Ethiopia was the most atypical of the countries we visited. Never colonized and long isolated by geography and its own culture, it lacks the educational and governmental infrastructure of the former French or British colonies. Thus, Ethiopia is making the most recent entry into the world of science of any country we visited. The newness of these efforts does not, however, diminish their enthusiasm or effectiveness. Haile Selassie I University, established in 1961, already has an enrollment of 4600 students, including approximately 2700 in the Faculty of Sciences.

Like universities almost everywhere, it is experiencing student unrest and was on an extended Christmas holiday at the time of our visit because of recent troubles. This circumstance did not interfere with our meeting with university officials and members of the faculty. The university has received extensive financial assistance from the United States and has enjoyed a close cooperation with Oklahoma State University in the area of agriculture. Its academic vice president, Frank Bowles, is sponsored by the Ford Foundation.

Like new universities throughout Africa, Haile Selassie I University sees as its highest priority the creation of educators, including those needed to reduce its own still heavy dependence on outside staff. Ethiopia has one of the lowest literacy rates in Africa; however, great strides are being made in overcoming this deficiency, and the country now has 600,000 students enrolled in elementary schools out of a total population of 22 million.

The scientific emphasis in Ethiopia is appropriately on agriculture and medicine, but an interesting and important geophysical observatory is located in Addis Ababa, founded and headed by Pierre Gouin, a talented Canadian. It takes advantage of the strong seismicity associated with the nearby Rift Valley and of the close proximity of the earth's magnetic equator. Dr. Gouin's seismic and magnetic measurements are, therefore, of worldwide interest, but perhaps the most important function of this excellent small laboratory for Ethiopia is the role it plays in teaching young scientists how a small research institution can do high-quality work with a minimum of resources. In fact, this lesson is one that Dr. Gouin could teach admirably to many of our own laboratories. The observatory manages, against difficult odds, to keep its instruments in operating order. Instrument maintenance, as I have noted earlier, is an ever-present problem in Africa, and Ethiopia shares the need for more trained technicians and for more spare parts.

Ethiopia is a focal point of tropical diseases. Partly for this reason, Addis Ababa is the site of the Naval Medical Research Unit No. 3 (NAMRU 3), the U.S. Navy's principal tropical disease research facility. Under the able directorship of Jack R. Schmidt, this well-equipped and well-staffed laboratory is doing outstanding research on tropical diseases of interest both within and outside Ethiopia. Here again, a valuable by-product of the effort is the opportunities created for the training of local scientists.

The majority of Ethiopians are highland animal raisers, and animal production thus presents some of the most critical problems in Ethiopia's largely agricultural economy. Lands are over-grazed, and the underfed cattle are highly susceptible to diseases. Veterinary medicine is seriously underdeveloped, with only about a dozen Ethiopian and 50 foreign professionals working in this field. Outside assistance would obviously bring large returns in this area.

In Ethiopia, we had our first discussion of a medical problem that plagues sub-Sahara Africa and is surprisingly akin to difficulties we face ourselves. This is the problem of bringing medical care to the outlying villages and rural areas where the vast majority of Africa's population lives. As in the United States, the professionally qualified physicians, native as well as foreign, congregate in the few cities. Most rural residents go through life without seeing a physician. Dr. Getachen, associate dean of the medical faculty, told us that Ethiopia is therefore actively developing programs for the training of paramedical personnel.

Kenya

The attraction of Kenya for its former colonial population and for today's tourist is evident immediately upon arrival at Nairobi, where the climate is surprisingly comfortable for a city within 2 degrees of the equator.

Seaborg with his wife Helen L. Seaborg touring Nairobi National Park (Game Park), Kenya, 10 January 1970. Reproduced by courtesy of the author.

Kenya has a close association with the neighboring republics of Uganda and Tanzania. Although the hoped-for political union of these countries as an East Africa Federation did not materialize, cooperation is still close, and many common institutions, including universities and research organizations, still exist.

The Kenyans are largely farmers and are deeply attached to the land. The British left them with an impressive scientific and educational infrastructure which they have preserved well and built upon significantly. All of these factors, physical and human, combine to make Kenya a land of great current attractiveness and future potential.

Kenya's vice president, Daniel Arap Moi, in common with many other African leaders we met, expressed a keen understanding of the role science could play in his country's development. He also pointed to the inspirational value for African scientists of contacts with scientists from abroad, as exemplified by our visit.

Few of the scientists we met on our visit matched the enthusiasm, eloquence, and ability of J. M. Mungai, dean of the Faculty of Medicine of the University College of Nairobi. Dean Mungai stressed the need for an expanded system of public health for the country. He described the modest beginning that has already been made in the form of regional clinics manned by paramedical personnel sufficiently skillful to treat common ailments and capable of referring cases to physicians in the city when necessary.

Kenya has 1200 physicians for a population of 11 million, but most of these physicians practice in the cities and only 20 percent are native Kenyans. Dean Mungai sees the current role of the physicians in countries such as Kenya as that of education, administration of a public health service, maintenance of satisfactory standards of health care, and, of course, treatment of cases beyond the capacity of paramedical personnel.

It was Dean Mungai who was most articulate in expressing to us the importance of allowing Africans to set their own development goals. Experts, he explained, will always be welcome to help in reaching those goals, but the objectives and programs themselves must be locally determined.

University College, with an enrollment of 2000, is located on an attractive campus that would do justice in natural and architectural beauty to any in the United States. Its principal, A. T. Porter of Sierra Leone, described the close relationship between University College and the University of California, an arrangement that we believe can be a model for "sister university" relationships between every African university and an American counterpart.

The science and engineering faculties of University College still serve the regional grouping of Kenya, Uganda, and Tanzania. The engineering faculty is almost entirely British, but the proportion of native Kenyans in the science departments is considerably higher. The dean of the engineering faculty, W. B. Palmer, on loan from the University of Leeds, believes that upon graduation the African students are as well qualified as their European counterparts academically but that they fall short in experimental and practical skills because of the absence of a technological tradition in Africa.

Outside Nairobi is located the East Africa Agricultural and Forestry Research Organization (EAAFRO), an institute dating back to the pre-World War I German protectorate. It is extensively supported under the U.S. technical assistance program and is directed by an American, Ordway Starnes. As a regional organization for Kenya, Tanzania, and Uganda, it attempts to fill gaps in national agricultural research programs rather than conduct a comprehensive research program of its own.

An interesting sidelight is that the laboratory has a herbarium with 140,000 species of plants which are being examined in the United States by the National Institutes of Health for anti-carcinogenic properties. However, its main programs are directed toward local needs, including the improvement of livestock and the development of grains suitable for use on the marginal lands that constitute much of the three countries' land area.

The EAAFRO publishes one of the few science journals produced in Africa, the *Agriculture and Forestry Journal*. An activity related to the journal is the East African science information service, initiated in 1966. Through this service, scientists in East Africa receive copies of tables of contents of journals and can subsequently obtain any desired articles upon request. The service distributes about 40,000 articles a month to about 113 locations in eastern Africa. The information service has approximately 934 journals in its library, and it calls on other libraries of universities in the area for assistance.

Aside from direction of the laboratory's research program, Dr. Starnes, like overseas experts throughout Africa, regards his principal task to be one of developing local scientists to replace himself and his expatriate colleagues. The laboratory's staff today numbers over 100 scientists and technicians.

Congo (Kinshasa)

In Kinshasa (formerly Leopoldville), we visited the impressive Lovanium University, which was created in 1954 and now has an enrollment of 3300 students. The present rector is Monsignor Th. Tshibangu. Here we met Luc Gillon, a driving force in the establishment of the university, first rector of the university, and still a very active administrator there. Monsignor Gillon, a nuclear physicist from Belgium, was instrumental in the establishment of the university's Nuclear Research Center. The director of the center and the government's commissioner of nuclear science is Felix Malu.

The Nuclear Research Center has a 50-kilowatt (thermal) Mark I Triga research reactor, which was placed in operation in 1959 and is the only nuclear reactor in Africa outside of the United Arab Republic and South Africa. It is soon to be upgraded to 200 kilowatts (thermal). The center has received much assistance from the International Atomic Energy Agency and from the United States. The general research program of the center includes the areas of agronomy, medicine, biochemistry, radiochemistry, and radiological protection. It is evolving into a regional center providing training for students from member states of the Organization for African Unity.

A United Nations hospital is located at Lovanium University and is staffed and operated by Danish personnel. Although the hospital's facilities did not appear to be particularly modern, the training and enthusiasm of the staff made it an impressive operation. As in other African countries visited, infectious and nutritional diseases are the most prevalent in the Congo, and outstanding research is being conducted by Jacob Raft and his Danish colleagues at the hospital.

On the flight out of the Congo, we were able to fly over the Inga Dam site on the Congo River. Construction is underway on a dam that will provide a modest amount of electricity to meet Kinshasa's short-term requirements. If fully developed, the site could be one of the world's largest and cheapest sources of electricity.

Ghana

Ghana, with its three universities and a comparatively long history of its talented students obtaining graduate degrees from British universities, has a very strong intellectual structure. Emphasis in current programs is on encouragement of graduate training in the sciences.

In Accra, I visited with Prime Minister K. A. Busia for discussion of Ghana's needs in the education field and the value of scientific visits such as ours.

At the University of Ghana at Legon, Alexander Kwapong, the exceptionally able vice chancellor, explained that the university now has a student body of 2500 and has produced 4000 graduates so far. The university has research institutes in the fields of animal science, irrigation, and forest products. Dr. Kwapong described the exchange relationships that the university has with numerous universities in the United States and the United Kingdom. He said that the exchanges of students and professors had been very beneficial but that increased numbers of American scientists, as well as up-to-date equipment, were needed for advancing education at the university. For example, the university has acquired an IBM 1620 computer and a grant to obtain a director for the operation of a computer center, but thus far it has been unable to locate a properly qualified man.

Dr. Kwapong noted that plans are being developed for a research program on environmental effects, with emphasis on the biological sciences. Also planned is a tropical biology center to be located in West Africa. American efforts have contributed already to the Volta Lake project, a scientific laboratory for the study of artificial lakes and their phenomena, which is of great importance to the future of Africa.

After discussions with Modjaben Dowuona, chairman of the Council for Scientific and Industrial Research (CSIR), we visited some of the laboratories of the CSIR, which is composed of a number of institutes relating to industrial applications, aquatic biology, and agricultural research. At present, approximately 70 percent of its funds are spent on agricultural research, with cocoa receiving the most attention because it is the major cash crop. Although we were unable to visit the Cocoa Research Institute, 105 kilometers north of Accra, we were told that it has an impressive Radioisotope Research Institute, which is well equipped for up-to-date research on ways to control the diseases that now threaten the important cocoa crop of Ghana.

Another laboratory we visited was the Radioisotope Applications Center directed by B. W. Garbrah, which falls under the jurisdiction of Ghana's Atomic Energy Committee, headed by J. A. K. Quartey of the department of chemistry of the University of Ghana. Research there is devoted primarily to the use of radioisotopes in the fields of medicine and agriculture. The center has fine facilities, but its work is limited by the lack of modern equipment. The laboratory is located at the site of an uncompleted research reactor, which the Nkrumah government had acquired from the Soviet Union. The decision of the new government not to complete this costly facility and to concentrate on radioisotope applications that are more in keeping with Ghana's needs and capabilities is indicative of the responsible approach of the present government.

Particular attention at the Radioisotope Applications Center is given to film badge monitoring for hospitals in Ghana and to the study of the effects of antimalarial drugs on humans. A whole body counter is being built, and renography and thyroid studies (using iodine-131) are to be undertaken.

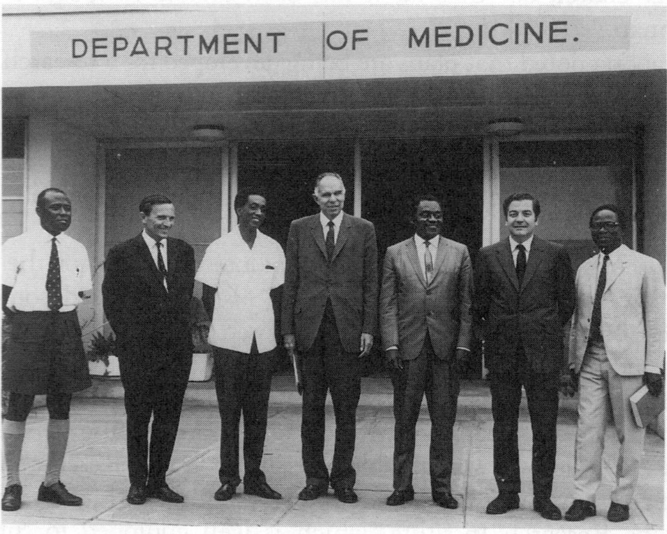

At the Medical School of the University of Ghana, Accra, Ghana, 13 January 1970. From L to R: Dr. F. I. D. Konotey Ahulu (lecturer in medicine), Professor B. Ringelham (chemistry pathologist), Professor Silas R. A. Dodu (Head, Department of Medicine of the University of Ghana), Seaborg, Dr. C. O. Easomon (Dean, Medical School of the University of Ghana), Dr. Henry N. Wagner, Jr. (Professor of Medicine, Radiology and Radiological Sciences, Johns Hopkins University, Baltimore, Maryland, USA), and Dr. Kwamina Ocran (research officer, Ghana Atomic Energy Commission). Reproduced by courtesy of the author.

At the Medical School of the University of Ghana, located at the Korle Bu Hospital, we visited with Dean C. O. Easmon and Silas R. R. Dodu, head of the department of medicine. Their work includes research on malaria and development of a family planning program through establishment of community houses.

Observations and Conclusions

I have already put forward the principal conclusion of our mission. Briefly, scientific talent is present in Africa and is making real and current contributions to the solution of Africa's problems, contributions that can be greatly enlarged with even modest outside support.

Despite the generally early stage of scientific and technological development in the countries visited, they are making good use of the small nucleus of the well-qualified and dedicated administrators, scientists, and doctors present, most of whom were educated in the United States or Western Europe, to help ensure the growth and independence of their respective countries.

This core of trained and enthusiastic professionals constitutes an irreplaceable asset to the African nations, which needs to be properly encouraged and supported in the development of scientific, educational, medical, and agricultural programs. The focal points for development of these programs are generally the universities, most of which have received assistance from the United States. Fortunately, government leaders usually recognize the important role to be played by these institutions and, within the limitations of available manpower and financial resources, provide support in maintaining and expanding them.

There was widespread agreement in the countries visited that improvements in agricultural techniques are imperative in order to provide more food, to cope with population increases, and to reduce the present high incidence of disease through improvement of nutritional standards. Almost equally important is the expansion of public health activities in the rural areas where most of the people live.The degree of progress to be achieved in resolving problems in these areas will depend, to a great extent, on how quickly the African nations can apply scientific techniques to these problems.

African government and university officials welcome the assistance and support that is needed from the advanced countries, but, to their credit, these leaders are anxious to proceed on their own initiative and would regard such assistance as being of an interim nature.

We found that technical assistance provided under United Nations development programs has been most valuable in meeting immediate requirements. In my own field of special knowledge, for example, in each of the countries we visited the International Atomic Energy Agency has provided equipment and experts, which are being used in the development of new techniques in such areas as agriculture, medicine, and mineral exploration.

Private foundations and universities in the United States have been particularly helpful in supporting and encouraging African education and medical research. We were often reminded of the importance of this assistance and the desire of African universities and specialized institutes for closer relationships with these American groups. We were encouraged to see that local representatives of the U.S. Agency for International Development are becoming increasingly involved with technical assistance to scientific and educational programs and are aware of the role of science and technology in the developing countries.

These are observations on the bright side. Clearly, African science is not without problems. We found that major communications gaps are present at all levels of African science. There is an inadequate and delayed flow of information between Africa and the United States and other advanced countries, among the African states themselves, and even among different institutions within each state. Members of our group repeatedly heard from members of the African scientific community that they felt cut off from their colleagues in the mainstream of scientific activities, particularly those in the advanced countries that have the most to offer. Owing to limitations in their own budgets, these African scientists are usually unable to travel abroad and to participate in international scientific meetings. Even the number of scientific journals and other publications that they receive is limited. Moreover, these journals are received many months after publication, with the result that they cannot keep current on research being conducted elsewhere.

The African states must cope with the problem of training an ever-increasing number of professionally qualified people in all fields, including the sciences. Further, there are very limited opportunities for graduate and postgraduate education at the universities today, and funds are limited for sending students abroad for advanced training.

There is a shortage of equipment in most of the universities and hospitals, as well as an acute shortage of skilled technicians to maintain the equipment that is already there.

Perhaps less demonstrable, but equally valid in our collective view, is the general conclusion that Africa has bright prospects for sound and steady development. Part of this feeling may have derived from the observation that Africa, in comparison with other developing areas visited by various members of our group, is by and large not yet overpopulated. It would be too much of a simplification to say that population problems are not present, but the crushing burden of people that is so evident already in parts of Asia and Latin America is still absent in all or most of Africa. But population growth rates are high, and a well-planned and determined effort is needed now if Africa is to avoid the morass of overpopulation.

Finally, we felt that we saw in the people a receptivity to ideas and advice, and to change itself, which should accelerate the development process.

If we are correct in these observations, Africa is deserving of all the help we can give it. Here are our views on what can be done.

Recommended Actions

The single most important step that we can take is to increase the exchange of scientists and technologists at the graduate and postgraduate levels between African nations and the United States. This exchange should take place in both directions and should cover all principal scientific fields, especially medicine, biology, and agriculture.

The current tight limitations on resources in the United States require that we devise new means by which these exchanges might be encouraged and supported. In the short term, an increase in any kind of communications between African scientists and their American counterparts could have salutary effects. Personal correspondence and newsletters from the science departments of American laboratories and universities would involve almost no funds and very little effort. Visits to Africa might be included on the itinerary of American scientists already traveling abroad, perhaps on vacation, at some cost savings in relation to trips made expressly to Africa.

In the longer run, I believe one of the most effective means for improving exchanges and communication between African and American science is by what is sometimes called the "sister laboratory" or "sister university" approach. This technique has been used to advantage in nuclear cooperation between laboratories of the U.S. Atomic Energy Commission and those in a number of developing countries and has also been used in other scientific fields. It involves the establishment of a permanent association between an overseas institution and an American one with similar interests. I believe that this arrangement develops an attitude of responsibility on the part of the "senior partner" — the American institution — for the progress of its junior partner, which brings about results that would not be possible through more casual and temporary forms of cooperation. Even where funds for salaries, equipment, and other forms of concrete assistance to implement these partnerships are severely limited, useful results can still be achieve through the continual guidance and consultation that these arrangements can provide at very little cost.

I believe the science community of the United States could set as a goal that within three years the science, agriculture, and medical faculties of every African university that so desires should have a "sister college" relationship with its counterpart department or college in an American university.

Another area where important help can be given at modest cost is in increasing the availability of scientific literature and documentation to African institutions. The addition of a few African institutions to the circulation list of the major American scientific journals would cost each scientific society very little. We should also try to find a way to establish one or more document centers in Africa through which scientific literature from abroad can be made readily available to African scientists.

We also need to increase substantially the number of scholarships available to Africans in science, medicine, and agriculture, and we need to find more Americans in these fields who can spend one year or more in Africa. I believe we should consider the establishment of a program by which young Americans would agree to serve a period of time abroad in exchange for government financial help in their education. We learn in

Africa that such a program is pursued by France, and other countries are considering making similar arrangements. Fellowships provided by private foundations, which are already being made available to a gratifying degree, should also be increased if at all possible.

The American concept of agricultural extension services to promote the education of farmers and the practical application of existing knowledge could make a tremendous contribution in Africa. Here, too, I believe the concept of the "sister" organization relationship may be applicable. I believe African nations would be encouraged to establish extension services or to expand present ones where they exist, if they knew that their organization would be "adopted" by and have frequent access to the advice of a specific U.S. agricultural extension service.

It was heartening to find in Africa that the major share of the development burden is being carried by the international organizations of the United Nations family. One obstacle to even more effective help from these organizations is the inexperience of many of the African nations in soliciting the advice and assistance that these bodies can provide. I would therefore suggest that each of these nations consider the establishment of an organizational unit, staffed by professionals experienced in the ways of the international organizations, to serve as a focal point for requesting technical assistance from them and from other available sources.

Beyond the ideas suggested above, there are many steps that could be taken, both by private organizations and universities and by individual scientists, to promote scientific cooperation with Africa. Our group believes that the nongovernmental role in African scientific and technological development could be greatly assisted if there were a focal point in the United States for promoting, coordinating, and following up these activities. Perhaps the most logical approach in this direction would be for a philanthropic foundation with existing ties in Africa to assume this responsibility. Another possibility is the organization envisaged by Robert E. Marshak, recently appointed president of the City College of New York, and by Roger Revelle, director of the Harvard Center for Population Studies, who have proposed establishment of an International Science Foundation that would assist scientists in the developing countries and that would not be limited to support from the United States alone. Either type of organization might well provide a timely mechanism to stimulate scientific cooperation with the African and other developing countries.

I believe there are few goals today that are more worthy of the serious support of the American scientific community.

Notes

1. Other members of the visiting team were Herman Pollack (Bureau of International Scientific and Technological Affairs, Department of State, Washington, D.C. 20520); Cyril L. Comar (New York State Veterinary College, Cornell University, Ithaca, New York 14850); William H. Taft, III (Bureau of International Scientific and Technological Affairs, Department of State, Washington, D.C. 20520); Henry N. Wagner, Jr. (Radioisotope Laboratory, School of Hygiene and Public Health, 615 North Wolfe Street, Baltimore, Maryland 21205); Norman P. Neureiter (Office of Science and Technology, The White House, Washington, D.C.); Myron B. Kratzer (U.S. Atomic Energy Commission, Washington, D.C. 20545); Justin L. Bloom (U.S. Atomic Energy Commission, Washington, D.C. 20545); John R. Totter (Division of Biology and Medicine, U.S. Atomic Energy Commission, Washington, D.C. 20545); James E. Ammons (Division of International Affairs, U.S. Atomic Energy Commission, Washington D.C. 20545); and Charles F. Baxter (Division of Space Nuclear Systems, U.S. Atomic Energy Commission, Washington, D.C. 20545).

Shaping the Future — Through Science and Technology

Remarks made at the 1971 Nobel Conference, at the Gustavus Adolphus College, St. Peter, Minnesota, January 7, 1971.

It is a pleasure for me to return for another of your outstanding Nobel Conferences. In the past these conferences have discussed such topics as "Genetics and the Future of Man," "The Uniqueness of Man," "Communications" and "Creativity." This year you have chosen to cover the very broad topic, "Shaping the Future," and you have selected a distinguished group of "futurists" to discuss its many aspects.

For my own contribution I have decided to speak on the subject "Shaping the Future — *Through* Science and Technology" and I have chosen this approach for several reasons. One is that this conference is conducted in honor of Alfred Nobel, a man who had great faith in science and technology as the means of molding the future for a better life for all men. Nobel had a particularly difficult time justifying and fostering this idea because (similar to the situation we face in the Atomic Energy Commission) he personally was struggling to develop, and then distribute, for beneficial uses, a form of power that was introduced destructively and bore the stigma of danger and death.

Another reason I have chosen to speak on shaping the future through science and technology is that we have many people today who doubt this can or will be done — and some who even doubt we have much of a future to look forward to, much less to shape.

Yes, times have certainly changed since Winston Churchill said:

> "If the human race wishes to have a prolonged and indefinite period of material prosperity, they have only got to behave in a peaceful and

helpful way towards one another, and science will do for them all that they wish and more than they can dream ..."

Today Sir Winston's faith in science would be looked upon as incredibly naive. And there would be some who would go even further and claim that it is exactly his kind of thinking that has got us where we are today. But let me take issue with this. Just where are we today? What has been the role of science and technology in getting us there, and what will it be in the future?

We face some very negative opinions today about science and technology in particular, about man in general, and about their relationships. Broadly speaking, we are told that man is a failure and that science and technology are responsible. This attitude is expressed in many ways by many groups and individuals. There are the bands of disenchanted youths who go off to reject modern society by living in countryside communes where they move closer to nature but still need electricity, still drive to town to pick up supplies with which to live, and still depend on the benefits of that society's laws, communications and public health and medical knowledge. There are those groups who, not having gone so far as rejecting all of society, reject all of science and technology on the grounds that some of it serves military, social or political causes with which they do not agree. Yet many of these are the very same groups that clamor for the uplifting of the people of underdeveloped lands, admitting that science and technology must play a role in their advances.

And there are those groups and individuals who are just as unilateral in their critiques of science and technology from other standpoints — some claiming they are leading us toward irreversible ecological destruction, others saying they are the source of an alienation that deprives us of our humanity, and still others seeing them as forces separating us from a biological inheritance that we should retain. Yet all of these will agree, after some reflection, that science and technology have contributed to our humanity, are needed to deal constructively with ecological problems and could be directed to establish a healthy relationship between man and his environment.

There is a strange mixture of tragedy and truth in all these outlooks — and, unfortunately, a strange distortion of past history and present reality also. The young hippie, for example — the new Thoreau — who goes off with his companions to live a communal life close to nature, no matter how noble his intentions or how strong his feelings, cannot be a model for his fellowman. In his total dedication to his values he can only become a parasite, reflecting a kind of pseudo-independence from the modern world. And should this aspect of his lifestyle be emulated by all his fellow youth, social and ecological disaster would eventually follow. Contrary to many romantic notions that abound today, nature alone could not support the billions of people on earth now (and the billions due even with population control). Should this be even partially attempted every hardship that man has suffered in the past would be visited upon him many-fold, nature and man would be devastated in constant conflict with each other, and we would see eventually, rather than the communal spirit and love projected by the young idealists, an aggressiveness and vicious competitiveness the like of which history never revealed. If you do not believe me, try projecting what the world would be like as each of the benefits of modern science and technology were removed and then all the people of the world were asked to retreat further and further into those supposedly ideal and idyllic yesteryears when men lived closer to nature.

But I dislike even suggesting such an exercise because there is a basic flaw in the "return to nature" idea — in the idea that anything that man does scientifically or technologically is somehow "unnatural." This is a powerful notion, a pervasive feeling, that runs through the thinking of so many people today as a reaction to some of the problems which science and technology have created. And while there is no doubt that we have an inherent love of the natural and a need to enjoy nature's attributes, that love and that enjoyment are enhanced and highlighted today because we have the luxury of enjoying nature as a friend and not a foe. The people of the U.S. who can drive to a campsite in a National Park and set up housekeeping in a camper or tent trailer with all the conveniences of home can afford to think and feel differently about nature than those people of East Pakistan for example, some of whom, as seen in the last holocaust there, have little choice but to live or die according to nature's whims.

My main point, however, is not based merely on this comparison of perspectives. It is based on the idea that modern man is neither acting "unnaturally" nor separating himself from nature. He is, in the broadest sense "doing what comes naturally." For if we reflect even for a moment we can realize how much science and technology are a part of nature's evolutionary process. How can this be so? The answer is not difficult to follow. Nature has endowed all forms of life with some protective or responsive means of survival. But what has she given man in his evolution? He is most poorly equipped in many ways. His vision is terrible compared with a bird's. His sense of smell is inferior to almost every four-legged animal. His natural means of locomotion are relatively clumsy. He benefits from no camouflage, no claws, no large incisors, and in his birthday suit he is ill-equipped to live in most of the world's climates.

What, then, did nature give man? She developed in him a brain and neural system, and a degree of manual dexterity, that allows him to amplify his limited powers through abstract thought, language and the use of tools and energy. His survival and evolution are based on a dynamic interplay of these, and on his ability to grow by his adjustment to the feedback from his errors and successes. And I contend that his evolution today, through the various civilizations and societies we have seen, and into the modern global one that is struggling to be born, is a *natural* evolutionary process. Furthermore, it is a process that is moving us, not toward self-destruction, but toward a higher form of life. It is forcing moral and social growth and preparing man for what John Platt has referred to recently as a "Hierarchical Jump" to a new level of man or mankind.

So let me reemphasize that, contrary to much that is heard today, science and technology are not unnatural or anti-nature, nor are they aberrations of human development that will lead to man's downfall. Anyone who merely compiles their errors and their human misuses in a one-sided picture to condemn and dismiss them as that, is totally lacking in vision. If man were to pursue the illusion that he could "adjust" to nature without a constant upgrading of his science and technology he would soon be extinct. Those who would depend for survival strictly on nature's ecological balance, devoid of man's intellectual equalizers, would witness the rapid decline of man and the ascendency of another species — most likely the insect, perhaps even the lowly cockroach who, we are told, has survived for millions of years and is still going strong.

But in stating my case in defending science and technology from this broad standpoint, I decidedly do not want to give the impression that I believe science and technology as used by human beings are unmitigated good, that all the criticism of them today is invalid or undeserved, or that much good may not result from the values of our youth's "counter-culture" on our "science culture." Feedback from such a "counter-culture" is absorbed and the best of it will have a good and lasting effect in our society. It simply is not true that our scientific–technological civilization is unresponsive to, or destructive of, our humanity. In total, it is a very human and humane enterprise that is guided by human drives and that tends to elevate, not debase, those drives. Let us dwell on another aspect of this point for a moment.

If it seems we are being told from all sides today that man is a failure, it is, ironically enough, because we are being judged in terms of a whole new set of standards in a world where almost everything seems possible, and now almost every want, every injustice and every wrong seems unbearable. Alexis de Tocqueville anticipated this over a century ago when he said: "The evil which was suffered patiently as inevitable seems unendurable as soon as the idea of escaping from it crosses men's minds." This holds true with a vengeance today. And I do not say that we should not maintain that impatience as long as it drives us to act, and act constructively. In fact, a countryman of de Tocqueville, Victor Hugo, made a very cogent statement on the constructive use of this impatience when he said, "Good government consists of knowing how much of the future to introduce into the present." There is no doubt now that science and technology have made this aspect of government increasingly difficult.

But because we can make instant coffee does not mean we can achieve instant population control; neither can we throw our environment into a washing machine, nor can we, like a response to a cry for "demand feeding," alter the relationship between our technological and economic systems overnight. And certainly achieving the single goal of placing a man on the moon — no matter how impressive a scientific and engineering achievement — is not to be compared with an ability to raise the "quality of life" of 200 million people. The important thing to remember is that the combination of this impatience, those values brought on by our new awareness of conditions as measured against potentials, and the knowledge and power that science and technology can provide, represents a force that could truly elevate man.

At this point I must stress our new awareness of our environmental and human condition because it is the basis of a whole new era of scientific, technological, social and political action ahead, and because it is actually quite remarkable. As an example of this, just let me point out a personal observation. In a little bookstore across the street from my Washington office they have compiled a reading list "in ecology and related subjects" for their customers. At last count that list numbered some 260 books, most of which were published within the last three years — and these are books just stocked by this small store. Remember also that these are only books. The profusion of technical journals, newspaper and magazine articles and television and radio programs related to these subjects now has reached the proportions of a torrent of information and expression of concern. And all this cannot help but have a profound effect on the direction of our society. Eventually it should be a very positive effect, providing we can maintain a balanced perspective and not allow the revelation of our problems to drive us to despair and defeat rather than the necessary constructive thought and action.

Let me make one further point on this new awareness. It seems to me that, among other things, civilization progresses through our use of a combination of hindsight, insight and foresight — with our advances depending largely on the extent to which we can emphasize the latter two. A review of all the popular material on environmental and social problems today shows that at the moment we are still very heavy on hindsight, although there is much good insight coming through now. The result is that we are still spending too much time and effort in seeking out scapegoats for today's ills and making historical comparisons which are unproductive and invalid.

In this same vein, we have many people who see the past and present through a strange pair of glasses, one that filters out our human advances as they approach the present and highlights only the "glories of the past," giving the impression that the total condition of man is regressing. Nothing could be further from the truth — all our environmental problems notwithstanding.

I will not dwell on this subject today, but instead I urge any who doubt our accomplishments and our potential, to read a remarkable book, "'Optimism One — The Emerging Radicalism" by F. M. Esfandiary. Mr. Esfandiary — who was born in Iran, lived in many parts of the world, worked for the United Nations, and now lives and teaches in this country — attacks the entire mythology generated by today's cynics. His ruthlessly realistic appraisal of days past and of how those in the underdeveloped areas of the world live — essentially still living in the past — as compared with the conditions, care and compassion of the modern world is indeed impressive. He speaks of our relationship with nature, of social conditions of the old world, of identity, alienation and violence and he speaks as one who has *lived* in both past and present.

If we want to shape the future — and particularly through science and technology — we must get out of the traps created by the despair of today's cynics or the frustration generated by unrealistic expectations. Neither condition allows us to develop the proper insight and move on to applying the necessary foresight which is so important today.

How then shall we move ahead? What should we expect from science and technology? What needs to be done to use them to the best advantage? And if this is done where might they take us? In other words, what will be the shape of the future they provide?

To answer these questions we must explore a basic dilemma of our times. That dilemma, stated oversimply, is: How can we have the growth to meet expectations already generated — global expectations that seem to make virtually unlimited demands — in a world of physical limitations, many of which are already being approached? The immediate reaction to this — and it is one shared by many people today — is to stop the growth; or to be more explicit, limit population, reduce expectations, lower production and pollution and concentrate on the redistribution of material things and the "quality of life." This is a very natural rational and human reaction. But, unfortunately, taken as single prescription, it is neither a very realistic nor imaginative one. Let me explain.

Man since his earliest days has always played what is known as the "zero-sum game." In a limited environment — limits enforced by physical, social or political boundaries or by his own ignorance — he had to compete for or share a limited amount of resources. But as he grew in his ability to manipulate his environment, through knowledge and tools which gave him advantages, both destructive and productive, he was able to expand the sum to be divided and extended himself to new limitations. Through history — the ages of exploration, of industrializations, of scientific revolution — he was able to expand both his frontiers and expectations. During such growth, the impression was gained that he was no longer playing that zero-sum game because different costs — human hardships and environmental changes — were not accounted for. But it should also be noted that each major expansion, each increase of the sum, was made possible by a major human advance in knowledge, in technology and in a new resolution of the human will to support it.

Today we have reached the end of another era of expansion, and in the eyes of many it is an end that is final and discouraging. Surveying population growth, resources depletion, environmental pollution and limitations and man's general social conditions these people tend to say, "The jig is up. We have reached the end of the line. There is only one way to go — freeze further growth and work within the limitations of this new zero-sum, this Spaceship Earth which our new perspective allows us to see in a new light."

When contrasted with the major feelings of only a few short years ago that the "sky was the limit," that economic growth was synonymous with progress and that both were essentially limitless, the new attitude I just mentioned seems far more sane. And I agree that if it were simply an either/or proposition I would immediately side with those who opt for the freeze. For the natural limitations, physical and otherwise, of uncontrolled growth are obvious. And rather than blindly moving ahead into their devastating effects — rather than letting natural laws put a harsh stop to our growth — we would be better off imposing our own limits, harsh as they also might seem to some.

But we are *not* faced with a strictly either/or proposition. Economic growth and ecological balance are not necessarily incompatible. There are areas where new growth and development are essential and there are those in which they should be leveled off or even cut back. There is also a new morality being introduced into the marketplace that will allow economic values to be assigned to environmental necessities, so that through a combination of regulations and incentives we can enjoy a type of human advancement neither tied only to a rising GNP nor bound to a harsh zero-growth policy. The key to this middle way, however, lies mainly in the wise development and application of science and technology. With our global frontiers seemingly closed, they hold the secrets to widening those frontiers, to radically expanding the new zero-sum to which we seemed to be confined.

Before explaining how they might do this, let me clarify what kind of growth I am talking about and why we must pursue further expansion in a rational way.

The idea of zero-growth is, surprisingly enough, one that appeals most to the educated middle-class American. For a combination of reasons, one of which might be his own feelings of guilt as he surveys his own abundance and environmental impact against the problems of the day, this citizen feels we must change our direction. He is sincere in his concern and willing (but probably only to a point) to make sacrifices to right the wrongs that have been perpetrated in the acquisition of his affluent society. But what he does not realize is that, for all its size and impact, if it were possible to distribute today's wealth evenly among the four billion people of the world, it would not go very far in meeting the needs and expectations that exist — not including those of the future already born.

And simply to compare the income and environmental effects of the average American and a citizen of a developing nation and use that as a guide to future action makes little sense either. To the hungry and the poverty-stricken ecology is irrelevant. Not having Society Security, or much other security, they breed for survival — to assure the perpetuation of their people and to have the necessary support should they survive to old age. So that while important programs of population control have been moderately successful they struggle against a natural instinct that could be adjusted by greater development — development that can take place through advances of science and technology without further environmental degradation and within some kind of rational economic framework.

The advance of the so-called "Green Revolution" — the creation and distribution of new high-yield crops as supported by the Rockefeller and Ford Foundations and carried out by men such as Nobel Laureate Norman Borlaug is an example of the kind of development I have in mind. Of course, the ability to lift a people above a marginal existence to a position where they have the strength and will to further advance must be supplemented by other means of development. Energy, education, improved transportation and communication systems, and appropriate industry must be brought in. But not in the way they were introduced into today's advanced nations, for then they would repeat and compound many-fold the environmental and social problems those nations face. And here is where advances in science and technology and their application can make all the difference, can shape a new future rather than disastrously imitate the past. There is no need for today's developing nations to relive the trials of a nineteenth century-type industrial revolution and arrive at many of the technologically induced dilemmas of the twentieth century. All these pitfalls can be avoided if a truly rational approach to development is used, and with such an approach the average citizen of a developing nation can arrive at a living standard satisfactory to him, in tune with his culture and lifestyle and not destructive of his environment. It does not have to follow, as some would have us believe, that because the middle-class American has fifty times the environmental impact of a peasant in India, that an Indian cannot achieve a life of material well-being and dignity without a comparable impact.

To understand how this is possible, we might look at the other side of the coin and see how the advanced nations might be "redeveloped" through new outlooks of science and technology.

The reason why our science and technology have, paradoxically, been both a success and a failure — why they have created the progress, expectations and problems they have — is because they and their applications have been developed and adopted in piecemeal fashion within a market economy. Until most recently in our scientific age we did not have the benefit of ecological or systems thinking — nor with expanding physical frontiers did we have the need or desire for them. Specialization seemed to gain the most productive results. Everything invented or developed was used to the limits of its profitability with its negative impact absorbed or written off, if noticed at all. And so we moved into what Simon Ramo has called the "Century of Mismatch" with unplanned or ill-planned growth, institutions incapable of understanding or directing that growth, and leading to our current era when new limits, new interfaces and all sorts of discontinuities have suddenly seemed to hit us just as we thought we had the "good life" really made. As I pointed out before, today we have a great awareness of this situation, and from this awareness a new set of truths is emerging and new guidelines for the course of society are being written.

One of these truths is that future development cannot come about through the direct exploitation of nature or man. It must be the result of using the capital of growing knowledge, of rethinking our values and revising our priorities, of learning to do more with less by increasing efficiency, by the maximum recycling of resources, by being more imaginative and less restricted by tradition in design, by learning to manage the greater complexity that is involved in the systems thinking and action we must employ.

The need for the systems approach — the ecological approach — is an important truth in itself. If we look around us, we see that almost all the problems of our age are problems at new interfaces or that take place because of a lack of integration between related forces both mechanical and human. For example, we no longer have merely a highway system, a railway system and an airways system — we have a total transportation system in which each of those components must flow together. If one ceases to function or becomes overloaded the backup immediately affects or even cripples the others. If we look one step beyond this, we can see that if our energy system — or any portion of it —fails, our transportation system is imperiled, our communications are affected, our health is endangered by the failure of other services and so on in a chain reaction. Similar critical relationships exist between the chain of resources, production and waste — and at the man-and-machine interface — as the role of man continues to shift from production to services. In other words, modern civilization has an ecology of its own, and maintaining its balances as well as the smooth functioning of its parts is now essential.

The human aspects of this ecology provide other examples of why discontinuities and mismatches must be adjusted. In a world witnessing an explosion of people and human activity at the same time that an implosion of these — through urbanization and other forced contacts — is taking place, enormous potential for conflict is generated at all points where there are disparities. These disparities may be harsh physical and economic differences, they may be differences in education and opportunity, they may also be psychological and cultural differences. And today on both a global and a national scale we are seeing violent reactions to these differences in the form of rising nationalism, intensified racial strife and increased conflict over ideological differences. The resulting turmoil caused by these polarizations cannot be wished or talked away. It cannot be ameliorated by retreating to irrationality or mysticism. It cannot be reduced by rationalizing that science and technology are responsible for it and therefore should not be allowed to interfere further, or, the opposite, that they alone should be responsible for correcting a situation they produced.

The solutions I see in resolving all these related problems of our age lie in recognizing fully both the organic nature of human civilization today and its inherent relationship with the natural environment that supports it. They lie in recognizing that science and technology, the major forces behind the growth and intensification of these relationships, must be used to gain the knowledge we need to fill the important gaps in our physical and social intelligence and to adjust our discontinuities and coordinate our mismatched relationships. They lie in building the social institutions (and perhaps restructuring some of our existing ones) to direct science and technology wisely, just as those institutions must extract intelligence and a certain wisdom from science and technology.

Actually, we are struggling with these solutions today — or beginning to. They seem so complex, so overwhelming at times that we wonder if we are not losing that "race between education and catastrophe." But I think we have at hand and are developing certain tools to help us win it. Some of these have arrived on the scene, historically speaking, virtually in the nick of time. I have often spoken of nuclear energy in this context because I believe that if we pursue its development carefully and apply it wisely it will provide a reasonably controlled population of reasonable demands with a virtually inexhaustible supply of power and at a time when we can anticipate the depletion of other sources of power. I emphasized the *reasonable* population size and its *reasonable* demands because there are those who upon hearing of inexhaustible energy blanch at the idea that it will only be used to support catastrophic growth. But if after all that we are learning these days we still believe that any technology will be used to its ridiculous extremes, or to support other obviously dangerous excesses, simply because it is there, or is temporarily profitable, then we face the greatest danger of all — literally that of lacking the collective intelligence or will to adapt to our own evolution. I refuse to believe that we are suicidal. I do not think we will go that route.

Nuclear energy used wisely will free man for eons from what Kenneth Boulding calls "the Entropy Trap." Abundant energy will allow us to save nature — not destroy it — by making a recycle civilization possible. In such a civilization, only energy will be depleted and natural resources, as we advance further in our knowledge of chemistry and physics, will become the building blocks of a world in which there can be endless variety without destruction. In such a civilization, we can expand the "zero-sum" so that all men can enjoy freedom from want, learn the true meaning of security and live in dignity with their fellowman. Here are the real challenges to today's anti-technologist, environmentalist, humanist, hippie and devotee of "Consciousness III" — help us build a world that can retain its natural beauty *and* be an endless source of human creativity.

Another technology that I believe is coming to the rescue of mankind — that may make mankind possible — is the computer. As the British cybernetics expert, Professor Stafford Beer, has pointed out, "Society has become a complex organism, and it needs a nervous system." We are now a global civilization that depends for its survival on a growing influx of data which must be processed into information, stored, distributed and eventually upgraded to knowledge and wisdom. Today the computer is the vital link in that system. And in addition to telling us where we are, it can help us shape our future by giving us the means to project and examine alternate futures. Through computer models we can "look ahead" to the consequences of various courses of action we may choose. And thus we may choose more wisely.

There are many other sciences and technologies that offer great promise for the future. The earth-orbiting satellite is one of these. Equipped in various ways, it has the potential of giving us long-term accurate weather forecasting that may save many lives and much property. It can give us vital information concerning environmental conditions, the health of crops, the state of the oceans, the atmosphere, the location of mineral resources and other natural phenomena. And it can unite the world through satellite communications, providing education and a means of sharing knowledge and culture.

Perhaps the most significant shaping of the future may yet come from the shaping of man himself — directly from the great advances now anticipated in the biological sciences.

Here again, and perhaps most forcefully, we are faced with the moral challenge that science and technology create. When life itself can be directly controlled, molded and even synthesized from basic chemicals, who will determine the nature, the direction and the ends of that life? Are we preparing ourselves for such God-like responsibility? In terms of cosmic time we are approaching it almost at the speed of light. We must think and plan and work and build our social institutions to manage this and other great responsibilities that our evolution is thrusting upon us. And we must do it now.

I am pleased that this conference has chosen to discuss "Shaping the Future." It expresses, in spite of all the despair around us today, an affirmation in that future, a belief that we have a future, and that if we will it we can be masters of that future, not its victims. Today I have expressed the belief that science and technology are essential to the shaping of that future. I hope we will turn to them with a growing care and dedication and use them with increasing wisdom. If we do, perhaps Sir Winston will not have been wrong and they will provide the human race with "all that they wish and more than they can dream."

From Man to Mankind — A Way to Go

Remarks made at An International Workshop: Conceptual and Attitudinal Pathways from Man to Mankind, Sponsored by the Center for Interdisciplinary Creativity, Southern Connecticut State College, New Haven, Connecticut, April 29, 1971.

In my remarks of one year ago to an International Joint Conference of the American Geographical Society and the American Division of the World Academy of Art and Science, I characterized the significance of the upheavals and crises of our time as follows:

> "What we are seeing today ... is not a forecast of doom. It is the birthpangs of a new world. It is period of struggle in which we are making the physical transition from man to mankind — a mankind that will be an organic as well as a spiritual whole on this earth."*

I submit that that is an idealistic statement. I am pleased that you considered it one intriguing enough to make the focal point of this symposium. Today, however, without appearing ungrateful or relinquishing my belief in that hopeful statement, I want to dash some of the cold waters of harsh reality in the face of such optimism. I want to emphasize that the birth of this new mankind will probably involve a long and difficult labor with many complications, one that calls for our vigilance, our encouragement and our care throughout a trying period.

Why should this birth of mankind be so long and difficult? The answer is that when we speak of mankind in the ideal we do not mean merely a

*See "The Positive Power of Science", pp. 266–274, this volume.

human race forced by growth alone to recognize its most common basis for survival. That much is imminent in time. From this standpoint we will eventually see a single mankind or none at all. No, hopefully we mean a breed of humans that has grown to recognize the need for a family of man, one capable of living in harmony with all its members and the earthly home that sustains it. As to whether we can achieve this, one must raise other important questions: Can we actually attain control of our growth and match our needs with an abundance through which we can "change human nature," developing human beings able to become what Abraham H. Maslow has termed "self-actualizing" people — people operating above the motivations generated by age-old insecurities? Can we achieve such abundance without some irreversible ecological effect, perhaps even by maintaining some ecological equilibrium that includes our man-made environment with its powerful technologies? And can we achieve all that without becoming a homogenized humanity, without losing our cultural diversity, our continued creativity and the desire to grow as individuals within a stable society?

These questions become our guide posts on the tortuous path to a global mankind. In terms of these guides let us take a few moments to analyze where we are today and what the journey ahead requires. Perhaps we can then see why it is such a difficult journey, why so many are so despairing so soon but, more importantly, why we must increase our determination and our efforts to prevail.

We idealists fortunate enough to live in this land so richly endowed with vast open spaces, with a wealth of natural resources, with a heritage of freedom and with a tremendous physical power brought by the pursuit and application of knowledge, how easily we speak of one world, of a "global village" and a "Spaceship Earth." And how we encourage, and are encouraged by, the impatience and fervor of our idealistic youth who want to see Utopia in their lifetime. Words like abundance and security and technology and development roll easily off our tongues. But their prevalence today creates something of a deception. They tend to emphasize that we live in a modern world of scientific and technological virtuosity, a world where a formula can be constructed to explain almost anything and a system conceived (and, if only funds were made available, assembled) to correct almost everything. Perhaps this is so, but it is also true we live in another world, one in which men still live almost as their ancient ancestors did, struggling from day to day to survive, knowing little more than a life of ceaseless toil and incessant hunger, and with scant hope that their children will rise to a better life. Strangely enough, these two worlds, and variations of them, exist almost side by side. In my travels around the world I have visited modern scientific laboratories where brilliant minds use the most complex tools to probe the mysteries of matter and life, while almost within sight a man unable to read or write works with a beast of burden much in the manner of his forebears of biblical days. Such is still the gap between man and mankind. And it does not close easily.

What are some of the dimensions of this gap and what kind of problems does it pose? A few facts help to reveal them.

A most startling one in a world where some people speak of our potential abundance is that two to three billion of the world's people suffer from hunger or malnutrition. The latest reports of the UN's Food and Agriculture Organization also indicate that the average person in the advanced areas of the world consumes four pounds of food per day compared with $1\frac{1}{4}$ pounds consumed by his counterpart in the poorer areas. Naturally, this problem contributes to other health gaps. For example, infant deaths per 1,000 live births are four times higher in the developing countries than in the advanced countries. And the life expectancy of a man in the developed nations is 40 percent longer than that of the average man in the developing ones.

But hunger and health are not the sole indicators of the disparities within mankind as it exists today. We still have some 800 million people in the world who are illiterate — the number has been growing — and most of them are in the poorer areas of the world. Also, it is estimated that about 20 percent of the world's entire male labor force is unemployed. This is reflected most predominantly in the developing areas and in general in a world where urban population is growing twice as fast as the number of urban jobs. The "Green Revolution" which I will discuss in a moment will aggravate this situation as will the spread of other technologies. Broadly speaking, this change in the relative productivity and "usefulness" of man, as we move toward an advanced, global mankind, will pose our greatest physical, psychological and philosophical problems — and in that order. For in a sense, as we increasingly solve the problems of "what" and "how" we will be more forcefully confronted with the questions of "why."

Returning though to our more immediate and "real" concerns, what other indications are there that we must intensify our efforts to uplift much of man to achieve true mankind? One such indication is delineated in a recent report to President Nixon from his Office of Science and Technology.[1] The report begins by projecting that by the year 2000 about three-quarters of the world's population, some five billion people, will live in the lesser developed countries. It admits to the gains in progress in these countries as shown by a rise in their GNP — a rise relatively greater than that in the developed nations, by the way. But then it elaborates on the implications of population growth in these countries by referring to a statement from the United Nations Association of the U.S. The statement relates the following:

> "In aggregate terms, the rate of population growth in the developing world renders difficult, if not impossible of attainment, rapid increase in per capita income. For example, with a GNP which is growing at an annual rate of five percent and a stationary population it takes 12 years to double per capita income. If population growth is not stationary but is rising, say at 2.4 percent (the projected average rate for the less developed world), it will take 27 years to achieve a doubling in per capita income.
>
> "If the annual growth in national income is four percent rather than five percent, doubling of per capita income will take 43 years. And since two-thirds of the world live in countries where per capita income is less than $300 per annum and often less than $100 per annum, even a doubling in per capita income is hardly likely to represent a satisfactory fulfillment of economic expectations."

That, I believe, sums up and dramatically highlights a "man to mankind" challenge we face — particularly when we see today what reactions result from the pressures of growth and unfulfilled expectations.

Of course, all this emphasizes the extreme urgency of attaining worldwide population control. It also emphasizes the urgency of narrowing the gap between the world's haves and have-nots. This can only be done by a major effort on the part of the haves. The developing nations can only lift themselves from poverty so far by their own bootstraps, but to lift themselves to the point where they can truly become self-sustaining, and then self-developing, they must have considerable assistance from their more advanced fellow-nations. The essence of this is that the path from man to mankind imposes a moral imperative on all those who would complete the journey. It is up to the stronger to see that the weaker do not falter or fall behind, to lighten their burden and to provide guidance and assistance all along the way.

Perhaps the greatest misconception about such a journey and the sharing it entails is that it necessitates that the have-nots be made over in the image of the haves. The spectre of this and its many implications, environmental and otherwise, is often raised. But we know that this is not necessary, nor is it possible, nor is it even desirable. A family of man can and must exist in which all members are healthy, all have their self-esteem and the esteem of each other and yet all are distinctive personalities whose difference contributes to the common interests and the creativity and vitality of the group. In fact, it is probably only through achieving such a family that we can attain a lasting global peace on this planet.

If I have made all this sound hopelessly difficult, let me inject a few encouraging signposts that we are at least on the right path. One is that during the past decade the average total growth rate (in Gross Domestic Product) for seventy low-income nations has been about five percent and some twenty of those countries have reach six percent. Even with an accelerating population growth eating into this gain, most of these countries have shown a favorable gain compared with the early experience of today's industrialized countries. During this period a great deal of development experience was gained which should provide valuable lessons for the future, both for the developing countries and those who seek to assist them.

Some of the most hopeful signs of development lie in agriculture. The so-called "Green Revolution" has, in fact, been almost spectacular in its results in some countries, particularly in India which is approaching food self-sufficiency. Because of its tremendous success in introducing and adding to its acreage of high-yield crops, that country's food supply is now growing at a faster rate than its population. And there are indications that its program of population control is meeting with some success. There is no doubt that the introduction and spread of new technologies — the new seeds, fertilizer production, the tremendous growth of the use of electrified tube pumps for irrigation — are all playing an important role in India's gains. We must lend our support and encouragement to this and similar development around the world. We must increase it.

This brings me to the second question I raised initially — in essence, can we increase such development, as well as that needed in our advanced nations, without devastating ecological effects? I think we can — within limits and providing we put great emphasis on the scientific, technological and organizational efforts necessary to meet global environmental requirements. There is no doubt that this is going to demand difficult economic adjustments and many value changes, particularly in the advanced nations. In the developing ones it will call for a certain vigilance, perhaps some restraints and hard choices, and no doubt some great technological ingenuity, if they are to bypass the environmental errors that the developed nations made in their industrialization. Many of the developing countries no doubt will want to emulate the industrial ways of the developed ones to give their people many of the amenities of life in those countries. And it may be difficult to convince them that in advising them to bear the cost of some improved technologies and methods of operation we are not asking them to make an unnecessary trade-off between more rapid progress and some environmental degradation. In an interview recently, John Kenneth Galbraith presented a different side of this question when he said:

> "My guess is that the greatest desire of the Third World is to have the same traffic jams and the same air pollution and the same environmental disharmony as the developed world has, and to call it progress. But these are matters of degree. Let's not worry about pollution in the Third World until we're sure that people have enough to eat and enough to wear and shelter. One knows that one can provide that without environmental damage. Let's not let up on that effort."[2]

Galbraith is probably right to the extent that we in the advanced nations will probably have to solve our environmental crises before "progress" becomes a real problem in the developing lands. But I think we are equipped to do so technologically if not yet economically and socially. As I have pointed out on many occasions, what we must aim for essentially is a "steady-state" world, one in which per capita consumption has virtually leveled off and the goods used are for the most part produced from recycled materials. This "steady-state" world would have to be an economic system with fixed trade-in values on all goods, all of "tagged" or known material content and preferably in products highly standardized with easily interchangeable parts. Literally nothing would be disposed of in the sense that we dispose of things today. Nature would not be the recipient of our "waste." Whatever was returned to nature's domain would first be reduced to its "natural state" and then returned in an ecologically acceptably way — to the right place, in the right amounts, at the right time.

This sounds like an impossible order but it is not. Achieving a "steady-state" society, particularly in a world of so many disparities and discontinuities in standards and values, will be extremely difficult, but it is logically the only way to go if we take stock of the world's resources and project our current growth. If we apply the current growth to the way we are operating today we can only end up with a totally degraded world in utter chaos. Resources would be unavailable to us because they would be depleted from their natural source, degraded and spread throughout the environment in a manner making them virtually unrecoverable to us. And this point could be reached only in theory because long before that the environmental impact of such a dispersion of materials would probably have been disastrous to all life.

It seems clear then that the rapid industrial growth man is experiencing now is part of a steep climb to what may be viewed as a "physical plateau of mankind," but one that is essentially a plateau of non-growth. M. King Hubbert recently expressed this idea in the following way:

> "... it is clear that the episode of industrial exponential growth can only be a transitory epoch of about three centuries duration in the totality of human history. It represents but a brief transitional epoch between two very much longer periods, each characterized by rates of change so slow as to be regarded essentially as periods of nongrowth. Although the forthcoming period poses no insuperable physical or biological difficulties, it can hardly fail to force a major revision of those aspects of our current economic and social thinking which stems from the assumption that the growth rates that have characterized this temporary period can somehow be made permanent."[3]

There are, of course, two factors that still must grow even when a near-state of equilibrium is reached in a "steady-state" world. The first of these is human creativity. And when we reach our "physical plateau" it will only be the continued growth of human creativity that will give us the innovation and variety to make life interesting. Perhaps I am wrong here, but I cannot image a "non-growth mankind," a global society that has stopped seeking new knowledge and new challenges — be they here on earth or elsewhere.

Of course a tremendous amount of knowledge and creativity will have to be generated and come into play just to reach that equilibrium. One need only imagine the problems in chemistry and physics to be solved in order to achieve a fine-tuned system of industrial recycle. Concurrent with this are also the problems of industrially and economically managing the residuals and by-products of the primary product.[4] Much technological assessment will also be necessary to determine whether the introduction of a product or service has any social validity in the first place.

The second input into a "steady-state" world that is non-renewable and will continue to be expended, though most likely at a decreasing growth rate, is energy. And we will need large amounts of it to manipulate matter to meet the requirements of mankind and nature in our "steady-state" world. The major problem related to this will probably not be the source of such energy — properly developed nuclear fission technologies and eventually controlled thermonuclear fusion should fill our needs for millions of years. More difficult than harnessing such energy sources will be the planning, in combination with associated technologies, and arrangements that will go into generating and distributing power so that its environmental impact is minimized. It will be possible to do this — to develop power technologies and locate them — so that man will have all the power he will need without any significant environmental problems. In fact we may see a day when power plants and their transmission facilities will literally never be seen by the average man because of the way they are designed and located — totally isolated from most human activity.

All things considered, I do not think that our path from man to mankind will be blocked by a lack of resources or a lack of the creativity to use those resources in an environmentally sound matter. What may be our greatest problem is finding ways to match and meld certain factions and functions within our evolving mankind so that mankind becomes a viable organic entity. A brief elaboration of this and a few recommendations are in order.

When we speak of a viable organic entity we are speaking about an organism in which all the parts are healthy, all serve a useful function and all are balanced in their relationship with the whole system. In order for a complex organism to be healthy, all its individual organs must be sound and functioning in harmony with each other. A weakness or disease in any one part usually affects others. There are symptoms — "referred pains" for example — often in areas remote from the diseased area. A fever or other general indication also warns of trouble as emergency activities spring into action within the system. The problems of evolving mankind are in many ways analogous to such a complex organism which is suffering because certain of its parts are poorly developed, and in general its metabolism is not well regulated. Troubles centered about these weaknesses affect widespread and often remote regions, and in general enervate the entire system. Also, if one looks objectively around the world one sees that evolving mankind's greatest dangers are much like those of a hypersensitive system which, in responding to the care directed to the wounds or diseases in local areas, runs the risk of a massive reaction to certain medication or treatment.

The reasons for both these local problems and our general hypersensitivity are related closely to the way we have grown. In a healthy organism, growth is carefully coded and regulated. Even then the organism may go through stages of maturing in which there seem to be "growing pains" — problems of awkward adjustment to itself and its surroundings. Evolving mankind seems to be suffering from a number of such growth problems. And, as I pointed out before, there are some kinds of development — in some parts of the globe — we must encourage and support. But also I think the feedback mechanism that tells us we must stop or adjust certain other kinds of growth is in operation — although some people question whether our responses will be fast enough. Perhaps the major difficulty lies in the way we have been growing and the actions and reactions involved. These bear a moment of consideration.

Mankind has been evolving through a process that seems to involve a sequence of explosion–implosion–displosion. This process is simply explained. The explosion factor relates to our exponential population growth and its increased impact via vastly increased technological capacity. The implosion aspect has to do with the phenomena of local, regional and global urbanization. This, in turn, causes the third and most symptomatic element of the sequence — that intense and disruptive reaction that some refer to as "displosion." Much analysis is done today of the first two factors. Data on population is compiled and studied in many ways and projections offer us some dire forecasts. The same is true of urbanization. On the other hand, we see far more manifestations of the displosion phenomena every day around the world, and though much is written about all the individual incidents — the violence, riots, wars, etc. — I do not think we have thought as deeply or as analytically as we might about their meaning — about their origins and what we might learn from them to help us open and smooth the path to mankind.

It is fairly obvious that technology has imploded humanity, has brought people together physically and in a wide variety of associations and activities. This has created all kinds of interfaces — direct meetings between peoples, institutions and ideas, more often than not of great dissimilarity. In most cases, these dissimilar forces at the interfaces tend to repel each other and are a source of agitation and conflict. They are the clashes of cultures, races, idealogies and values we are seeing today all around the world that our technological implosion — our "world shrinking" technology — has intensified over the past few decades. They are also manifestations of a striving for identity, individualism and self-esteem in a world that ironically both desires and resists greater social unification. Both as nations and as peoples we all need each other desperately, and yet we struggle to be ourselves, to maintain our individual identity.

On a global scale it is interesting to reflect on this struggle. For example, following World War II there was much talk of "one world," of international community, even of world government. And this continued even through the "Cold War" era. At the same time, however, a wave of nationalism swept the globe, as after 1945 more than 60 new nations gained formal independence within a period of 15 years. Even within these new nations there was much unrest and many upheavals due to further efforts at self-identity on a regional or tribal level and this continues in many areas today. The same became true in the older nations as various factions sought to increase their stature and power within the existing systems. One indication of the intensity of this long and painful struggle, aside from the lives it has taken, and destruction and suffering it has caused, is the fact that we have more than 16 million refugees in the world today — people who have been uprooted and are still homeless because they were the victims of these clashes at our human interfaces.

But let me return to, and conclude on, a note of optimism and hope. There are many ways available to man to solve the problems that exist at these interfaces, even as they continue to grow around the world. To employ the language of science, there are catalysts, there are common bonds and there are weak but significant interactions, unifying forces that must all be recognized and put to work — not to destroy all differences but to allow a human mix in which those differences are subordinated to the most vital common interests. When such a mix is reached those differences then take on a positive effect by adding variety and interest to the total scene. In the long run I think it may be crucial to mankind to maintain such variety and interest to prevent what could bring about the decline and fall of even a utopian, steady-state mankind — simply boredom.

But with the challenges ahead I believe we have a long way to go before that becomes a major concern. What are some of those catalysts, bonds and interactions to which I referred to help us achieve a unified mankind? Let me touch on a few of them. One of them, to which I admit a personal inclination because of my own experiences, is the greater exchange of scientific and technological information. The flow of such information often does more than provide the basis for the self-development of a people. If managed properly it may provide them with an important amount of self-esteem as well as focal points from which to stimulate broader educational growth. There is a significant amount of such exchange taking place today and the good it accomplishes rarely receives the credit it deserves. About 10,000 international conferences are now held each year, of which a major portion are scientific and technical. I know there is a great bond through science in bringing peoples together, directly and indirectly. I experienced this last year on what I like to refer to as my "Scientific Safari to Africa," a trip during which I visited six African nation.† I experienced it again only last week when I accompanied a delegation of Soviet scientists on a tour of U.S. nuclear laboratories. And I am sure I will experience this once more in even greater measure when, as head of the U.S. Delegation and President of the Conference, I attend the Fourth International Conference on the Peaceful Uses of Atomic Energy in Geneva this September. This conference will see more than 100 nations contributing and sharing information on uses of the atom that can benefit, and help unite, all mankind.

Perhaps even more important than sharing information is exchange on a people-to-people basis, particularly over an extended period of time. The Report of the Commission on International Development points this

---

†See pp. 275–294, this volume.

out when it speaks of the 100,000 experts and volunteers working in the developing countries under official programs and the 80,000 students and trainees studying in the industrialized countries. The report states:

> This two-way flow of people has powerfully stimulated change, introduced new management skills, raised educational standards and helped to create much infrastructure and expanded industry."[5]

I would also venture to say that such exchanges do far more than is indicated in that statement to advance the cause of mankind. The human contacts made are extremely important. Of course, this kind of bond is made through more than technical fields. International contacts through the arts and sports are vital links between peoples of the world. Apparently no contact or exchange is so small as to be insignificant if it is at the right time in the right place. Even the force of a ping-pong ball seems to be strong enough to open a wedge through which the greater forces of more common interests may flow, eventually breaking down seemingly impenetrable barriers. In every case, we must use the common interests, small as they may seem, to broaden contacts and work towards proving that cooperation must overshadow competition and conflict as we move toward the realization of mankind.

As man and his influence grow, and as his planet "shrinks" proportionately, so do the areas and fields of cooperation grow and multiply of necessity. International aid and trade must be fostered and balanced properly. Our global environment becomes a global concern that must be studied and managed cooperatively. The necessity of education to link peoples and nations, to bring man up to the level of knowledge and understanding that a universal mankind demands, requires tremendous cooperative efforts.

One key to eliciting the degree of cooperation that will be demanded in all these areas — and ranging from the community level to the international level — seems to be the emphasis on active participation. It has become increasingly obvious in recent years that man, both as a species and as individuals, recognizing the extent to which his destiny now resides in his own actions, wants to take a more active control of his life and not merely leave its future to the forces about him. While the resurgency of this participatory attitude is the source of some conflict in society today, it is a vital force toward ultimately creating better men and a finer mankind. A man who seeks to control his own destiny must inevitably see how closely that destiny is linked to the well-being of all. He will soon realize the true meaning of mankind and will be the most dedicated worker towards its perfection.

If I have indicated in my remarks today that we still have a long, difficult way on the path from man to mankind, I did not do so to be discouraging. We must move on with a sense of urgency, but time is *not* running out for the human race. We may feel the waves of adversity rising up all about us, but the tide is *not* against us. And if we see in today's upheavals what seems to be senseless conflict, destruction and waste, we might recall the words of Alfred North Whitehead:

> "It is the first step of wisdom to recognize that the major advances in civilization are processes which all but wreck the society in which they occur."

Truly, this is the time for such recognition. We are building a better world. Spread the word. Share the work. And help in every way to see that man completes this arduous but rewarding journey to mankind.

References

1. Report on "Population and The World Environment" Office of Science and Technology, Jan. 14, 1971.
2. *New Scientist and Science Journal*, "Abrasive Optimist," Feb. 18, 1971, pp. 375–377.
3. M. King Hubbert, U.S. Geological Survey, speaking at a Symposium on Environmental Aspects of Nuclear Power Stations, International Atomic Energy Agency, Aug. 1970, N.Y.
4. An interesting approach to such problems is discussed in *Economics and the Environment — Material Balance Approach* by Allen V. Kneese, Robert U. Ayres and Ralph C. D'Arge (Resources for the Future, Inc., Washington D.C., 1970).
5. *Partners in Development — Report of the Commission on International Development* (Praeger Publisher, Westport, CT, 1969).

The State of Knowledge

Remarks made at the Annual Dinner for the American Committee for the Weizmann Institute of Science, New York, N. Y., November 3, 1971.

I feel deeply honored that you have invited me to be with you this evening. To be in the company of men and women who are steeped in the legacy and love of learning always gives me a warm feeling and one of hope in this troubled world. I have had that feeling before — and noticeably so during my visit to the Weizmann Institute of Science during a trip to Israel.

I had the great pleasure, during one of my many trips abroad as Chairman of the Atomic Energy Commission, of visiting Israel in September of 1966, and while there of being invited to lecture at the Weizmann Institute. I recall that I gave my lecture on the transuranium elements to a full house in the large auditorium of the Institute, to a very receptive and responsive audience. It was a gratifying experience, and I was also most pleased when some weeks later I received a copy of the lecture I had given reproduced in Hebrew. (Although I must confess that I could only recognize it as my work by the chemical symbols printed in English.)

Following that lecture, Meyer Weisgal honored me by hosting a luncheon for my colleagues and me in his home. It was one of those rare occasions at which the good food was excelled only by the fine conversation. Later that day and evening, conversations of equal calibre were conducted in our meeting and dinner with Prime Minster and Mrs. Eshkol in a beautiful dining room in nearby Caesarea. That entire visit to Israel, which included much historic sightseeing and a tour of Soreq Nuclear Research Center, was one of the most intriguing and enlightening visits I have made in some 800,000 miles of travel to more than 60 countries. There were many reasons why this was so. I am sure you know most of them. But what is significant on this occasion — and which relates to what I want to discuss this evening —is what I witnessed of the pursuit and application of knowledge that prevails in the nation of Israel, surely a major factor in determining that nation's destiny and, as an example to others, the destiny of much of the world.

Israel *is* a living example of a country built and surviving on knowledge. It is in a sense a "State of Knowledge" where learning, scientific and cultural, is held in the highest esteem — but more than that, where it is applied daily to the business of living and growing. Knowledge is Israel's greatest capital, a major asset for its future development and an important item of export. In short, Israel knows the value of knowing. Its people have known it traditionally for hundreds of centuries but today they can, in their own state, see and feel its effect more precisely.

This is not generally true around the world today. And it is on this subject — the state of knowledge in the world today, where it stands now, people's attitudes toward it and what we might do to improve its present position and future possibilities — that I would like to focus this evening. This, of course, is a subject of enormous proportions, and the best I can hope to do in the few minutes I have with you is touch on a few aspects of it. But it is a subject so vital to our well-being — to our very survival — that it deserves our attention.

What we know now is that knowledge is the key to human evolution — and perhaps of all evolution. The transfer of knowledge from the individual man to his immediate tribe or community, to his larger society such as the nation-state, and today to global man — to what I have often referred as an evolving global mankind — has been a force of change far outstripping any natural process in recent centuries. Some thinkers today go so far as to postulate that essentially "natural evolution," as we have thought of it in the past, is no longer taking place, and that because of man's predominant influence now the effects of human activity have taken over the evolution and fate of all life on earth. This may be an exaggeration — but not by much. For surely what man knows today and what he does with that knowledge will determine the future health and well-being of this planet as well as that of his own future offspring.

Just what is the state of that knowledge? (And from here on I will use that word synonymously for scientific and technological knowledge although it applies more broadly.) How much do we know? How fast are we learning? And with what effect?

The explosion of the acquisition and transfer of knowledge in the world today is nothing short of phenomenal. A few facts serve to characterize what is taking place. We are told that the amount of scientific information published around the world every 24 hours would fill seven complete 24-volume sets of the Encyclopedia Britannica. This means that reading around the clock from day to day one man would need about 1,000 years to cover one year's output. Where is such information appearing? At present it is being offered annually in some 300,000 articles published in some 6,000 scientific and technical journals in the U.S. alone, and on a worldwide basis by some 2,000,000 articles appearing each year in about 35,000 journals written in about 50 languages. It is being further disseminated by the direct contacts of some 100,000 scientists and technologists meeting in about 2,000 international conferences that take place annually. I recently attended one such conference — the Fourth United Nations International Conference on the Peaceful Uses of Atomic Energy — in which scientists of 49 countries, one of them Israel, and nine international organizations presented more than 500 papers to some 4,000 conference attendees. I believe some of you were present at that conference, and many others of you have been attending similar conferences throughout the years.

But there are great paradoxes in all this explosion and exchange of information — many of them at the roots of the grave problems of the world. It is worth taking note of these paradoxes which we seem to face daily in so many ways. The first has to do with the nature of knowledge and the way it grows. The process starts with the accumulation of data which we assemble as "information." The information properly related to other information becomes a body of knowledge, which in itself may be highly specialized or fairly broad. On the basis of this, plus previous experience with similar and related knowledge we develop what I will loosely refer to as "wisdom." The problem is that this is a process that is wildly out of kilter today and one extremely difficult to correct. We are effective in gaining specific information, yet each conquest reveals new ignorance and the need for more specific data. This, of course, is nothing new. Neither is the problem of acting on the basis of insufficient data. As John Donne pointed out more than two centuries ago, "who are a little wise, the best fools be."

The other part of the problem is that even as we increase our specialized knowledge we tend to neglect the interdisciplinary knowledge needed to bridge specific areas. We have discovered that we are very poor "bridge builders" and are trying now to overcome this. That is why one hears so much emphasis today on subjects like systems, symbiosis, synergetics. That is why we have "discovered" subjects like ecology and macro-economics. For we are being plagued not only by the gaps in our knowledge and the rapid obsolescence of our expertise, but by the discontinuity of that knowledge and its effects when applied. So while our first paradox is that "the more we learn the more we need to know," the second one becomes "the more we apply the more we seem to need to apply — but correctly." That, we might say, is Catch-22 and one of the reasons why we are so caught up in our concern with what has been called "technological assessment."

The problem here is that the power of our applied knowledge, its scope and far-ranging influence within shrinking frontiers, has become so all-pervasive and effective that it must be treated with greater rationality and controls. When a statement such as this is made, much confusion seems to arise. On one hand there is the notion that science and technology must have complete freedom to be creative and innovative. On the other is the idea that a "free market" approach to the support and use of science and technology is no longer possible. Neither of these concepts is wholly true, nor is the conflict between them as harsh as some would have us believe. Technological assessment is not the equivalent of thought control and should not necessarily stifle scientific creativity. In fact, it may be great incentive to enhance it.

In the past, our freedom to innovate and develop technologically with only economic restraints has allowed us to carry forth each innovation to a saturation point. We became addicted to the belief that all "goods" were good *ad infinitum* and that growth in itself was a prime good. The reaching of many saturation points in recent years — many concurrently and with unsuspected impact — has changed that. The backlash of traffic congestion, pesticide-immune pests, accumulating nondegradeable waste and the effects of other products and by-products of a "free market" approach to research and development has led us to realize that we must now create a new outlook and framework for operating a scientific society. The fact that this outlook which demands the utmost in systems thinking and an ability to deal with dynamic, ever-changing forces in nature and society, calling for a kind of super-rationality brings up the next paradox we have to face.

This paradox revolves around the fact that at the very time we need this degree of rationality we are faced with a turn to irrationality and primarily by the very people whom reason needs most — the young. The "Counter-Culture" group who speak of the "myth of the objective consciousness," the "Consciousness III" advocates who believe that they can only "green" America — and the rest of the world — by doing their own thing, these are among the best of our youth who should be embracing knowledge and specifically the kind of knowledge they can apply to build the better world they say we have failed to give them.

As I pointed out a number of years ago, in a talk titled "Youth — A Vote of Confidence,"* history reveals that most of the great discoveries, accomplishments and artistic creations were made by relatively young people. The early years are years of immense creative energy and fresh insight. Too many of those years are being wasted today. Of course we in the scientific and academic communities must share some of the blame for this. On too many occasions we have been less than inspiring in words or deeds in winning over the young. Many of our institutions have also failed in their limited outlook and actions to capture, no less captivate, youth. Murray Gell-Man put this even more strongly when he said recently:

> "I believe that narrow rationality, pervading government, universities, industries, and other parts of our national and international life, is provoking a wave of insufficient rationality. Youngsters tired of the tyranny of badly programmed computers, and of people who act like badly programmed computers, are turning to tarot cards and charlatans."

Fortunately, I believe this is not a problem that plagues Israel today, as most of its youth see less of this misguided use of rationality and see much more directly how science and technology are sources of survival for their homeland and the basis for a very bright future in the years ahead. I believe they also see — and this is a point I have made many times — that there is not necessarily a conflict between the sciences and the arts, between the thinking man and the man of emotional sensitivity. This is an old myth that is taking on new life today — and it is a dangerous one.

---

*See pp. 202–210, this volume.

It is dangerous because, even partly believed, it takes on the character of a self-fulfilling prophecy, tending to keep many talented, sensitive people away from scientific and technological fields where they might narrow the gap between the "Two Cultures" as well as make considerable contributions in "humanizing" both the means and ends of technology.

Another paradox of knowledge growth involves what we might call "the rich get richer" syndrome. This, of course, has to do with the educational and technological gaps between advanced and developing nations and as a result much of the gap in living standards. Volumes have been written by people such as Barbara Ward, Harrison Brown and Aurelio Peccei dedicated to warning us of the impending disaster that could result if the advanced nations do not make a concerted effort to narrow these gaps. Let me read to you how Harrison Brown characterized this situation before a Congressional committee not long ago. Speaking in terms of the world's advanced nations as "the culture of affluence" and of the developing nations as "the culture of poverty," Professor Brown said:

> "The culture of affluence is now made up of some 1,000 million persons having an average per capita GNP of about $2,300. The culture of poverty is made up of some 2,500 million persons having a per capita GNP of about $180, or thirteenfold less than that characteristic of the rich countries. Within the rich countries per capita GNP is increasing about 4 percent yearly; within the poor countries the lot of the average person is improving much more slowly — about 2 percent per year.
>
> "As an illustration of the impossible situation which would develop were the present growth rates to continue, 130 years would be required for the poorer countries to reach the level of per capita which is now characteristic of the richer countries. But by that time, at present growth rates, the population of the poorer countries would reach 150 billion persons. At the same time the per capita GNPs of the richer countries would approach $1 million per person."†

Now I submit to you that that is what is known as becoming a millionaire the hard way — and, we know, the impossible way because the world

†Harrison Brown — "Science, Technology and the Developing Countries." Paper delivered at the 12th Meeting of the Panel on Science and Technology of the Committee on Science and Astronautics, U.S. House of Representatives.

could not tolerate the situation Harrison Brown describes. Disaster, in many irreversible forms, would overtake us long before that point. But that projection serves to dramatize the dynamics of a global process that must be changed. And it can only be changed by the effective sharing and application of knowledge. This goes far beyond the concept of "foreign aid" as most people think of it today. Education, science and technology — and of the kind that is totally compatible with a people's culture and values and goals — must become a part of the developing nation. As Rene Maheu, Director General of UNESCO once said, "The development of a nation cannot be achieved if science and technology do not cease to be an imported magic, to become a custom of its people." A proper "climate of knowledge" — a receptiveness to new ideas and change — is essential for this to take place. As senator Humphrey so cogently pointed out at this gathering in 1965, "Knowledge cannot merely be given. It must be received."

A complicating factor in this process today is that, while we wish to see the poorer peoples of the world raise their standards of living through development, we do not want to see them reap the burden of the environmental and social problems of development with which we in the advanced countries are now struggling. This makes even greater demands on our pursuit and sharing of knowledge — demands complicated by important psychological and ethical factors of which you are well aware. Nevertheless, we must use every instrument of national and international policy and every worthwhile institution — and develop new ones when necessary — to share knowledge with our fellowman. For never was there a time when ignorance was farther from bliss or knowledge more essential to human survival.

By way of conclusion, let me touch on my own feelings about Israel's stake in the growth and sharing of knowledge. As I said earlier, one might characterize Israel as a State of Knowledge. It is a state built on the proposition that knowledge is a prime resource, that it is a lever with which to perform the mammoth tasks of building a thriving human society of free men out of the sands of the Sinai and the rocks of Judea. I imagine that it is fashionable for those who have visited Israel and seen the miracle that men have wrought there, to say that nowhere have so few done so much with so little. But we know that this is far from the whole truth. Overlooked or undervalued in such a statement have been those great forces and catalysts of the human mind and spirit. Israel offers a living testament to their effectiveness, to what maximum human resources can do with minimum natural resources. There is a great lesson for many parts and peoples of the world in this — some not too far from Israel's own borders — and one would hope that that lesson could be quietly absorbed.

Of course much more remains to be done in Israel, and much of that is only possible when a lasting and secure peace has been established. Eventually — and we pray that it is soon — all men in the Middle East will find the wisdom and the greater courage needed to lay down their arms and turn to the task of recreating an Eden in those historic lands.

It has been my hope, and that of many of my colleagues in the nuclear field — men like Alvin Weinberg, Phillip Hammond, James Ramey, and particularly Lewis Strauss — that nuclear energy used for the large-scale desalting of seawater could someday play a significant role in that task — in bringing abundant fresh water and peace to the Middle East. And I want to tell you that, even though the political and economic difficulties still loom large in this matter, we still share that hope and feel that we will see the atom used as such a force for peace and plenty in that part of the world.

During my tenure as Chairman of the AEC, I was asked on numerous occasions whether I thought Israel was a nuclear power — or less euphemistically, did she have the bomb? My usual reply was that she was among those nations that had the knowledge to build one if she wanted to, and I speculated no further. Now in retrospect I often wished I had said "Yes, she is a nuclear power, the kind that knows of, and makes use of, the atom's power for peace. She is a member of the International Atomic Energy Agency, her nuclear laboratories do advanced research in most nuclear fields — nuclear physics, solid state physics, reactor physics, radiation chemistry, inorganic chemistry, nuclear chemistry. She uses radioisotopes in medicine, in agriculture and food preservation, in industry and in other areas of research. Her professors in nuclear sciences and engineering are excellent and their students enthusiastic and enlightened. Of course Israel is a nuclear power — the kind that more nations should strive to be." I recommend those or similar words now should such a question be asked.

Thank you for inviting me to share these thoughts with you this evening. As you all know, I will now be returning to the academic world — to my former role as teacher and researcher at the University of California at Berkeley. From there I will continue to watch the state of knowledge around the world and hope that I may add a little to it. And even if the Weizmann Institute does not wish another lecture on the transuranium elements, I hope to return to Israel someday to see much of what I missed on my past visit and all the new progress that must inevitably take place in that incredible land.

A Journey to China

Article written for Science News.

I have just returned from a 17-day visit to the People's Republic of China as a member of a delegation organized by the Committee for Scholarly Communication with the People's Republic of China (jointly sponsored by the National Academy of Sciences, Social Sciences Research Council and American Council of Learned Society). Throughout our visit, the 21 people in our delegation were the guests of the People's Republic of China through its Scientific and Technical Association. We visited research institutes, universities, hospitals, communes and factories, as well as many other points of historical and cultural interest.

A primary objective of our visit was to negotiate an agreement with our Chinese hosts for the exchange of visiting groups of scholars in the natural sciences, the social sciences and the humanities.

A high point of our visit was a two-hour meeting with Premier Chou En-lai in one of the rooms in the famous Great Hall of the People. Many of our host Chinese scholars were also present. The Premier showed a remarkable and detailed familiarity with the proposed exchange agreement, indicating which of our suggestions he found acceptable.

In Peking, I visited the Institute of Geophysics, the Institute of Agricultural Applications of Atomic Energy, the Institute of Chemistry, the Institute of Physics and the Institute of Atomic Energy, as well as Peking University and Tsinghua University. At each place we were given informative tours, and our questions were freely and completely answered. Since the Cultural Revolution (1966–70), the work at these and at the

Reprinted by permission from *Science News*, **104**, 14 (1973),

Chinese Premier Zhou Enlai greeting Seaborg. Helen Seaborg and Dr. and Mrs. Albert Feuerwerker are in background. Beijing, China, 27 May 1973.

other places we visited has taken on a very practical aspect, with less emphasis on theory than is found in comparable establishments in the United States or in Europe.

At the Institute of Geophysics, the emphasis is on seismology and the problem of predicting the location, magnitude and time of earthquakes. China has maintained earthquake records for 2,000 years, somewhat fragmentary for the earliest years, but in enough detail to include estimates of magnitude as well as records of place and date during the last 500 years. These records are being analyzed to show alternating periods of high and low seismicity; such data should be of value to seismologists in the United States.

At the Institute of Agricultural Applications of Atomic Energy, the emphasis is on production of new strains of wheat and other crops, using selective breeding and radiation-induced mutations.

The work at the Institute of Chemistry is focused on high polymer chemistry, particularly photosensitive polymers with applications in the printing industry, cross-linked block polymers that have both plastic and elastic properties, polymers with photoconductive properties that might lead to the development of a chemical method for the conversion of solar energy to electrical power and nylon-like polymers with unusual strength. They are also working on organic silicon compounds used for lubrication oils, rubber, insulation in electric generators, resins for printed circuits, water repellent coatings and rust resistant varnish.

At the Institute of Physics, research is carried on in plasma physics, magnetism, lasers, crystallography, low-temperature physics, high-pressure phenomena, acoustics and theoretical physics. The X-ray crystallography program has concentrated on the structure of insulin, in collaboration with the group at the Institute of Biochemistry in Shanghai, which achieved the first synthesis of insulin a few years ago. Of special interest are their efforts in controlled thermonuclear reactions (nuclear fusion). They are investigating three approaches — two using magnetic confinement and one using the concept of laser-induced implosions. For magnetic confinement they have apparatus of the theta pinch and strong-focus type, and they have observed the production of neutrons (which could be evidence of fusions taking place) with both of these machines. They readily conceded, however, that these could be the result of nuclear reactions produced by deuterons accelerated in the electric and magnetic fields in the apparatus. The work on laser-induced fusion is of a preliminary nature and has not reached the stage of testing for the production of neutrons.

The Institute of Atomic Energy is located some 40 kilometers (an hour's ride) from the center of Peking. Associated with this institute is a group in Peking concerned with theoretical physics, including elementary particle physics and cosmic rays, and a cosmic-ray laboratory in the mountains (3,200 meters altitude) of Yunnan Province in Southern China. The institute is organized into divisions of reactor engineering, accelerator engineering, neutron physics, nuclear reactions, theoretical nuclear physics, elementary particle physics, nuclear electronics, nuclear detectors, radioactive isotope production and stable isotope production. Since 1958, they have had a seven-megawatt heavy-water and a 120-cm cyclotron, both of Soviet design. There is also a 2.5-million volt Van de Graaff of Soviet design. The cyclotron delivers 24-million-electron-volt (MeV) helium ions, 12.5 MeV deuterons (with an external beam of 80 microamperes) and, as a result of a recent modification to include a 3-sector magnet, variable energy protons of 3 to 19 MeV energy. Present emphasis at the cyclotron is on deuteron stripping reactions.

A neutron crystal spectrometer has been used with the reactor to measure neutron absorption cross sections, including the fission cross sections of uranium-233 and uranium-235, over the energy range of 10^{-2} to 10 eV. No work with plutonium is done here. A neutron diffraction apparatus is used to measure crystal structures and disorder in alloys. The reactor is also used for neutron activation analysis and, together with the cyclotron, as a source of isotopes for processing in the radiochemistry laboratory. In the radiochemistry laboratory such radioactive isotopes as P-32, S-35, Cr-51 and I-125 are separated, purified and synthesized into compounds for use in Chinese hospitals.

Of special interest are the two 180-degree magnetic separators for stable isotopes — one with a 220-ton magnet, 6 kilogauss field, and orbit radius of one meter; the other with a 280-ton magnet, 3 kilogauss field, and orbit radius of 1.7 meters. They separate the stable isotopes of such elements as Cu, Fe, Yb, Ce, Mg, Ni, Cr, Se, Li, Sr and Rb with average yields of the order of 10–20 milligrams per hour, for use in research laboratories throughout the country.

In Shanghai, I visited the more recently established Institute of Nuclear Physics in the company of Victor Weisskopf (Massachusetts Institute of Technology) and Robert Sachs (director, Argonne National Laboratory), fellow members of our delegation. We had the impression that we were the first Americans to visit this institute. Here there is also a 120-cm cyclotron (a copy of the cyclotron at the Institute of Atomic Energy in Peking), built by themselves in the period 1960–64, after the break with the Soviet Union (which occurred in 1958–60). Here they are measuring the fast neutron fission cross sections of Th-232, U-238 and Pu-239. The cyclotron is also used in a program of charged particle activation analysis. This institute includes divisions of nuclear physics (including a theoretical group), instruments, radiation and isotopes applications and radiochemistry. In the later division, C^{14}-labeled amino acid and tritium-labeled steroids are synthesized for use in other institutes such as the Shanghai Institute of Biochemistry. Especially noteworthy is their work on nuclear batteries, in which they plan to use Sr-90 with thermoelectric conversion to electricity of the heat of radioactive decay.

I gave talks in my specialty — the transuranium and superheavy elements — on three occasions, each lasting at least a couple of hours including time for translation. I spoke to a combined meeting of the Chinese Physical and Chemical Societies in Peking, to the science faculty at Nanking University in Nanking, and to a group representing numerous research institutes and educational institutions at the Science and Technology Center (Station) in Shanghai. In Shanghai, my presentation in the morning lasted two and a half hours and the discussion was pursued for two hours in the afternoon. At each place, the questions and discussion following my talk were carried on at a very informed and sophisticated level.

We did not see any direct work on atomic power during our visit to the People's Republic of China. We were told that China has adequate supplies of fossil fuels, so the development of atomic power is not an urgent matter. We received indications, however, that they are beginning to move toward the development of atomic power, with preference for the enriched-uranium water-cooled reactors (the pressurized-water reactor (PWR) was mentioned) rather than the natural-uranium graphite reactors.

Members of our delegation were interested in visiting Chinese hospitals and to learn about medical practices. Acupuncture plays a central role in Chinese traditional medicine, which has been revived by Chairman Mao since the Liberation. It has been used for thousands of years for curative purposes, which may be of limited value, but was not expanded for use in anesthesia until the 1960's. Its use in anesthesia seems to have definite value and should be introduced into Western medicine. Many Western doctors are planning to visit the People's Republic to learn this technique. Although I was not able to do so several members of our delegation saw caesarean sections and abdominal operations, performed under acupuncture anesthesia, throughout which the patients were conscious and perfectly content. We saw graphic movies of such operations.

All of the universities and colleges were closed down throughout the Cultural Revolution (1966–70). Peking and Fu Tan Universities reopened in 1970, the others that we heard about in 1972. During the Cultural Revolution, these institutions retained their teaching personnel, who spent the time reflecting on Chairman Mao's teachings and how they could reform their own teaching to better serve the masses. They usually also spent time working in the factories (especially the scientists and engineers) or fields (the social scientists and humanists). The post-Cultural Revolution philosophy emphasizes the practical; it downgrades theory, even in science, downgrades social science, and most of all, literature and the arts. The entrance requirements for students are entirely changed. Important in this process is their standing with "the masses" and their peers, and their acceptance of Chairman Mao's philosophy. Because the teaching faculties were retained during the Cultural Revolution and the schools have only recently reopened, the ratio of teachers to students at this time is usually one to one. This will decrease as successive classes arrive (up to the total of three for the uniformly three-year course), but apparently the ratio is not expected to drop below about one to five. These figures include assistant instructors and what we would call teaching assistants.

Our reception was very friendly everywhere by people in all walks of life. Our scientist hosts were uniformly glad to see us, volunteered information on their education in the United States, and asked about the state of health of their friends in the United States. The people in the streets were always friendly, without exception, and often stood in groups to welcome us with clapping of hands. Children almost invariably welcomed us with clapping of hands, individually and in groups as we entered or left our cars, hotels or points of visit, and even as we rode past in our cars. Often we were objects of curiosity, and people on foot or on bicycles would turn around to catch another look at us after passing. We were afraid someone on a bicycle would have an accident as he or she looked back. Sometimes little children were frightened by the strange sight of seeing us, and would hide behind their mother's legs when we looked their way, but would invariably smile back and look joyfully at their mothers when we welcomed them with a smile.

Besides laying all the essential groundwork for the exchange agreement, I believe our delegation contributed to increased understanding between the scientists, and more broadly, between the people of the United States and the People's Republic of China. I personally found our journey to China a very interesting and rewarding experience.

New Directions in Development

Address given at the Nobel Symposium, Norwegian Nobel Institute, Oslo, Norway, September 18, 1973.

We traditionally have held that objectivity is a fundamental aspect of science and scientific research. But perhaps nowhere is it more difficult to be objective these days than in analyzing the roles of science and technology in international development. In fact, the entire matter of development is one filled with strong subjective beliefs, with varying opinions on what goals should be set and the best means of achieving them.

Discussing development is a difficult matter for a number of reasons. First of all, in considering development we are dealing fundamentally with issues that determine human well-being or human deprivation, human happiness or human misery, and perhaps in the long run human survival altogether.

At the same time that we are gripped by these large issues, we are also torn between feelings of international altruism and a degree of ingrained national self-interest. We still live and work on these two levels.

In addition, as scientists, specialists, and members of various organizations, we hold, and tend to remain loyal to, certain points of view which are bound to influence our outlook on the total picture, large and complex and multi-dimensional as it is. We may be Neo-Malthusians,

---

Reprinted with permission from *Coordination in the Field of Science and Technology — The Role of the Specialized Agencies of UN* (*Proceedings of the 26th Nobel Symposium*, Oslo, September 17–19, 1973), eds. A. Schou and F. Sollie (Universitetsforlaget, Oslo, 1974), p. 67.

ISBN 82-00-2333-8.

advocates of economic adjustments, believers in the "technological fix," or champions of a regulatory approach.

But also, most of us have lived and worked in the so-called "real world" long enough to be its captives intellectually and emotionally. That is, we are realists; we know the forces of politics, of economics, of human habit, of all the factors that inhibit necessary change or exact their toll when rapid change is thrust upon the world either through new ideas or new technologies.

As a result of all this, a certain pattern seems to take shape whenever men of science gather in international meetings to discuss the role of science and technology in development. One can usually sense an extended fervor in the air because of the very nature of the subject and its implications for mankind. We know what is at stake. We know the potential power of science and technology for good. We have seen it work in the past. We see it at work today. And we want to see it work in the future. But our background, our interests, our "realism" tend to shade our vision and our ability to shape the alternatives that must take root in the world today. We resort to a considerable degree of linear thinking, to looking at and extrapolating from historical trends and examples, and to projecting futures based on extensions of some, rather than of building the kind of new world that the projections might indicate is necessary.

But there is hope. In recent years some of our projects, particularly those involving resources and environmental problems, have shocked us into newer modes of thinking, and these are beginning to affect our ideas about development. One can hear rumblings of these effects. They followed the publication of the study "The Limits of Growth," supported by the Club of Rome. They surfaced at the Stockholm conference and in the planning of the UN Environmental Program, in terms of possible conflicts between the needs of development and the protection of the environment. We are bound to hear more on this score as efforts are made to internalize the cost of a healthy environment in the world's industries. We are also going to hear more on matters of population and population distribution, land development and tenure, food, resources, trade, aid, the seas and space — to name a few of the topics that will provoke new thinking in the light of development needs and trends.

In my brief remarks here I would like to construct a type of framework that might be conducive to some of that newer thinking, and within that framework offer some suggestions as to ways that science and technology might serve development more rationally and constructively.

Let me begin by referring to a concept of global development upon which I speculated a few years ago in speaking to an international conference on "Environment and Society in Transition."[1] In my remarks at that time I expressed the philosophical belief that we — the human race — were evolving from a condition of tribal man to a truly organic mankind, a global civilization that would eventually live in harmony with itself and its environment. I stressed the point that a global civilization could exist but momentarily on this earth if it were to act as a parasite or cancer, that it had to establish a non-exploitive, closed-cycle way of life, and that this could only be done by the formulation and application of a new scientific outlook and new ecological–technological relationship.

In the intervening years, all that I have seen and heard in the world has reinforced my belief in this idea. And it is particularly applicable to the matter of development and the relationships between the developing and developed nations. Development must take the form of a globally integrated process, in which the natural and human resources of each nation become part of a system through which each will benefit and fulfill its potential in a way that allows the entire system to thrive. Hence, an organic mankind — a system of nations — will be much the same as a healthy individual who enjoys good health because all his organs and parts are functioning in a balanced way, contributing to each other and to the whole man.

This biological analogy can be alluded to in many ways relating to our world civilization. One way, of course, is to think in terms of a global metabolism into which industrial production must be worked so as to assure that the flow of materials and energy in and out of manmade systems does not interfere with the cycles of natural systems. The environmental movement has made us fully aware of this requirement and the kind of activity that must take place to meet it.

Another way is to think of emerging mankind as one might conceive of an unborn individual, perhaps in the stage of an early fetus. At this point the brain and the nervous system precede the development of the organs and skeletal structure of the body. One could liken this to the global situation in which our lines of communication first reach out to the various parts of the world and their peoples. This is followed by early international economic activity and trade which acts as a growing circulatory system helping to feed and build the individual organs. Pursuing the analogy still further, we might compare technological growth to increased hormonal activity, and economic growth through industrialization to muscular and metabolic activities.

Of course these analogies are far from perfect. But I think they can help to illustrate, if only in a crude way, something which is occurring in the world of which we must be aware — namely, that the formation of a global system bearing some resemblance to an organic one has been taking place.

But there are now indications that its development is entering a new phase where the system must regulate itself so that its parts mature in a way best suited to their own health and that of the system. In realistic terms, nations cannot and should not all grow in the same way, attempting to become copies of other nations in different regions with different resources, backgrounds, cultures, and needs. Each must develop in a manner, at a pace, and towards goals best suited to the needs of its own people while contributing to the overall needs of the world.

How is all this to come about? Do we have in our global system the equivalent of a DNA or RNA code destined to regulate our growth?

I believe that, broadly speaking, it may exist in science, and that a major role — if not *the* major role — of science in the coming years is going to be the direction that scientific thinking can give to development. This direction should take place in many forms and on many levels.

At the highest level it should determine and recommend a system of integrated development based on economic readjustments — combining trade, aid, monetary measures — designed to balance better the growth of developing and advanced nations. Part of this balance would probably be achieved by setting up conditions that would favor the production and export of goods from countries best suited to produce them, in terms of resources, environment, labor and other conditions. Developing nations badly need this kind of production and trade advantage. As a recent report by the U.S. Secretary of State indicates: "Trade is more important than aid for developing countries. Export earnings account for nearly four-fifths of their total foreign exchange resources."[2]

Economics must be used as a basic and applied science to accomplish this end, just as it is being used as a means of instituting environmental changes. And as economic forces do not operate in a vacuum, the social and political sciences must be used in conjunction with them. Of course this implies that we foster an international commitment that somehow supersedes many of today's national interests and nationalistic feelings. Ironically, these seem to be on the rise today at a time that begs for closer international cooperation. Perhaps, therefore, the entire key to the future rests on whether the nations of the world — and particularly their leaders — can be made to see the folly, the shortsightedness and, ultimately, the disaster of a divided world. And this is true not only in terms of rich and poor, of ideological division, but in the matters of advantages of natural and human resources, including scientific and technical knowledge.

(Let me indicate at this point that there are many organizations and individuals in the world who have strongly advocated this point of view, or ones similar to it, for some time. I think their ranks and influence are growing. Foremost among them are, of course, the specialized agencies of the United Nations — particularly the United Nations Development Program (UNDP) which assists projects in over 130 developing countries. The individuals are too numerous to mention but a few that come to mind because of their tireless efforts over the years are Barbara Ward, Aurelio Peccei and the late Lester Pearson.)

In dealing with this difficult matter of integrated international development, I believe that in addition to the new work of established organizations such as the UNDP, we will see important scientific contributions coming from new organizations such as the International Federation of Institutes for Advanced Study, the UN Environmental Program, and the International Institute for Applied Systems Analysis being established in Vienna. The work of such organizations should establish a better scientific basis for the type of development compatible with environmental needs and the technological change taking place in the world.

This leads me to the second level of the role science should play in directing development. If the first was helping plan global integration, the second must be working on regional and national levels to see that each country and each area develops in a way that is best for its own people and environment. This, unfortunately, is bound to be a matter of some controversy over the years. There are several reasons. A basic one lies with the very notion of development as various people conceive it. To many, development has become synonymous with industrialization, particularly with the process of industrialization through which the Western World went in the 19th century. We know today that technologically speaking much of that process can be bypassed, and also that many of the conditions that made it possible in Europe and the United States do not exist in a number of areas that are striving for industrialization today. The kind of industry that most developing nations could sustain today does not have the labor requirement that would absorb the large influx of workers being forced to leave rural areas because of changing agricultural practices. A whole new outlook toward the development process is necessary, and is, in fact, seriously being studied and discussed throughout the world.

In essence this outlook emphasizes the need to stem the influx of population to already-crowded urban areas by a combination of two related efforts. One is new rural development — building the need for rural employment based on agriculture-related needs. The other is the development of small-scale industry in outlying areas, existing villages or new towns, calling for less capital-intensive and more labor-intensive production. Much of both of these — agricultural and industrial work — would involve what is coming to be known as Intermediate (or Appropriate) Technology: relatively light machinery and equipment that allows the productive employment of more people but eliminates the backbreaking labor and drudgery historically associated with labor-intensive production. Intermediate Technology should also be of the type to take advantage of local conditions and resources. It should be able to harness the natural energy most available in the area — solar, wind, water, or geothermal power. It should be designed for simple operation, service and repair. And it should be low in cost and maintenance. The creation of such technology offers a great challenge to science and engineering today.

The reasons we have only recently begun to emphasize this kind of technology are many. Barbara Ward stated a major one when she wrote several years ago:

> "The technology which is dominant in the world today is not always appropriate to the needs of the developing countries — and this for a very simple reason. The whole weight of economic research and of investment in further research is virtually confined to the developed countries and has, for 50 years and more, taken the form of trying to find labor-saving methods of production."[3]

This, of course, is a difficult pattern to change. Viable, profitable, markets must be created, literally overnight, if this new kind of technology is to be introduced. And its introduction calls for considerable other social and economic change in a country, even if we assume that the country is willing to recognize the need for it. This in itself poses major problems. Many countries believe they can be successful in following the conventional process of industrialization and that any efforts to convince them otherwise are an infringement on, or an insult to, their national sovereignty. Some are already in the midst — and the agony — of that process, with the result that many have a rate of unemployment and underemployment ranging up to more than 30 percent. No one knows what the outcome of this will be, but it seems likely that much social upheaval and human misery will take place before it becomes clear.

In speaking of this new type of development involving the effort to maintain productive and healthy populations away from the large urban areas, to redevelop rural regions and locate small-scale industry there with the help of intermediate technology, let me emphasize that there are examples of this under way. One example of how a single item of intermediate technology can establish a linkage between agriculture and small-scale industry has been the tubewell. As one development economist has pointed out:

> "In the last decade entire towns and sections of towns have grown and developed almost entirely around the domestic manufacture of diesel engines for private tubewells, other tubewell parts, and of course the drilling installation and repair operations. The agricultural investment in private tubewells have encouraged additional investment in small-scale manufacturing, very little of which actually gets measured for national account purposes."[4]

Walter Elkan, the economist who wrote that statement, emphasized the importance of establishing in developing countries small-scale industries that have what he calls "backward and forward linkage" — that is, which stimulate related industries that supply materials and services to the central item being produced, and also create future needs. Such an industry would be building construction, for example, which calls for raw and finished materials and, when completed, stimulates the need for furnishings and household goods which might be manufactured within the country.

In the future we should see, with the help of good international cooperation in science and technology, some productive combinations of intermediate and high technology. One of these might be the use of inexpensive television receivers in remote rural areas, and small villages and towns in developing nations, to make use of the telecasting of educational and technical information from communications satellites. This combination could bring to people — who are otherwise isolated or lack the means — education for their children and themselves, information to aid them in their farming and the marketing of their products, important weather information, and general news and entertainment. India will provide a prime test for education via satellite television when in 1975 the United States makes available to her its Applications Technology Satellite. This satellite, to be positioned in an equatorial orbit, will be available to India for six hours a day for at least one year under a current agreement. India is now preparing to provide direct telecast from the satellite into receivers in 150,000 villages. Much of the success of satellite television of this type will hinge on the ability to produce and distribute very inexpensive but reliable television receivers. The problem is obvious when we realize that the people who need these sets the most may have an average annual income of about $100. Eventually, we may see a system of nuclear-powered, synchronous satellites in use at about 22,000 miles above the Earth which would allow direct telecast to battery-operated television sets anywhere on earth. At that time it would be to our great advantage to have available a television set comparable with the cheap battery-operated transistor radios so common around the world.

Up to this point, I have spoken of science and technology helping to direct international development from two rational approaches: that of planning and controlling the exchange of resources so as to narrow the gap between rich and poor, and that of encouraging a redirection of development within the developing areas to emphasize employment and dispersed growth based on different style of industrialization. But science and technology must work, as they always have, on the level of dealing directly with individual problem areas in a disciplinary and interdisciplinary manner. And it is to this level that I want to devote the remainder of my remarks, emphasizing some key problems and some new directions in dealing with them.

Foremost among the problems facing man, and particularly the developing nations, are still those associated with food and health. This is true in spite of the fact that we have made enormous strides in these areas during the past decade, thanks to such men as Norman Borlaug, Carlos Chagas, and some of those participating in this Symposium.

The Green Revolution must be pursued to its fullest potential throughout the world, and the problems that its initial success has highlighted must receive the fullest attention of science and technology. One of the most important of these has to do with man's use and control of water. In many parts of the world, including those where the Green Revolution has been most successful, food production is highly dependent on weather patterns that have been erratic from year to year; accordingly, we find ourselves fluctuating from years of surplus to years of famine. Periods of droughts or floods seem to be the rule rather than the exception in many areas. While we certainly cannot control the world's weather, we must at least develop a better understanding of it — enough to avoid having the fate of a nation hinge on the coming of a monsoon that never materializes. We also need to intensify our understanding and control of the water cycle throughout the world, to know more about underground water and how better to use it, to increase our knowledge and ability in flood control, to further our efforts in water recycling and desalting technology, and to pursue the possibilities of weather modification.

As recently estimated, our global water demands will increase by about 100 billion cubic meters per year just to keep up with the population increase. We will therefore have to make great efforts in the years ahead in water conservation and technologies for tapping new sources for fresh water.[5]

Another problem area associated with food that demands the best efforts of science around the world is the destruction of food by pests. Up to 50 percent of the world's food that is produced is destroyed before it can be consumed by human beings. It is devoured and spoiled by insects and rodents or is attacked by micro-organisms. It has been estimated that in Africa alone the amount of grain lost annually could feed 55 million people. We clearly need better means of pest control and food preservation and storage throughout the world. There is much promising work being conducted to meet these needs. Some examples are the work of the International Rice Research Institute to breed rice varieties resistant to certain pest, a multi-university research program on integrated methods of pest management being coordinated in the United States by the International Center for Biological Control at the University of California, Berkeley, and programs of the UN's Food and Agricultural Organization and the International Atomic Energy Agency. Most recently a new international organization, the Group for Assistance on Storage of Grain in Africa, has been formed to devote itself to the prevention of post-harvest crop loses on that continent. Its members are conducting research and trying to improve traditional methods of storage to reduce the huge destruction of food caused by rodents.

The scientific community around the world must continue to increase its investigations of the seas, particularly as they relate to marine resources and food supply. A better knowledge of such phenomena as coastal upwelling and other natural conditions that influence the growth and movement of fish populations may help us to regulate our exploitation of certain species which seem to be declining just at a time when we are turning increasingly to the sea as a greater source of protein. The fact that international competition in fishing has intensified, to the point of open conflict over off-shore territorial limits in some areas, indicates even more strongly the need for a new era of cooperation in our understanding and uses of the sea. Much has been said over the years about "farming the seas" but we still remain primarily sea hunters, even to the extent of sometimes accusing each other of poaching. Next year we will have a major United Nations Conference on the Law of the Sea which may settle some of the legal questions concerning our expanding use of the sea's resources. But the time is now overdue when we must develop a true science and technology of the sea, emphasizing aquaculture on oceans and inland. This would be of particular benefit to many of the developing nations, some of which depend heavily on fishing as a source of food and revenue. It has even been suggested that international agreements be made that would favor the developing nations as the world's fisheries because the fishing industry is labor-intensive and can keep large numbers of people employed.

Closely related to the food problems of the developing world is the matter of health. It is this matter that has always been a major focus of the world's science community and continues to be, with new problems taking hold as older ones are solved. In my visits around the world to more than 60 countries, I have always been impressed by the dedication of the medical profession and those engaged in health and biological research, particularly in the developing countries where they seem to be working against such overwhelming odds.

Many of the health problems throughout the developing nations will improve slowly as conditions related to poverty — poor nutrition, sanitary conditions, and education — are improved. Great strides have been made in reducing such major scourges as malaria, tuberculosis, yaws and the major communicable diseases. Credit for much of this success is due to the World Health Organization (WHO), which coordinates an extensive network of medical science activities around the world on a pitifully small budget. As Martin Kaplan, Director of the Office of Science and Technology of WHO, has pointed out, WHO's activities "are carried out with a total effective working budget that the sanitation services of many large cities in the United States would consider grossly inadequate, some 100 million dollars annually."[6] Over a period of the last 15 years WHO's research has made a significant contribution to medical science and technology in, among other things, evaluating the effectiveness of various vaccines and serums, in providing insights into malnutrition, in conducting studies on the biology of many insect vectors and their resistance to insecticides, and in developing nearly 300 biological standards and working preparations to provide baseline references for all countries. Dr. Kaplan sees the unfilled priority areas of WHO's work in science and technology to lie in such matters as: the development of a worldwide health information network, an increase in fundamental and applied research in parasitic diseases, a better understanding of the biology of human reproduction, more knowledge of the effects of toxic chemicals (especially at the molecular level), and improved technological aids for diagnostic and survey procedures. Such aids would be invaluable in mass campaigns against communicable diseases, for population studies in genetics, nutrition and metabolic disorders. Less expensive technological procedures are also necessary for dealing with environmental and sanitation problems.

In outlining these priority areas of research and development, Dr. Kaplan stressed two major requirements for their success; one was increased funds, the other was greater participation of leading scientists and advanced laboratories. It is on this matter of participation that I want to conclude my remarks. And rather than focus on the role of national and international institutions with which you are familiar, I would like to discuss briefly a new organization whose work we hope will increase the participation of scientists in meeting the needs of developing nations. I say "we" because of my personal involvement in its initial planning and formation together with Roger Revelle, Robert Marshak, Maurice Levy, Sven Brohult, and many others. I shall also describe briefly another proposal I have made for promoting the contributions of scientists and technologists all over the world to international development.

The organization to which I refer is the recently formed International Foundation for Science (IFS), an organization designed to "support and encourage scientific research in developing countries by seeking out and supporting promising young scientists." IFS was formed by a consortium of national scientific academies, research councils, and equivalent organizations in developed and developing countries. The consortium consists, at present, of 21 members, almost equally divided between developing and developed nations.

A major purpose of IFS is to fill a large gap that remains between the efforts of national and intergovernmental organizations in assisting the growth of science and technology in the developing countries through "institution building" and providing training and advice by outside experts, and the need to support research by individuals and small groups in the universities and research institutions of the poor countries. IFS will focus its efforts on the latter, particularly on supporting the work of younger faculty members and their graduate students in the natural and social sciences. The foundation hopes to ameliorate the plight of young scientists in the developing countries, who now have little opportunity to develop their own potential, by providing small grants that would enable them to undertake specific research projects, which they have proposed themselves, in their home countries. The emphasis of the grants would be on allowing the scientists to maintain the continuity of their research over a period of years, and on providing them with technicians, instruments and scientific literature.

Among the criteria for the grants would be the relevance of the proposed research to the needs of the country. The foundation hopes to encourage cooperation among scientists in the poor countries by seeking proposals in the same field from different countries.

The ideas that went into the formation of IFS were tested over the past few years by contacts — through correspondence and personal visits —

between members and associates of the foundation's Interim Board and officers of national scientific bodies in Asia, Africa and Latin America. These contacts verified the great need for the type of support IFS will offer.

When the Interim Board of IFS met in Stockholm last February, several kinds of research that the foundation might support were discussed. Representative of these were: comparative studies of soil fungi that seem to be essential to the successful introduction of new tree crops in areas where they had not previously grown; the microbiology of nitrogen fixation for rice, soy beans and other crops; the chemistry of natural products, including those which may be of pharmaceutical value; geophysical methods of prospecting for and appraising underground water; the biology of reproduction of fresh water and brackish water fish species that are suitable for fish culture; and studies of social and demographic changes related to the introduction of new agricultural technologies.

The importance of research projects such as these to development is obvious. Furthermore, they have several common characteristics that make them desirable for IFS support. They represent relatively inexpensive research that could be undertaken by individuals or small groups in universities and existing research institutions. Their results contain foreseeable applications. There are competent individual scientists in the developing countries eager to do such research. The training of graduate students could be an integral part of this research. And similar projects in other countries could provide opportunities for cooperative work.

The International Foundation for Science has been launched in the firm belief that there is a substantial reservoir of scientific talent in the developing countries that must be supported and encouraged to grow. Each country should have the opportunity to develop its own cadre of competent young scientists who have, in addition to their scientific and technical knowledge, a firsthand understanding of the problems of their country and an ability to make the most of the outside expertise provided to them by the regional and international scientific organizations and through the scientific exchange programs. We look forward now to seeing the ideas behind the IFS take root, to being able to increase its support and the scope of its work.

In devoting some of my remarks to this new organization for supporting research in the developing countries, I have not meant to play down the important role of the major national and international scientific organizations, several of which I have mentioned and all of which deserve mention and praise. They have done and continue to do outstanding work.

But there has been no widespread activity among individual scientists and engineers throughout the world directed toward the solution of practical contemporary problems and the promotion of human welfare on an international scale. It was with the thought of moving toward the mobilization of such action that I proposed[7] at the recent conference "Science and Man in the Americas" the formation of a true *American* Association for the Advancement of Science to be followed in due course by a worldwide joining together of national associations of science in an International Association for the Advancement of Science (or perhaps an International Association for the Advancement of Science and Technology).

The many Associations for the Advancement of Science in the countries throughout the world are well suited for action directed toward the application of science and technology in the promotion of human welfare. They are in general noncommercial, nonpolitical associations of scientists and engineers and of other people interested in science and engineering. They are nationwide in scope and include all of the natural and social sciences. In each case the membership is in general open to all without restriction as to scientific qualifications and thus can be referred to as the "Lower House of Science." Also, typically, each such national association has organized affiliations with many other associations and societies in its country.

Thus these associations are, in varying degrees, representative of organized science in the countries in which they operate and can act with the great authority inherent in their broad membership. There are at present some 40 associations of this type with a wide variation in their names. I believe the effectiveness of their member scientists in promoting the contributions of international science and technology to the forging of an organic mankind will be greatly enhanced through some worldwide unification of the present individual efforts.

The work for all of us in science and technology has only begun, in terms of what we might — and what we must — accomplish in world development. The forces that our past accomplishments have set in motion must be controlled and given new direction. The knowledge uncovered by our past efforts raises new questions that demand answers. In dealing with development at this state we face what may be man's greatest task — essentially taking over where natural evolution has left off in determining the destiny of our species, and perhaps that of the entire planet. Man will be called upon for new wisdom and leadership in meeting this awesome challenge. To not rise to this challenge would be not only to fail that portion of mankind seeking to lift itself from deprivation and despair, but to deprive the world of its first — and perhaps only — chance to unite in a global civilization, one in which all can live in dignity and harmony. I believe we will succeed in our efforts to help build that kind of world.

References

1. Glenn T. Seaborg, "The Positive Power of Science," *Environment and Society in Transition* (Proceedings of the International Joint Conference of the American Geographical Society and the American Division of the World Academy of Art and Science, April 27–May 2, 1970), *Annals* of the New York Academy of Sciences, Vol. 184, 1972, pp. 682–687, see also pp. 266–274 this volume.
2. *U.S. Foreign Policy, 1972: A Report of the Secretary of State*. Department of State Publication 8699, General Foreign Policy Series 274 (Superintendent of Documents, Washington, April 1973), p. 42.
3. Barbara Ward, "The Decade of Development," *Reshaping the World Economy*, ed. John A. Pincus (Prentice-Hall, Englewood Cliffs, 1968), p. 27.
4. Walter Elkan, *Introduction to Development Economics* (Penguin Books, Harmonesworth, England, 1973), p. 115.
5. Jose P. Peixoto and M. Ali Kettoni, "The Control of the Water Cycle," *Scientific American*, Vol. 228, No. 4 (April 1973), p. 46.
6. Martin Kaplan, "Science's Role in the World Health Organization," *Science*, Vol. 180, No. 4090 (8 June 1973), p. 1028.
7. Glenn T. Seaborg, "Science, Technology, and Development: A New World Outlook," *Science*, Vol. 181, No. 4094 (6 July 1973), pp. 13–19.

Doing More with Less

Commencement address at Kent State University,
Kent, Ohio, August 30, 1975.

I am privileged to have this opportunity to speak to the graduating class of Kent State University.

It is traditional in a commencement address to try to predict to a graduating class something about the future that you will face, your role in it and the needs you can fill, and the problems that will confront you.

And the speaker often handles these topics in very broad, philosophical ways. I would like today, however, to focus on a specific matter that will profoundly affect our quality of life individually and as a society.

You are familiar with the energy crisis that we face today and for the indefinite future. I believe that the energy crisis may be the forerunner of a number of crises in the supply of natural resources — crises that only can be averted by sound planning. As a nation with 6% of the world's population we have increased our demands to the point where today the United States is a net importer of minerals, consuming about 30% of the world's production. For some — for example, chromium, cobalt, nickel, tin, manganese, tungsten, platinum and platinum metals — we import all or nearly all of our supply. We are dependent upon substantial imports for a disturbing total of some 30 or 40 minerals.

Rapidly as our consumption is rising, it is rising even faster in many other parts of the world. As economic growth and industrialization accelerate over much of the world, the competition for mineral supplies will increase and the developing countries will exert more control over the disposition of their mineral resources. This situation has the seeds for crises if we plan no better here than we have in the energy area.

Science and technology can do much to postpone our day of reckoning. Chemists and geochemists, working with geologists, can help us understand the origin of minerals and hence can lead us to the discovery of concealed deposits. Chemical engineers can develop methods for use of lower grade minerals, improve the efficiency of mineral recovery and reduce the waste of valuable by-product materials. And with the help of science we can develop more substitute alloys and materials that can shift the burden from substances in short supply to those available in greater abundance.

But the world's mineral resources are finite in magnitude. We will simply have to slow down our increase of consumption, learn to conserve, to do more with less. And we will have to learn how to recycle our supply.

We will eventually — in a few decades — have to create a "recycle society," a society in which virtually all materials are reused indefinitely and our virgin resources become primarily the "make-up" materials to account for the amounts lost in use and production and needed for new production to take care of any growth that would improve the quality of life.

In such a society the present materials situation is literally reversed; all waste and scrap — what are now called "secondary materials" — become our major resources, and our natural, untapped resources become our back-up supplies. This eventually must be the industrial philosophy of a stabilized society and the one toward which we must work. It first involves a shift in industry to the design and production of consumer goods that are essentially non-obsolescent. This means that products will be built to be more durable, easily reparable with standardized replaceable parts; new parts would be easily inserted, and goods could be repaired with simple basic tools.

In the recycle society, all products and parts might be labeled in such a way that their use, origin and material content could be readily identified; and they might have a regulated trade-in value. Many items of furniture, housewares, appliances and tools — in addition to low-maintenance qualities — will be multi-functional, modular and designed for easy assembly and breakdown to be readily moved and set up in a different location when necessary. Their design and construction will also allow for their reassembly and redesign into essentially new products when their owners have different uses for them or seek a change. When a consumer (or, more applicable, a "user") wishes to replace an item or trade for something better or different, he can return the old item for the standard trade-in price.

The industrial processing aspect of the recycle society may be the easiest to achieve as there are already several major companies that claim the ability to recycle totally the waste products of selected manufacturing plants without any economic penalty. We also should see extensive recycling of organic material from agriculture and forest industries. Animal waste will find many uses as fertilizer, fuel and feed. Protein will be grown on petroleum waste and extracted from otherwise inedible plants and agricultural products.

To make the transition to an economic recycle society, much will be required in the way of new legislation, regulation, tax incentives and other measures that will make the use of secondary materials more economical than that of virgin resources. The opposite situation prevails today.

In addition to the setting up of new methods of marketing and management required to operate a recycle society, a long-term consumer education program will be needed. An entire society reusing and recycling almost all of its possessions, especially after an extended era of conspicuous consumption and waste, will take a great deal of pride in a lifestyle that is extremely creative and varied and based on a new degree of human ingenuity and innovation. The Recycle Society of the 1990's will be better off than the Affluent Society of the 1970's.

Several types of assets will accrue from the movement toward the kind of society I have been describing. One lies in the fact that it will be less energy-intensive. For example, recycled steel requires 75% less energy than steel made from iron ore; 70% less energy is used in recycling paper than in using virgin pulp; 12 times as much energy is needed to produce primary aluminum as to recover aluminum scrap. A society set up to reuse most of its resources systematically and habitually could effect substantial energy savings.

Perhaps even more important would be the reduction in the environmental impact of a recycle society. This would be true for several reasons. Such a society should develop an environmental and conservationist ethic due to the scarcity of resources as well as the new value placed on land, water and air. We can see this ethic already in the making today. There is some fear that because of our energy crisis we will severely compromise this ethic or even abandon it. I do not think this will be the case. Rather, I believe we will be making substantial sacrifices in the years ahead to change our lifestyle in order to match our economic and environmental needs.

Our industrial and power systems will be much more efficient users of energy; hence they will not be rejecting as great a percentage of waste heat to the environment. In fact, most systems probably will be planned and designed to make maximum use of waste heat, using it for space heating or possibly agriculture or aquaculture. Waste water and sewage water also will be recycled in industrial and perhaps even municipal water systems. What water is returned to the environment will be as clean, if not cleaner, than it was when it entered the manmade system.

It would be foolish to believe that even with a recycle society we will not still be drawing on a substantial amount of new resources. Even with population growth leveling off and economic growth cooling off, we can expect substantial growing demands for new materials and resources well into the next century. What this means is that by that time we will have to develop a new level of ingenuity in materials substitution and in what Buckminster Fuller calls "ephemeralization" — the process of doing more with less. Fuller uses as an example of this the Telstar satellite which, while weighing only one-tenth of a ton, outperforms 75,000 tons of transatlantic cable. Communications, of course, have offered the best examples of this "more with less" phenomenon, as in electronics we have seen the size of basic devices reduced by a factor of ten roughly every five years. Today, a single chip of silicon a tenth of an inch square may hold microscopic units that perform the functions of as many as 1,000 separate electronic components. Furthermore, silicon is one of the most abundant substances in the earth's crust.

Substitutions not only in materials but in functions — the way we conduct our business and personal affairs — may vastly alter our lives, effecting many savings in energy and time, and therefore affecting how we otherwise spend our energy and time. For example, communications as a substitution for transportation can effect such savings to a great extent. Shopping by telephone and having good home delivery is a very old example of this, although largely out of style today. But if one could survey local supermarkets and department stores via videophone and a computer to do some quick comparative shopping and then have the selections delivered, one would have more time, money and energy (personal and automotive) left for other things. Another aspect of this type of shopping involves a considerable saving in space. A simple warehouse with a small fleet of trucks could service thousands of customers, eliminating the paving-over of large parking areas and the operation of an elaborate market.

A society that exercises this option of using communication in place of transportation in many of its activities — whether in shopping, business or educational activities — can conserve many resources. But it must be one that has learned to substitute other activities for the social and entertainment value that we have come to find in our more random way of life.

It is possible to speculate that we may find a situation in which our working world will be served mainly by communication and public transportation, and the savings from this will allow us to use private transportation in a limited way for recreation and vacations.

Much of what I am projecting here is likely to take place. In addition, I see some of the following as changes that may be initiated or well under way during the coming decades:

(1) A large shift toward clustered, attached and high-rise housing surrounded by community-owned open lands. This planned housing would rise as "new cities" and as neighborhood communities within larger urban areas.
(2) Increased use of underground space made possible by advances in excavation technology: underground shopping centers, warehouses, recreation and entertainment complexes, rapid mass transit lines, and power and communications cables.
(3) Offshore power plants with extra-high voltage, superconducting transmission cables carrying electricity greater distances inland. In the next century they might include both nuclear- and solar-powered electric and hydrogen-producing plants.
(4) Integrated industrial complexes planned to concentrate energy sources, materials and manufacturing in single locations to reduce long shipments of fuel and material resources, make more efficient use of waste heat and materials, confine and control environmental impact.

My concluding thoughts about the prospective changes in our lifestyle center on global cooperation. In the midst of an energy crisis aggravated by the withholding of oil as a political weapon, this does not seem to be a popular topic. In our immediate reaction to strive for energy self-sufficiency, we should not over-react, not delude ourselves into believing that self-sufficiency in energy and other resources is the total solution to national security and well-being. In fact, self-sufficiency in mineral resources is probably an impossible goal. And much of the developing world faces a food problem that can be solved only be means of international cooperation.

We live in a highly interdependent world, one in which hard bargaining will continue but in which accommodation grows increasingly important. My travels over the world in the past dozen years to more than 60 countries — and recently to the People's Republic of China — lead me to believe that all nations of the world need each other, that all have something to offer and all could benefit by a greater exchange of human and material resources, of knowledge and goods.

Over the next twenty years we will have to make enormous strides — together — in controlling population, increasing food production, managing our environment, investigating and controlling the resources of the seas, conducting global research, developing methods to reduce the human impact of natural disasters, and generally uplifting the economic conditions of a large number of the world's peoples. There are no alternatives to these measures — except a tremendous increase in human misery that ultimately will affect all the world's peoples.

I haven't meant to exhaust you with thoughts of the work to be done even before we've received today's benediction. But I hope that I have succeeded in conveying to you some thoughts about the future that you face. I have tried to do this in a realistic manner — for the problems and difficulties are indeed complex and real.

I hope you also have been able to absorb the posture of optimism which I also believe is realistic. These problems can be solved: we *can* find ways sufficiently to do more with less. It will take determination and planning by you and by the other young people who are now entering the mainstream of society and are ready to show the rest of us how to do it.

My congratulations to you in the graduating class on your success in completing your work at Kent State. At this commencement of the new chapter in your life, accept my warm wishes for your continued success and for the triumph of your own imagination and courage.

Knowledge and Survival

Address delivered at the 500th Anniversary Celebration of Uppsala University, Uppsala, Sweden, September 29, 1977.

I am honored by your invitation to participate in this celebration of the 500th anniversary of Uppsala University, one of the world's oldest and most distinguished institutions of higher education. My remarks today, I believe, are in keeping with the spirit of this occasion. The message is a simple one, but one I believe is already being universally recognized and acted upon. It is that the extension and dissemination of knowledge, perhaps now more than at any time in history, hold the key to human survival.

The growth and spread of knowledge have long been a major aspect of our social evolution, largely determining humanity's place in the scheme of things. Now we are at a stage when new and pressing demands are being placed upon us to expand that knowledge and to use it to make wise changes that could affect the lives of generations to come. This will require, among other things, a revitalization of our academic institutions and of the spirit of learning and research. It may also require changes in their operations and the ways in which they are supported.

Some see this as particularly difficult at this time because of a prevailing pessimism in many quarters. That pessimism stems, in part, from the great demands now being made on our environment, our resources and our social and economic institutions as a result of our current style of growth and pressures to meet the needs and expectations of a major portion of mankind.

Those who subscribe to this pessimism are suspicious of all growth. They see only disaster ahead for the human race because of the spectre of uncontrolled growth and its various physical, economic and social ramifications. But others, myself included, see the possibility of controlling that growth through expanded knowledge, and, perhaps more important, of using new knowledge as the principal means of making what growth is necessary, acceptable and advantageous to the number of people who will inevitably inhabit the earth over the next few decades.

Economists are particularly cognizant of the fact that it is human creativity and innovativeness that deal effectively with limits and the pursuit of new knowledge is central to such a process. As Kenneth Boulding has stated: "... human welfare is a race between growing social potential — properly disseminated and organized knowledge — and the constant tendency for things to be used up." Progress today, more than ever, involves us in such a race — in the pursuit of ways to do more with less.

Knowledge today can also be used to anticipate directions the future might take and help us to avoid disastrous circumstances. With theoretical knowledge and the use of computer simulation, for example, we can model various futures and provide ourselves with options as to which direction we might best go.

Today, a great deal of our knowledge-gathering and use — what is usually referred to as research and development — is wasted, just as we waste so much of our natural resources. I am thinking, in part, of the $30,000 million being spent annually around the world on military research and development. While many view that as necessary to their national security, those resources and the human intellect and creativity they employ (a third more than goes into civil R&D) could have a major impact on world security — on the fulfillment of human needs — if released to the pursuit of peaceful progress.

Whether or not we can make a shift of that kind to any degree, it will be essential that we expand our knowledge in other areas to meet a number of major problems ahead. Let me take the next few moments to elaborate briefly on some of those challenges and how they might be met.

Aside from the ever-present need to avoid armed conflict, the world faces three broad challenges that could reach crisis proportions. They are:

- the need to match global natural resources — the use of energy, materials, food and water — to the requirements of a growing humanity;
- the need to understand more fully the environment and ecology of the planet and to conduct our human activities both in harmony with them and in ways conducive to our survival and satisfaction;

- and, the need to fulfill as well as possible the human potential — the desire of individuals to be useful and rewarded members of their society.

These are the three broad criteria for a world of peace, economic and social stability, and general health. They are so closely related that defects in any are felt deeply in the others.

Let me explore briefly some aspects of these challenges, particularly in terms of the requirements for expanded knowledge and its rapid application.

In recent years, much has been discussed and debated concerning the possible limitations of the earth's resources available to humanity. While there can be no argument that the growth of human population, and its demands on resources, cannot grow indefinitely (or even at its current rate for very long without disastrous consequences), the resource limits we face over the next few decades and well into the 21st century are due more to our lack of knowledge and political and social organization than to any real physical shortages.

For example, the world abounds in potential energy resources — from untapped fossil fuels (particularly coal and shale oil deposits), from nuclear fuels (both uranium and thorium), and from the sun (through direct solar energy, bioconversion and the heat stored in the oceans). But, we are held back from the fullest utilization of these resources by the lack of various bits and pieces of knowledge necessary to develop and implement energy systems that would use these resources efficiently, economically and in environmentally acceptable ways. And, in some cases, where the scientific and technological knowledge is available, the state of the public knowledge is inadequate for the political unanimity that would allow its timely application.

We see these knowledge shortcomings in such areas as the large-scale burning of coal: to what extent will its release of carbon dioxide to the atmosphere affect global temperatures? What effect will this have on climate, on plant life, and consequently on human life? We need to know much more about such environmental matters, as well as the technological matters involved in the safer mining of coal and ways to use it more efficiently in various forms.

In the case of nuclear energy, a combination of scientific, technological and sociological shortcomings inhibit the fullest use of this enormous resource. There are a number of ways that the nuclear fuels can be used

— many alternative fuel cycles — and some countries have come a long way in the development of a few of these. But, there remain a number of unanswered questions regarding the extent of uranium resources, the choice of reactor technologies, the recycling of fuels and the management of radioactive waste. If these can be resolved soon, and widespread public acceptance gained, along with the technological solutions, nuclear energy could make a major contribution to satisfying global energy needs, particularly in the industrial countries.

The use of solar energy, which should play a major role in meeting global energy demands in the next century, also poses a great many interesting questions that must be answered. We can harness the energy of the sun in such a variety of ways that a multitude of technologies, both large and small, need to be investigated and developed. These range from relatively simple systems for solar space and water heating, cooking and drying and industrial processing, to the bioconversion to fuels (liquid and gas) of vegetation and organic waste, and to the production of electricity by thermal–electric methods and photovoltaic systems. The large-scale production of hydrogen via photo-conversion of water is also an important possibility.

The knowledge to do almost all of this is near at hand or attainable in the not-too-distant future. The bigger question might be how we develop and disseminate these technologies so that they best fill the economic needs of people in different areas of the world. There is currently taking place among those concerned with global energy policy a debate over the principal approach that the world must take to satisfy its energy needs. One side believes that we must continue to develop large-scale energy technologies and depend to a great extent on central power systems. These would include coal and nuclear systems, and fusion power, when and if it becomes feasible. It would not preclude solar energy in any forms that become economical. The other side believes that we must decentralize our energy systems, depending primarily on smaller-scale technologies under local and even personal control. Their emphasis would be on the use of renewable resources and a more labor-intensive, less capital-intensive, way of life, with the thought also of avoiding what they consider a greater vulnerability to the failure of large systems.

Both sides present some strong and convincing points. But, as in the case of the "Limits to Growth" debate of a few years ago, there is a weakening of idealistic positions. What will probably happen in the real world is that, after much trial and error, we will see different areas of the world adopting the energy systems that are best suited to their needs and that they are able to develop economically. In all these cases, the rate at which we generate and apply new knowledge will determine the kind of progress we make, whether we choose large systems or the so-called "intermediate" or "appropriate" technologies.

Agriculture and the production and distribution of food constitute another area where we are in a race between the attainment and application of knowledge and possible catastrophe. At present, we live in a world where it is estimated there are 450 to 1,000 million malnourished people. Most of these people live in the developing countries, which in the next two decades will add approximately another 1.8 billion people. To meet their food needs, as well as those of the rest of the world, we face the monumental task of at least doubling, and perhaps tripling, global agricultural production within a generation. And, most of this increased production will have to take place in the developing areas, on land now considered marginal for agricultural use, or other uncultivated regions, some in arid lands and much in the humid tropics.

But, as difficult as this challenge seems, it is far from impossible, and there are some reasons for optimism based on current thinking and new approaches in agriculture. What will be required is an enormous and concerted global effort of research and development, concentrating on such subjects as genetic improvement of plants and genetic variety to meet different weather, soil and pest conditions; research in photosynthesis to improve plant production efficiency and variety; methods of biological nitrogen fixation to reduce synthetic chemical fertilizer requirements and their associated energy and environmental impact; and, improved knowledge of the management of tropical soils, which, if handled properly, could produce high-yield crops on some 700 million acres of presently uncultivated land in the humid areas of Africa and South America. Great efforts will also have to be made in land management and farm production systems, in the control of soil erosion, in the management of water resources — to cultivate arid and semi-arid areas and improve drainage and avoid logging and salinity problems in others — and in food distribution throughout the world. Much can also be done to reduce the huge losses of food, before and after harvest, taken by bad weather, insects, birds, rodents and the spoilage resulting from inadequate means of preservation during transport, storage and marketing. In some areas of the world, the total losses due to these causes run higher than 50 percent of all the food produced.

Difficult as it may seem, we can make enormous progress in all these areas, and much work is already underway on them. Success in any number of them will have a significant impact: the additional annual nitrogen that might be fixed biologically if we could achieve this process in cereal crops at an efficiency of just 30 percent of that of soybeans would be the equivalent of 36 million tons — more than five times the total amount of nitrogen fertilizer now being used in the developing countries. With intensive research in photosynthesis, we might produce plants that offer 50 to 100 percent increases in agricultural yields. In the tropics, success in dealing with soil problems could result in annual production 150 to 200 percent of that achieved per hectare on temperate zone soils.

It is essential that all this work be done cooperatively, between the advanced and developing countries, and that the latter achieve self-sufficiency in food production. The current consensus of belief is that in addition to the research results, an important requirement will be that the developing countries maintain highly productive and socially acceptable agricultural systems and other rural programs that are more labor- than capital-intensive. The importance of this lies not only in the need for adequate, economic food production, but to keep large numbers of people from moving into the cities and creating problems of overcrowding and unemployment there. Some hope that all of this can be done is reflected in the fact that the People's Republic of China, with one-quarter of the world's population, essentially supports herself agriculturally on 11 percent of her land.

Because of its consequences affecting global agriculture, as well as a great many other resources, we need to increase our knowledge of climate. There is new stimulation of interest in this subject because recent research seems to support the idea that we may be entering an era of climatic change that could contain significant weather variability. Such variability brings with its fluctuations in temperature and precipitation — freezing, shifts in wind and rain patterns, and periods of drought — that could play havoc with food production, energy needs and other economic conditions. Some of these weather-related problems have already been felt in my country, and in other areas of the world. We know something of their causes, but a great deal more must be learned if we are to have the ability to forecast climate and weather changes with enough accuracy to develop the best protective measures. For example, in addition to the need to change soil management techniques to meet drought conditions, and to devise better irrigation methods, plant genetic research is needed to develop more drought-resistant crops, and those with different growing seasons to meet changing temperature patterns.

Climate and weather changes make new demands on our management of livestock, in the ways they are bred, fed, confined and transported. Changes in weather also affect the life cycles and movement of pests, including shifts in the swarms of locust in Africa and the migratory grasshopper in North America. We could avoid great damage by these and other pests if we could predict their movement more accurately and take the proper defensive measures. All these matters of new and improved knowledge will bear on our future and will affect the lives of billions of people.

Our interest in climate is but one aspect of a new sensitivity to environmental conditions the world has experienced over the past decade or so. The pressures of population, of economic growth, of industrial and agricultural expansion — all with their increased demands on resources and their greater impact on the environment — have demonstrated that we must better understand and control the environmental effects of our activities, lest we seriously degrade or even destroy the life support systems of this planet. The advances of science and technology, with their instruments of great and growing sensitivity, with their ability to survey large areas of the earth, and with their capacity for assembling and analyzing enormous quantities of data, are giving us fair warning that we must change many of our ways. We will have to continue to improve and use these technologies, such as the earth-orbiting, remote-sensing satellite systems, and we will have to develop the knowledge and the means to correct the problems they indicate, as well as follow-up on the opportunities they offer.

An improved sensitivity to the role of environment has made us aware not only of its broad ecological relation but of that directly related to human health. It is interesting, and important, to note, as a result of epidemiologic studies, how the leading causes of death and disease in the advanced nations have, over the past several decades, shifted from the communicable diseases, such as tuberculosis, cholera, diphtheria, dysentery and other bacterial infections, to those based on different, man-made, environmental change. While many of the developing areas of the world still suffer from bacterial and parasitic disease, the leading causes of death among us in the Western World are now heart attacks, arteriosclerosis, cancer, diabetes and accidents, all of which are related to different environmental factors. Particularly alarming are the recent indications that 60 to 90 percent of all human cancer has an environmental cause. This gives us good reason to learn more about, and to control better, the chemicals we produce and release to the environment. In controlling our other degenerative and man-made diseases, we obviously need to learn a great deal more about our nutrition and our work and social habits, which both directly and indirectly have an important bearing on these.

In the developing areas, malnutrition and parasite-borne disease represent a related, and the greatest, threat to human health. We must learn the best ways to help the people of those regions to help themselves combat those causes of debilitation and early death. This is another case for knowledge for survival.

In conclusion, let me add a few brief comments regarding what may be among our most important reasons for expanding and successfully applying our new knowledge. We need to find new and improved ways of increasing human productivity and economic development — both from the standpoint of providing the physical necessities of life, and of providing more human beings with employment, with a livelihood and a useful, satisfying role in their society. As was stated at the last Bellagio Conference, "None of our hopes for mankind can be realized without continued economic development ... In a capital-short world, the most powerful sources of productivity increases are new technologies, and the effective engagement of well-trained people. Research, education and engineering are major sources of these productivity-enhancing capabilities."

We need to generate the human innovations that will permit us to make better use of all our resources, to produce more of the kinds of goods and services that fill the greatest needs of the greatest number of people. And, we must do this in ways that will provide employment for the greatest number. The latter challenge cannot be overemphasized. Unemployment, particularly among our youth, is high today, and presents a troublesome problem throughout Europe and the United States. In the developing world, it presents an even greater challenge of providing additional, productive employment opportunities for 1,000 million people by the end of the century — for 700 million new entrants to the labor force, as well as for the 300 million who are presently unemployed or inadequately employed.

Much of what I have related today appears overwhelming to many people. It drives some to pessimism and despair. It makes other doubt the future of mankind, unless we retreat to simpler lives, and even to the ways of our ancestors. What these people fail to realize is that we cannot go back to those ways and those days. Furthermore, for all our difficulties, life today is far better for more people, and the possibilities for the future can be brighter than ever, if we develop not only new knowledge, but a greater faith and confidence in the human mind and spirit. The development of that spirit, of that faith and confidence, as well as new knowledge, is a major responsibility of our universities throughout the world. It will do us little good to bring forth new ideas and technologies if the world suffers from a general failure of nerve to use them, to venture forth to new frontiers, to accept some risk to achieve greater gain.

As we celebrate the 500th anniversary of this great university, we should reflect on the five centuries of human history that have taken place since its founding and realize the great distance humanity has traveled. We have come this far because there were those who dared to defy human ignorance, and because there were others who had a vision of a better world. Together, they taught us and led us to new ways and new opportunities. Such ways and opportunities still exist, if we are willing to work for them. If we do, we will not only survive, we may find ourselves entering a new and great age for all of humanity.

A Second Visit to the People's Republic of China: Some Comparisons

I participated in a three-and-a-half week visit (May 17 to June 11, 1978) to the People's Republic of China, during which I served as chairman of a delegation of chemists and chemical engineers. The twelve-member delegation included eight chemists covering a wide range of specialties, two chemical engineers, a China specialist and a staff member of the United States Committee on Scholarly Communication with the People's Republic of China — the organization which sponsored our visit. Members of the delegation were: John D. Baldeschwieler, California Institute of Technology; Jacob Bigeleisen, State University of New York at Stony Brook; Ronald C. Breslow, Columbia University; James A. Ibers, Northwestern University; Thurston E. Larson, Illinois State Water Survey, Urbana, Illinois; Yuan Tseh Lee, University of California, Berkeley; Alan Schriesheim, Exxon Research and Engineering Company, Linden, New Jersey; Richard S. Stein, University of Massachusetts; Richard P. Suttmeier, Hamilton College, Clinton, New York; James Wei, Massachusetts Institute of Technology; Robert B. Geyer, Committee on Scholarly Communication with the People's Republic of China and me.

A group of chemists from the People's Republic of China visited the United States a year ago (April–May, 1977) and we had the opportunity to see each of the ten chemists from that delegation during our travels in China. Members of that delegation were: T'ang Ao-ch'ing (head of the delegation), Kirin University; Wu Yueh, Kirin Institute of Applied

---

Reprinted with permission from *THE HEXAGON of Alpha Chi Sigma*, Vol. LXIX(1), pp. 3–6, Autumn 1978.

Members of Pure & Applied Chemistry delegation, Hsi'ling Guest House, Hangchow, 4 June 1978. L to R: John Baldeschweiler, James Ibers, Jacob Bigeleisen, Alan Schriesheim, James Wei, Richard Suttmeir, Ronald Breslow, Seaborg, Yuan T. Lee, Robert Geyer, Thurston Larson, and Richard Stein. Reproduced by courtesy of Jacob Bigeleisen.

Members of chemistry delegation from the People's Republic of China in Berkeley, California, 26–27 May 1977. From L to R: Su Feng-lin, Chou En-lo, Wang Erh-k'ang, Hsia Tsung-hsiang, Wu Yueh, T'ang Ao-ch'ing, Huang Wei-yuan, Chang Lo-feng, Chiang Ping-nan, Yin Yuan-ken, and Hsu Mao. Reproduced by courtesy of the Lawrence Berkeley Laboratory.

Chemistry; Yin Yuan-ken, Lanchow Institute of Chemical Physics; Chiang Ping-nan, Talien Institute of Chemical Physics; Huang Wei-yuan, Shanghai Institute of Organic Chemistry; Hsu Mao, Institute of Chemistry, Peking; Chou En-lo, Kirin Institute of Applied Chemistry; Wang Ehr-k'ang, Kirin Institute of Applied Chemistry; Chang Lo-feng, Talien Institute of Chemical Physics; Hsia Tsung-hsiang, Shanghai Institute of Organic Chemistry; Su Feng-lin, Foreign Affairs Bureau, Science and Technology Association of the People's Republic of China.

This was my second visit to the People's Republic. I was privileged to be a member of a delegation that visited China five years ago that had the responsibility for negotiating the terms under which these exchange visits are taking place. At that time we had the unusual opportunity to meet with Premier Chou En-lai to discuss the conditions of future exchange visits, which would take place under the provisions of the Shanghai Communique of 1972. During our one-and-a-half-hour meeting, the Premier gave us a wide-ranging view of pertinent Chinese historical background and on this occasion told us he would accept nine out of the twelve subject areas we had requested to be covered by our exchange agreement.*

Our delegation travelled widely throughout China and visited, as a complete delegation or in smaller groups, a total of 45 installations — research institutes, universities, colleges, industrial plants, factories and some non-scientific sites. Prior to our visit we submitted a list of research institutes, universities, industrial plants, etc. that we wanted to visit; and this request was honored in all detail. Our itinerary took us to four areas of China — (1) Peking and environs, (2) northeast China (the former Manchuria), (3) Shanghai and environs and (4) western (actually central) China including Sian and Lanchow. The portion of our trip to northeast China included the cities of Talien (the former Port Arthur, also known as Dalien and Darien), Shenyang (the former Mukden), Fushun, Changchun (the former Hsinking), Harbin and the Tach'ing oil fields; and, during some of this portion of our trip, our delegation was divided into two groups.

---

*Reported in detail in "A Journey to China," pp. 331–336 of this volume.

One of the interesting features of our visits to the various research institutes and universities was the opportunity for the members of our delegation to give talks on their specialties. In many instances such a talk, together with the following question and answer period, would take half a day. In some cases a delegation member would give such a talk in the morning and another talk in the afternoon of the same day. We were impressed with the quality of the questions following our talks, indicating an extraordinary knowledge of the scientific literature by the members of our audience. We were also impressed by the high degree of motivation of the young scientists that we met on these occasions and during our visits to the institutes and universities. I had occasion to give a number of such talks, including a full day of talks.

We were received everywhere with great cordiality and warm hospitality, which did much to obviate our very strenuous schedule. Our hosts in each city entertained us with a multi-course banquet, punctuated with the traditional Chinese toasts, in which mai tai was featured and which always included welcoming remarks by the evening's host at the beginning and a response by me near the end of the meal. The theme of these remarks always included emphasis on friendship between the scientists and the peoples of our two countries.

Audience at Seaborg lecture in Library Building, Institute of Atomic Energy, Peking, 22 May 1978. Reproduced by courtesy of the author.

Oil shale processing unit, Petroleum Refinery Unit No. 2, Fushun, China, 29 May 1978.
Reproduced by courtesy of the author.

Cockroft–Walton accelerator, Institute of Modern Physics, Lanchow, China, 8 June 1978.
Reproduced by courtesy of the author.

Early during our visit, in Peking, we had a three-hour meeting with Tsien San-tsiang, Vice President of the Chinese Academy of Sciences (and an old friend whom I met in 1973 in his role as Director of the Institute of Atomic Energy), and other Chinese officials to discuss Chinese science policy and plans. He briefed us thoroughly on their National Science Plan, announced by Vice Premier Fang Yi at their National Science Congress early this year. This plan focuses on eight priority areas. The first three — agriculture, materials science and energy resources — are designed to meet the most immediate needs. The next three — computer, space and laser sciences — are meant to bring Chinese science and technology abreast of that of the advanced countries. The final two areas — high energy physics and genetic engineering — are selected to help Chinese science catch up with the advanced countries in prestige areas. Their energy plan includes emphasis on nuclear power and they have plans to build — with Chinese engineers — and to import nuclear reactors during the next decade.

A remaining task of our delegation is to prepare a comprehensive report in which we will summarize our findings and try to make an evaluation of the status of chemical and chemical engineering science in the People's Republic of China. It would be premature to try to anticipate these conclusions, but a few general statements can be made. The emphasis until now has been on applied rather than basic research. This is a consequence of the undeveloped state of science and technology at the time of "Liberation" in 1949 and the requirement to give first priority to the immediate practical needs of the people. When the time seemed ripe for a turn toward basic research in the 1960's, the Cultural Revolution intervened, which led to a further delay in the pursuit of basic research. They have, nevertheless, made good progress in a number of particular areas, such as: chemistry of natural products, polymer science, catalysis, and areas of analytical chemistry, including gas chromatography. Our delegation was particularly interested to find an oil shale conversion plant in full operation at Fushun — this and two other plants in China are probably the only such plants in the world operating as production units.

We were impressed by the high degree of motivation exhibited by the students and young scientists. They were anxious to learn, do good research and serve their country in whatever task they might be assigned. They were very familiar with the foreign scientific literature. I spoke to several of the scientists in positions of responsibility of the value to their country of setting up a relationship with Chemical Abstracts so that they can better interact with scientists throughout the world; there are good indications that they will move in this direction.

I was particularly interested, of course, to visit three nuclear research institutes — the Institute of Atomic Energy near Peking, the Institute of Nuclear Research near Shanghai and the Institute of Modern Physics in Lanchow. Each of these institutes has an operating cyclotron and other low energy accelerators, radiochemistry laboratories and an interesting program of research in nuclear physics and nuclear chemistry. The Institute of Modern Physics, in 1973, converted its 1.5-meter cyclotron to the acceleration of light heavy ions (carbon, nitrogen and oxygen ions). They have received authorization for and are planning construction of a tandem cyclotron accelerator system with improved capability for the acceleration of heavy ions. Their 1.5-meter cyclotron will inject into a 6.15-meter sector-focused cyclotron. This will produce 50 MeV per nucleon carbon ions and 6 MeV per nucleon up to xenon ions. For a second phase, with a better injector, they hope to accelerate uranium ions to 10 MeV per nucleon.

Yuan T. Lee and Seaborg at the Great Wall, near Peking, 21 May 1978. Reproduced by courtesy of the author.

Mao Mausoleum, Tien An Men Square, Peking, China, 11 June 1978. Reproduced by courtesy of the author.

Figures in museum at site of the Tomb of Emperor Ch'in-shih-huang, near Sian, China, 6 June 1978. Reproduced by courtesy of the author.

Of especial interest is the fact that we were the first foreign delegation to visit Lanchow. This made us objects of curiosity to an even greater extent than during the rest of our visit. Following a dinner in Lanchow Restaurant, our delegation found a crowd of some five hundred curious people waiting for us when we emerged from the restaurant. They were smiling, and very, very friendly, but obviously we were a strange species.

Our hosts managed to squeeze in visits for us to a number of interesting scenic and historic spots. We visited the Great Wall, the Ming Tombs, the Forbidden City, the Summer Palace, the Temple of Heaven, the Mao Mausoleum on Tien An Men (Red) Square — all in Peking. We also visited picturesque Hangchow with its famous West Lake; the Palace of the Ch'ing Dynasty (Manchus) and the Northern Tomb in Shenyang; the Hot Springs in Sian, where the Sian incident (capture of Chiang Kai-shek in December, 1936) occurred. Also in Sian we visited the extensive memorial site of Ch'in-shih-huang, the first emperor of the Ch'in Dynasty (221–206 B.C.) and the first to unite the warring states in a unified China.

It was interesting to note differences in attitude and life styles in various parts of the country. The life style and general method of dress seem to be most conservative in Peking. Here is evidenced an almost universal adherence to the familiar gray and blue Mao suits. There seems to be very little activity after dark and no night life. Shanghai, of all the cities we visited, has a life style and dress closest to the "West," although still distinctly Chinese; here Huai Hua Street is full of life until about 7.30 p.m. and displays some activity later than this, and there are perpetual line-ups for admission to the cinema theaters. (The Acrobatic Show here was fantastic, but I believe this is true of other cities as well.) Many people dress colorfully in Shanghai, and the young people are open in their courtship, although very proper. In the northeast we found the life style and dress intermediate — between that of Shanghai and Peking. Hangchow is a fantastic city that almost defies description — scenic, historical, with a good climate. Sian is another extraordinary city, with a 3000-year history, 2000 years as the capital of China; and our specially arranged visit to the memorial site of Emperor Ch'in-shih-huang, the first emperor of the Ch'in Dynasty and the first to unite China, was a privilege and a beautiful experience; this memorial, now under excavation, apparently contains thousands of life-size figures. In some ways I found the visit to Lanchow the high point of our visit. In this semi-desert area Lanchow reminds me of Albuquerque, New Mexico. Lanchow has not yet been "spoiled" by the entry of foreigners. We were the first official foreign delegation to visit, and the first Americans to venture beyond the protected areas of the hotels.

It is impossible to refrain from comparing conditions in 1973 (my first trip) and in 1978 (my recent rip). The most apparent difference results from the overthrow of the "Gang of Four" and the complete termination of the "Cultural Revolution." Five years ago there was much talk about the end of the Cultural Revolution in the sense that universities and colleges were apparently being reopened, and there was talk of starting graduate education again. Apparently this was a false hope, and there was a setback extending into 1974–75, due to the actions of the "Gang of Four." Every briefing at the start of our welcoming sessions made reference to conditions before and after the "overthrow." All universities are starting to take advantage of the national entrance examination, are expanding back to the four year (from the three year) curriculum, and all the major universities and research institutes are now starting to admit graduate students on the basis of the national examination augmented by additional local examinations. An important consequence of this reform is the new turn to basic research. We were told that emphasis on applied research was, of course, necessary during the years before the Cultural Revolution, but this would normally have been followed by a turn to basic research in the 1960's had this natural course of events not been interrupted by the Cultural Revolution and/or the "Gang of Four."

Our visit came at a propitious time from the point of view of the real changes that have occurred and are occurring following the Cultural Revolution. We suggested to our hosts, in Peking and at the other places we visited, that our two-country exchange agreement be expanded to include bilateral symposia on selected subjects, exchange of graduate students and other scientists for extended periods of time, and other means of increasing cooperation. We had the impression that our hosts have a great desire to move in this direction and events that have occurred since our return home give me much encouragement that we shall indeed see much more of our Chinese friends.

Preparing for the 21st Century

Presented to a Regional Conference of the American Federation of Teachers, San Francisco, California, October 15, 1983.

We all recognize that we live in a rapidly changing, increasingly high-technology world. I have characterized our present age as that of the Third Revolution. The Revolution of Independence gave birth to our nation and established the democratic principles on which our classical concept of "equality of opportunity" — largely through education — is based. The Industrial Revolution rewarded the American spirit of inventiveness and made us leaders in the world's economy, blessed with an extremely high standard of living. The Third Revolution, the Revolution of Science, has already transformed how we understand our world — through the remarkable expansion of knowledge in a few decades — and is radically altering almost every aspect of our lives. Our response to the challenges of the revolution in science will, quite simply, decide our future. The strength of our technological and scientific enterprise will determine our economic well-being, our security, and our health and safety.

Our most valuable resources are our intelligence and ingenuity. As a nation, we pride ourselves on our history of pioneering new technologies; in the future much will depend not only on that capacity for innovation but also on our general preparedness to participate in the practice and production of those technological advances. There can be no doubt that scientific literacy — a solid understanding of science and mathematics — is now more important than ever before. And there is irrefutable evidence that the skills of our youth are not only not progressing with the increasing demands — but actually are deteriorating at an alarming rate.

The United States owes its current preeminent position in the world and its high standard of living in considerable part to science and technology. We have been leaders in development of new knowledge and ideas and in their application to human needs and desires. This leadership has brought us unprecedented prosperity, health and national strength. Recently, however, our lead has been challenged by other countries in what are now all too familiar areas, such as automobiles, consumer electronic products and even steel. The United States now imports 28% of its cars, 55% of its consumer electronic products and 18% of its steel. Imported goods account for 19% of American consumption, up from 9% in 1970. Certainly, low wage rates overseas contribute to the competitiveness of foreign products; however, one cannot underestimate the significance of more widespread technological know-how and sophisticated production techniques.

Simon Ramo, one of our great leaders in the application of science and technology, believes that the United States today is a giant but slipping technological nation, where concern is growing that its innovative entrepreneurial spark is ebbing, as in area after area its productivity leads are closed by other nations. Ramo is convinced that science and engineering are the keys to re-establishing our competitive position and reversing our declining economy. Scientific research can teach us to apply the laws of nature more effectively and engineering can lead to economically advantageous products and techniques which could create new jobs and increase exports to restore the U.S. trade balance. The economist Edward Denison found that in the period 1948–73, during which national income per person employed increased 83%, more than half of that 83% increase was the result of new knowledge, consisting of better technology and management techniques.

The conclusions of the Panel on Advanced Technology Competition and Industrialized Allies, published earlier this year by the National Academy Press, emphasize the need to "elevate, in the scheme of national priorities, efforts to strengthen the nation's capacity for technological innovation." I quote, "The United States must act now to preserve its basic capacity to develop and use economically advanced technology. This innovative capacity is essential for the self-renewal and well-being of the economy and the nation's military security. Trade in advanced technology products and services will contribute enormously to our economic health."

Innovative scientific research and development can help us to acquire new knowledge. However, the acquisition will mean little to our future economic well-being without its application. An outstanding example of this is the fact that although industrial robots were first developed in America, it is estimated that only about 6,500 industrial robots are in use in this country, while in Japan as many as 75,000 may already be employed. I trust that Nobel Laureate Wassily Leontief's prediction that "the human worker will go the way of the horse" is an exaggeration — but I do not question the certainty that robots and computers will do much of the work in the future that human beings are now doing.

I believe that those relatively few students who have elected to pursue science as a career are studying and learning as much science and mathematics as ever. However, the vast majority of students are ending their study of these subjects at earlier ages and performing less and less well on tests of their knowledge. The National Commission on Excellence in Education (on which I served) found that half of our high school students take no math or science courses after the tenth grade. If this situation continues, it is apparent that the pool from which future scientists and engineers might be drawn will decrease and also that today's secondary school students may know too little to function effectively in an increasingly scientific and technological society.

We must improve general science education for all of our young (and not only for those who plan to continue their education and become professional scientists, mathematicians or engineers) because we need a large number of scientifically literate, trained nonprofessional workers with the understanding and skills to manufacture, operate and repair increasingly complex technological equipment. Future employment opportunities, necessary to replace jobs lost in our declining "smokestack industries," will be in areas requiring technical sophistication and will depend on a work force prepared to adapt to changing needs.

Our schools now offer a range of vocational programs designed to prepare students to enter the job market immediately after high school. However, we run a serious risk that skills which these students are working hard to master now may soon be obsolete in today's unpredictable and rapidly changing technological economy. A technical memorandum, issued by the Congressional Office of Technology Assessment in March of this year, "Automation and the Workplace," points this out. The chapter on "Education, Training and Retraining" begins, "At various stages in U.S. history, changes in workplace operation and procedures in all sectors of the economy have resulted in changes in education, training, and retraining requirements for those employed and preparing for employment ... For example, new production techniques introduced during the Industrial Revolution had much to do with the creation of a system of free public education, since large-scale production and continued industrial expansion required a literate work force capable of functioning on production lines, supervising manufacturing operations, keeping administrative records, and performing other functions." They note that "The application of programmable automation in manufacturing operations has the potential to trigger widespread changes in education and training requirements. Robots and other forms of programmable equipment and systems may change the organization of the manufacturing process, the character of the production line, the occupational mix, and the human–machine relationship." We cannot afford too narrow a focus on vocational training.

Dension, whom I quoted earlier, found that the increase in the amount of education received by employed persons is the second largest source, after new knowledge, of growth in productivity. In his view, educational background decisively conditions both the types of work a person is able to perform and his/her proficiency in any particular occupation. Education not only enhances the skills of individuals within an occupation — it also permits beneficial shifts in occupation and heightens awareness of job opportunities, thereby improving the chance of employment in new areas as the need arises. An educated work force is better able to learn and use more efficient production practices. Experiments with "quality circles" in the automobile industry have demonstrated also that thinking workers can make a substantial contribution to improving production techniques and products.

It is very difficult to forecast detailed changes in occupational employment. However, we can state with confidence that future opportunities in the job market will be highly colored by the high technology society of the Third Revolution. At the turn of the century almost one-third of America worked on the farm; now barely one in thirty works the land. The mechanization of the American farm led us to heavy industry in World War II. Our economic development and industrial growth relied for a long time on energy-consuming products like the automobile. Now that oil is no longer cheap and in ample supply, the employment trend has shifted. Today more than one-third of America's workers are in service occupations. In 1974 over 20 million Americans were employed in manufacturing and slightly less than 13 million worked in service. By 1982, only eight years later, manufacturing jobs had dropped to just over 18 million while service occupations accounted for over 19 million jobs. We can anticipate an acceleration of this change in occupational opportunities as robots take over more of the jobs in manufacturing and computers perform many of the functions of white collar office workers. But the servicing of robots and the operation of computers and wordprocessors will require millions of technicians by the year 2000.

I don't think it is too much of an exaggeration to argue that our economy is changing from heavy industry based on the availability of an affordable supply of oil to an industrial base structured around a new type of raw material — the microchip. The rapidly developing solid state technology allows us to provide a large number of complex logical functions on a single semiconductor chip about 1/4 inch on a side. Such chips can now provide storage space for 64,000 binary bits of information or a complete microcomputer. The capacity of chips has about doubled yearly since 1960 while their price has remained roughly constant. The resulting reduction in cost has encouraged the rapid adoption of computer-controlled equipment in factories, offices and homes. Already it seems impossible to image airline reservation systems or banks operating effectively without the aid of computers — and computers are being introduced into almost every other area of business management and daily life at an amazing rate. I think we can confidently predict that the largest area of demand in the next 20 years will probably be for computer programmers, operators and technicians.

However, there are also many other new technologies which will surely require a huge new labor pool with sophisticated technical skills by the year 2000. We have only just begun to understand the possible applications of developing laser technology, which will replace machine and foundry tools on a large scale. The challenge to develop and produce substitutes for oil as a source of energy will employ highly trained technicians, as well as scientists and engineers, as we learn more about producing synthetic liquid fuels from coal and oil shale and electrical power from solar energy, breeder reactors, nuclear fusion and geothermal energy. And most of these energy-producing methods will give rise to waste materials, whose safe disposal will require the development of new methods by highly trained people. Housing construction techniques and materials will change as we perfect the heating and cooling of buildings using sunlight and as our understanding of energy conservation methods improves.

We can also predict that there will be a growing need for people to administer our health and medical services at a new higher level of training. Advances in the development of new tools to diagnose disease suggest that there will be a great demand for the services of paramedics. There is also bound to be an explosive increase in the need for what may soon be called "geriatric social technicians" to care for the mental, physical and social needs of our nation's aging population. Who knows where such fields as bionics and genetic engineering may lead us?

The point really is that our future success in these explorations will be based on more than a simple inventory of acquired skills; it must be based on a solid foundation of scientific literacy and a people endowed with a practice in learning and the flexibility to adapt to a society constantly changing. Although we think of the demand for new knowledge and skills as being especially evident in such high-tech industries as microelectronics, it is pervasive in all fields of productivity. Think of the farmer, in the past prepared to produce food with a basic knowledge of soil and seeds and rainfall, who now needs a workable knowledge of insect physiology, chemical fertilizers, weather modification techniques, hybrid strands, animal breeding, mechanized and computer support systems, price supports and parity, and international commodity agreements and trade patterns.

In addition to the need for trained scientists, mathematicians and engineers, and nonprofessional workers with an understanding of complex technological equipment, we need widespread understanding of science among the general population. Support of the critical need for scientific literacy is best expressed by Thomas Jefferson's famous dictum, often-quoted and profoundly true: "I know of no safe depository of the ultimate powers of the society but the people themselves; and if we think them not enlightened enough to exercise their control with a wholesome discretion, the remedy is not to take it from them but to inform their discretion." It is fundamental to the effectiveness of our democratic system that our citizens be able to make informed judgements on the more and more complex issues of scientific and technological public policy. In recent years a growing number of referenda have concerned such issues as nuclear power, laetrile, recombinant DNA facilities, fluoridation, chemical waste management. Decisions must be made which are of critical importance to our health and safety.

The scientific community has been growing more and more apprehensive about the public's ability to understand the issues and make informed judgements. And the public, I believe, is growing more and more distrustful of the experts on whom they are forced to rely, whose opinions vary and are often expressed in inaccessible, incomprehensible language. As a scientist, I am distressed by recent movements to limit scientific inquiry which result from ignorance and fear of loss of control. Scientific inquiry is founded on the faith that the world is orderly, and that if we can discover the laws which govern our physical universe, we can understand cause-and-effect, forecast and, to some degree, control our future. A lack of understanding of the most basic scientific principles can lead to the acceptance of "pseudoscience" — creation science, astrology, miracle cancer cures, fad diets — as science, which confuses the issues even more and contributes to a general deterioration of public confidence in scientific information.

The determination of science policy is not only a matter of making laws; it is also a question of deciding what science to pursue and how to support it. Public *mis*understanding of science can result in such destructive policies as wide fluctuations in science funding, setting research priorities based on political rather than scientific criteria, unrealistic demands for quick, practical results, and misguided regulations and accounting procedures. The success of our scientific research and development efforts depends in no small part on the support of the public and its confidence in the value of the work. The temptation to rely on specialists to direct our future development is great; however, we cannot afford to abdicate responsibility to experts or to fakirs, posing as experts. The stakes are too great.

The body of scientific knowledge already at our disposal is vast and expanding at an almost unimaginable rate. We have progressed further in the past several decades than in all previous history and progress in the lifetime of each of you here today will likely exceed that of all time past — of all time previous to 1983. What, then, must a student master in order to become scientifically literature? And, how well are our schools performing their obligation to prepare our young to participate fully in our economy and our democracy and in future adventures of discovery?

The most basic and commonly used definition of literacy is the ability to read and write. More than 20 million Americans are functionally illiterate by the simplest tests of everyday reading, writing and comprehension. About 13% of all 17-year-olds in the United States can be considered functionally illiterate and that figure may be as high as 40% among our minority youth. There is every indication that more and more young people are being shunted through our schools without being taught basic skills and that, for the first time in our history, the educational skills of this generation of high school graduates will not surpass, not even equal, those of their parents who received a high school education. The Department of the Navy reported to the National Commission on Excellence in Education that 25% of its recent recruits cannot read at the ninth grade level, the minimum necessary to understand simple written safety instructions. Our first priority must be to teach our students to read, and to understand, evaluate and *use* what they read.

The most elementary understanding of scientific concepts requires a familiarity with the basic vocabulary of science. One cannot begin to explore the study of physics, for example, without a grasp of such terms as velocity, acceleration, force, energy and gravity. However, too often science is taught in the old "tell, text, test" pattern, feeding students "dictionary definitions" and settling for regurgitated responses, lacking any true understanding of the principles operating or how we arrived at them. Scientific concepts are the product of human intelligence and imagination. A solid foundation in mathematics and science must include the understanding that a scientific concept involves an idea first and a name afterward; understanding does not reside in technical terms themselves.

In addition to introducing students to the concepts, laws and processes of physical, biological and mathematical sciences, we must teach them the skills of deductive reasoning. Science education must make clear the methods of scientific inquiry and enable the student to distinguish between observation and inference, and understand the difference between the occasional role of accidental discovery and the deliberate design of forming and testing hypotheses. Newton's revolutionary theory of gravity was not the result of his being bopped on the head by an apple — but of what he inferred from the experience. One study of 17-year-olds showed that nearly 40% cannot draw inferences from written material and 35% cannot solve a problem in mathematics requiring several steps. A particularly appalling example of this lack of mathematical problem-solving skills is presented in the April 1983 National Assessment of Education Progress report. They found that 48% of a representative sample of 17-year-olds incorrectly answered the following question: "A hockey team won 5 of the 20 games it played. What percent of the games did it win?"

Clearly, we are failing in our obligation to train students to exercise their intelligence, to reason independently, and to test their conclusions. A scientifically literate person must be able to understand how theories are tested and validated. Ignorance of the rules of evidence invites acceptance of such arguments as the creationist dismissal of the theory of evolution as "just a theory."

A study of the history of science lends much to an understanding of the creative process. An educated person is aware of at least a few examples of ways in which scientific knowledge has had a direct impact on intellectual history and on his/her own personal view of the universe. One should understand not only specific examples of the practical applications of scientific knowledge to everyday life, but also have some notion of how the development of scientific knowledge has influenced our human condition and of the moral, ethical and social implications of those scientific developments.

I doubt that any of you would argue with my conviction that few students are endowed with these understandings in the course of their pre-college education. Evidence that our students are not developing the basic skills or absorbing fundamental information is indisputable. Average achievement of high school students on standardized tests is now lower than 26 years ago when Sputnik was launched; scores on Scholastic Aptitude Tests (SAT) declined steadily from 1963 to 1980. Just between 1975 and 1980, remedial mathematics courses in four year colleges increased by 72% and now constitute 25% of all mathematics courses taught in those institutions; at community colleges, 42% of the classes cover only secondary level material. A study of 15 developed and 4 developing countries conducted by the International Association for the Evaluation of Education Achievement in 1970–71 demonstrated that American children are falling far behind children of other industrialized nations. Although the United States had by far the largest percentage of students still in school at age 18, our 18-year-olds scored *lowest* in achievement and achievement gain. In 19 academic tests, Americans had the lowest achievement scores in seven subjects. Children of major economic competitors, like Japan, scored dramatically higher on the tests.

I believe that the need for educational reform is now recognized as a top priority on the national agenda. Across the country, people have begun to explore possible solutions to the crisis in our schools which the National Commission on Excellence in Education identified as a "rising tide of mediocrity" resulting from "an act of unthinking, unilateral educational disarmament."

I recently reread a speech I delivered in 1959, during the post-Sputnik era when there was so much concern about our ability to compete in such fields as national security and exploration of space. In that speech I warned, "The question is whether science, with all its bounty and mercy, will mock us by placing chains upon the individual. Serious students of our society cannot dismiss lightly such nightmares as those suggested by George Orwell ... But unless we attain a wider foundation of knowledge of the general nature and principles of science, we may lose democratic control of our destinies. Through ignorance and only through ignorance, we could take the dark road to *1984*."* Here we are, a few short months from 1984. Orwell's worst imaginings are not fulfilled and neither are our needs for a scientifically literate and well-educated citizenry. We now understand that it is not only our democracy that is at stake, but also our place in the technological economy which supports our high standard of living. Whatever the expense of improving education, it is an investment in the future we must make. Excellence costs. But in the long run mediocrity costs far more.

---

*See "Science and Liberal Education in the Space Age," pp. 33–42 of this volume.

Chemistry in the Third World

I believe that we scientists fortunate enough to work in developed countries, with comparatively rich resources in education, facilities and funds, have a special obligation to share these resources and our energies with scientists in less developed countries and to work together to seek solutions to the world's most pressing problems. Among these problems are many areas (for example, improved control of disease and increased production of food) whose solutions may well be reached through the field of chemistry.

The International Organization for Chemical Sciences in Development (IOCD), for which I serve as President, was founded in 1981 with the main objective of involving chemists from Third World nations in the search for solutions to the urgent problems of their countries. To do this, IOCD uses two main channels — initiation of research programs and provision of services. Because scientists from industrial countries serve as scientific advisers and monitor IOCD research projects, technology transfer and the strengthening of institutions in Third World countries is also facilitated.

The IOCD has already made an impressive start in accomplishing some of its objectives in the brief period since its founding. A program of chemical synthesis designed to uncover new drugs for the treatment of tropical diseases has been initiated. Eleven laboratories (located mainly in developing countries) are participating in this program. IOCD is planning a similar chemical synthesis program in the area of food sciences and investigating possibilities for making contributions in the field of human fertility regulation. IOCD is conducting a survey to identify principal investigators and groups throughout the world who are involved in research

---

Reprinted with permission from *The Chemist* **62**, 21 (1985).

relevant to such activities. In particular, the organization is making an attempt to identify scientists from developing countries who are working in the chemistry of natural products.

In addition to organizing facilities for analyzing samples from developing countries, IOCD is initiating a program for maintaining and repairing instrumentation and equipment in chemical laboratories in these countries. Eventually there will be a comprehensive biological screening program in Third World nations for compounds prepared by synthesis or isolated from natural sources. Scientific and technical advice will be provided free of charge to managers or governments in countries attempting to build chemical research programs or industrial facilities.

The administrators and staff of IOCD are all volunteers who work directly with scientists from developing countries. Its small infrastructure and low overhead give IOCD a great deal of independence and flexibility.

Start-up funding from UNESCO enabled IOCD to take some significant first steps toward accomplishing its major objectives. This year, IOCD moved its Secretariat to Mexico City, at the invitation of the Mexican government. It has initiated several joint programs, some sponsored in part by such organizations as the World Health Organization, the Walter Reed Institute for Medical Research, the National Institutes of Health, the United Nations Fund for Population Activities, and others. In the future, however, IOCD must obtain funding from governmental organizations and private sources, including foundations and companies in industrialized countries as well as in the developing world.

Responses from chemists around the world have been encouraging and have confirmed the conviction that IOCD has an important role in aiding the development of Third World countries and in assisting them in improving the quality of their life through chemistry. It is gratifying that chemists in developed nations now have a way to collaborate in their areas of expertise with colleagues in developing countries. So far, only the groundwork for facilitating international communication and collaboration has been laid; now a great deal more work must be done.

The Role of Basic Research in Society

Remarks at a media conference concerning the Superconducting Super Collider at the Lawrence Hall of Science, July 30, 1987.

We are gathered here in the Lawrence Hall of Science, a national science center devoted to the promotion of the public understanding of science, the teaching of science to people of all ages and especially to students and teachers in and of grades K-12, and to the development and dissemination of science teaching materials. The LHS is a memorial to Ernest Orlando Lawrence, inventor of the cyclotron, the accelerator which might be considered the first ancestor of the Superconducting Super Collider. Here in the Lawrence Memorial Hall you may view the actual accelerating chambers of the first cyclotrons. The five-inch diameter, pill box-shaped chamber of the first (1930) cyclotron might be contrasted with its descendant, the proposed 20-mile diameter, doughnut-shaped SSC.

We all recognize that we live in a rapidly changing, increasingly high technology world. I have characterized our present age as that of the Third Revolution. The Revolution of Independence gave birth to our nation and established the democratic principles on which our classical concept of "equality of opportunity" — largely through education — is based. The Industrial Revolution rewarded the American spirit of inventiveness and made us leaders in the world's economy, blessed with an extremely high standard of living. The Third Revolution, the Revolution of Science, has already transformed how we understand our world — through the remarkable expansion of knowledge in a few decades — and is radically altering almost every aspect of our lives. Our response to the challenges of the revolution in science will, quite simply, decide our future.

The strength of our technological and scientific enterprise will determine our economic well-being, our security, and our health and safety.

Our most valuable resources are our intelligence and ingenuity. As a nation, we pride ourselves on our history of pioneering new technologies; in the future much will depend not only on that capacity for innovation but also on our general preparedness to participate in the practice and production of those technological advances.

The United States owes its current preeminent position in the world and its high standard of living in considerable part to science and technology. We have been leaders in development of new knowledge and ideas and in their application to human needs and desires. This leadership has brought us unprecedented prosperity, health and national strength. Recently, however, our lead has been challenged by other countries.

Basic research is research driven by intellectual curiosity, rather than by immediate practical problems. Its results are usually unpredictable, and can't be programmed. It is the engine that drives applied research and technology. Scientists are explorers. However, we do not stumble upon innovative new technologies in a barren wilderness. New ideas take root from knowledge already accumulated. Basic research is the fertile field where grows the understanding which is the stock of applied research. Incremental scientific advances, as well as major discoveries, result in new technologies of great commercial importance. They can give us entire new industries, as in the case of advances in molecular biology. They can give us whole new ranges of products, as in the case of polymer chemistry. They can revolutionize other technologies and industries, as has been the case for the transistor and the laser. Incidentally, here in the exhibit halls you will see a new exhibit that explains laser technology and some of its uses.

Basic research leads to the creation not only of new products, but also new industrial processes and manufacturing systems. These can greatly increase industrial productivity, reduce costs, and improve the quality of products. For example, advances in microelectronics are aiding the production of automobiles, steel, and many other manufactured goods. Discoveries in biology are influencing the processing and production of pharmaceuticals, foods and chemicals.

Basic research conducted with accelerators has made a broad range of contributions to society. These range from the most fundamental contributions to our understanding of atomic and nuclear structure to the huge reservoir of results obtained from the application of accelerator-produced radioactive tracers to research in physics, chemistry, biochemistry, biology, medicine, physiology, agriculture, zoology, botany, engineering, genetics, nutrition, environmental problems, and many other fields. Activation analysis is another large field of investigation.

To be more specific, let me mention some of the results of basic research achieved with accelerators here at Berkeley. These have included the synthesis and identification — i.e., the discovery of — the transuranium elements neptunium (atomic number 93), plutonium (94), curium (96), berkelium (97), californium (98), mendelevium (101), nobelium (102), lawrencium (103), rutherfordium (104), hahnium (105), and unnamed element 106. The transuranium elements americium (95), einsteinium (99) and fermium (100) were discovered, not through the use of accelerators, but by the use of intense neutron sources. (Unnamed elements 107, 108 and 109 have recently been discovered with accelerators in a laboratory in West Germany.)

The profound impact which the discovery of a transuranium element can have on society is vividly demonstrated by the use of plutonium in the production of nuclear weapons. On September 10, 1942, my co-workers weighed a sample of plutonium for the first time. Using a quartz fiber balance at the wartime Metallurgical Laboratory at the University of Chicago, we weighed 2.77 micrograms of the oxide. This sample is on display here in the Lawrence Memorial Hall. The more beneficial use of plutonium-239 as a nuclear fuel is well-known. Another isotope of plutonium, plutonium-238, is used as an energy source for heart pacemakers and to provide energy for small thermoelectric power units. Nuclear isotopic power in the thermoelectric generator units, utilizing plutonium-238, made possible the operation of instrumentation to send back information from the moon, furnished power for the Viking unmanned lander and for the Pioneer and Voyager space probes to Jupiter, Saturn and beyond. Americium-241 is useful for a wide range of industrial gauging and oil well logging operations, and for a number of other practical applications. A number of radioactive isotopes of lighter elements discovered in basic research utilizing cyclotrons are used to an extraordinary extent in nuclear medicine. The Berkeley 37-Inch Cyclotron, used for the discovery of many of these radioactive isotopes, now stands as a monument in the entrance area outside the Lawrence Hall of Science.

Let me go on to describe more broadly a few of the many discoveries resulting from basic research that have had an enormous impact on society. A recent issue of the now-defunct AAAS publication, *Science '85*, listed two dozen discoveries that could change our lives — designer catalysts, composite materials, curious quarks, replacement genes, biology of birth, number theory, plant power, physics in nine dimensions, egg development, versatile ceramics, ultrasmall microchips, space butterflies, cancer-causing genes, superfast computing, brain drugs, evolution's revolution, commuting to space, experimental math, biochemical control of behavior, talking by light wave, the space telescope, substitute body parts, living drugs, and 21st century software.

A good starting place for a more specific discussion might be early this century, with Ernest Rutherford's discovery of the atomic nucleus and — with the contributions of Niels Bohr — the structure of the atom, which changed our whole understanding of nature. This led to a far-reaching surprise — namely, that the arithmetic that had served basic science so well throughout history had to be replaced by a new arithmetic, quantum mechanics, now necessary to describe the phenomenon of the microcosm. When combined with electromagnetism to yield quantum electrodynamics, quantum mechanics has yielded the most precise theory of matter yet known.

During the same time period as Rutherford's and Bohr's discoveries relating to atomic structure, we received the theory of relativity from Albert Einstein — perhaps the greatest intellectual achievement of this century. This led to the formulation of the famous $E = mc^2$, the basis for the release of nuclear energy.

The release of nuclear energy, made possible by the discovery of the nuclear fission reaction in 1938 by two German chemists, Otto Hahn and Fritz Strassmann, not only gave us nuclear weapons and the promise of electricity from nuclear power, but also a bevy of life-saving radiopharmaceuticals and radiation sources and radioactive isotopes for other uses.

Close to 20 million nuclear medicine procedures are performed annually in the United States. More than half of all patients are either diagnosed, treated or monitored using such techniques. Most of these administrations use iodine-131, cobalt-60 or technetium-99m, discovered here at Berkeley in the 1930's through the use of cyclotrons, or iodine-123, iodine-125, thallium-201, all of which are now manufactured in the required quantities in nuclear reactors and accelerators. Today technetium-99m is the workhorse of nuclear medicine, accounting for more than seven million diagnostic procedures per year in studies of liver, bone, lung, thyroid, heart and brain.

Turning to the cosmos, our theoreticians have deduced that our universe started in a cataclysmic explosion about 18 billion years ago (or is it 11 billion years ago, as recently reported?). We have re-created in our laboratories conditions that existed tiny fractions of a second after the beginning of the universe, and this is one of the regions of exploration that will be possible through the operation of the superconducting super collider. It has been estimated that our universe consists of an assembly of some 10^{80} atoms. The SSC will also help unravel the mystery of the quarks and gluons that apparently account for the inner workings of the nucleus of the atom. Violent collisions of quarks may provide a window on the internal structure of quarks.

A goal of basic science is to discover the rules that govern matter at the most fundamental level, and basic to this objective is the need to unify all the known fundamental forces of nature — gravity, electromagnetism, the strong and the weak nuclear forces — into a single theory, a unified field theory. The newest approach, known as the superstring theory, holds that the universe has not just three spatial dimensions, but nine. Achievement of a unified field theory would surely lead to the most basic understanding of matter and the Superconducting Super Collider could make a contribution toward this objective.

Turning to some other results in the field of physics, the discovery of the transistor, the development of microchips, and other advances have led to a revolution in communications that has converted our planet into a global village with the aid of satellite relays for television signals. Now the availability of solid state lasers and of optical wave fibers is giving us a new revolution in communications. These fibers each carry the equivalent communication load of thousands of coaxial copper cables. These developments have led to a revolution in the handling of information through the availability of enormous amounts of fast computer memory.

The attention of the entire world has been captured by the recent discoveries of superconductivity in which this phenomenon has been observed in a number of ternary oxides containing rare earth elements at temperatures as high as above the boiling point of nitrogen, with some indications that it may be possible to reach room temperatures. The ultimate effect on such everyday applications as the economic transmission of electricity, the levitation of superfast trains, and numerous other applications, are such as to boggle our imagination. A recent symposium in New York drew so many people it took on a "Woodstock" quality and a symposium in Berkeley in June was attended by 400 people when only 40 had been expected.

And, I have not even yet come to the results of basic research in my own field of chemistry, which is perhaps the most utilitarian of all the experimental sciences. Chemists will synthesize millions of new compounds tailored for a wide spectrum of practical uses. Nuclear chemists will be involved in the synthesis of additional chemical elements, hopefully in the region of the superheavy elements predicted to exist in the "island of stability."

A recent (1985) study, "Opportunities in Chemistry," was chaired by George Pimentel of our Berkeley Chemistry Department for the National Research Council. This so-called "Pimentel Report" suggested that priority in basic research in chemistry should be given to the following frontiers: understanding chemical reactivity, chemical catalysis, chemistry of life processes, chemistry relating to our environment, and chemical behavior under extreme conditions. The report pointed out that advances in chemical synthesis and catalysis will be a key to maintaining our current position in world leadership; that the interdisciplinary field of chemistry of materials will play a vital role because advances will depend upon ability to tailor new substances including polymers to replace and outperform traditional or scarce materials; that chemistry will play a central role as we seek to clarify and detail the chemistry of biological life sciences at the molecular level, which will lead to the production of more food and better health; that the remarkable progress made in recent years by molecular biologists and biochemists has been built upon basic chemical principles; and that analytical chemical techniques will play an increasingly important role in protecting the environment.

Our endless supply of solar energy could be put to practical use through chemical processes yet to be discovered — direct catalytic conversion to electricity, splitting of water to produce hydrogen fuel, or widespread bioconversion of vegetation and waste products. Chemists will increase the efficiency of extraction of new sources of minerals, and material scientists will synthesize substitute materials from more abundant supplies.

Central to the growth of biochemistry has been the elucidation of the functions and characteristics of proteins. Some early scientists believed that living matter contained "vital" properties which made it indescribable in terms of existing chemical laws. The evolution of our knowledge about proteins shows how mistaken they were. We have progressed from the postulate in 1902 that proteins are polypeptides to the point where the amino acid sequences and crystal structures of complicated proteins can be determined routinely.

Only a century ago, the term "enzyme" did not exist. Today, we carefully can probe the reactions and properties of these biological catalysts made of protein. As a result, tremendous progress has been made in understanding physiological processes such as the Krebs cycle, glycolysis, electron transport and oxidative phosphorylation, biosynthesis, contraction and movement, and the maintenance of concentration gradients across membranes. Our understanding of membrane-bound enzymes may someday lead to the design of artificial membrane systems capable of performing sequential reactions or transporting materials across membranes. And through the manipulation of whole organisms or enzyme systems alone, we may be able to create new energy sources, such as the production of hydrogen through the biophotolysis of water.

In the field of nutrition, chemistry and biochemistry also have contributed enormously — particularly in vitamin studies. The importance of these chemicals was well-established in 1911 when Casimir Funk coined the term "vitamine." By 1948, scientists were beginning to recognize the participation of several of these chemicals in enzymic reactions. Now, an impressive array of vitamins has been identified, including vitamins A, D, E, K, C, B_1, B_2, B_6, niacin, biotin, folic acid, and B_{12}.

Most progress will be made in the chemistry of life processes — in biochemistry, molecular biology, and related areas concerning the study of proteins, enzymes, nucleic acids, and other macromolecules. Chemical and biological investigations at the molecular and cellular levels, aided by enormously efficient computers, will elucidate the origin of life and perhaps lead to the artificial synthesis of life. Biochemical genetics will give us a great deal of control over the genetic code and, beneficently applied (which will pose a real challenge), should result in a substantial reduction of genetic defects.

Immunochemistry, computer-aided molecular medicine, and chemotherapy should lead to the alleviation, treatment, cure or prevention of our major ailments, including "mental" illness, and to a slowing of the aging process. We will understand the structure, mechanism, and functioning of nerve and brain, leading to control of our memory, through investigations in neurochemistry and the related fields of neuroanatomy, neurophysiology, statistical biology, and experimental psychology. Biochemical engineering should make available implantable (microcomputer assisted) artificial hearts, kidneys, eyes (instruments to permit the blind to "see"), ears (instruments to permit the deaf to "hear"), and other bodily parts and organs.

The use of lasers in chemistry should make it possible to reach the ultimate limit of single-atom or single-molecule detection. I referred earlier to the exhibit on lasers here in the LHS. Laser techniques are used to study the microchemistry of surfaces, a prerequisite for understanding such practical processes as catalysis, adhesion, and corrosion. Laser fluorimetry and multiphoton ionization makes possible the detection of impurities present in part-per-billion or part-per-trillion ranges. Lasers can be used in the separation of isotopes, including the separation of fissionable uranium-235 from uranium-238. Physicians now use lasers for applications ranging from eye surgery, and vaporizing tumors, to removing warts and tattoos. Masers and lasers were discovered and developed by Charles Townes and Arthur Schawlow in the 1950's and 1960's.

The development of powerful synchrotron radiation sources, a by-product of accelerator research, has provided one of the most important new tools for the study of materials. These sources provide radiation many orders of magnitude more intense than has been available in the ultraviolet and X-ray wave lengths. They have enormous importance for probing the nature of solids and of surfaces, and work with the sources has provided new insight into the mechanisms of catalysis and the nature of chemical bonds — both matters of major industrial importance. With these sources much more sophisticated imaging techniques become possible for use in diagnostic and clinical medicine.

The superconductor technology associated with the SSC has an important potential for the development of very high field magnets needed for medical science studies on the causes of arteriosclerosis, cancer and Alzheimer's Disease.

In these references to chemistry, synchrotron radiation, and lasers, I have mentioned medical applications. Let me develop this theme further. Another technique applied to medicine, furnished by the physicists, is that of nuclear magnetic resonance (NMR), discovered by Edward Purcell and Felix Bloch in 1946. NMR spectroscopy has already revolutionized large areas of biological and clinical research and the diagnosis of disease.

Another powerful export from physics to biology and medicine is positron emission tomography (PET), which technique has been of particular use in elucidating brain function. Tom Budinger of the Berkeley Donner Laboratory developed the first high-resolution PET instrument. The use of accelerated charged-particle heavy-ion beams in clinical medicine has led to increased success in the treatment of specialized cancers, including, here at Berkeley, encouraging results in the treatment of otherwise intractable pituitary and pancreatic cancers, and, in the research of Jacob Fabrikant, the treatment of arterial venous malformations.

Medicine underwent a revolution in the middle of this century when it became possible to cure some infectious diseases with antibiotics. Now medicine is undergoing another revolution based on the applications of molecular biology — the mapping, cloning and study of thousands of human genes to understand the body's normal functions at the molecular level, made possible by the unravelling of the role of DNA by James Watson and Francis Crick in the 1950's. A consequence of this will be gene therapy, one of the most exciting ramifications of molecular biology. The first successful gene therapy in a mammal was reported in 1984, when researchers injected a growth hormone gene into a fertilized mouse egg to overcome a genetic deficiency of growth hormone. The power of recombinant DNA technology lies in the opportunity it affords, for the first time, to produce virtually unlimited quantities of practically any protein and the promise it gives for the production of new drugs.

A recent request has gone into the Department of Energy to support a large multi-year, multi-disciplinary technological undertaking to order and sequence the human genome. (I must confess that "genome" was not a part of my vocabulary a year ago. Seeking its definition in Webster's *New Collegiate Dictionary*, one finds "one haploid set of chromosomes with the genes they contain." Pursuing this further, one finds the definition of haploid as "having the gametic number of chromosomes or half the number of characteristic somatic cells." Not giving up, one finds for gamete "a mature germ cell possessing a haploid chromosome set and capable of initiating formation of a new individual by fusion with another gamete.") Actually, a genome is the complete set of instructions guiding the development of a living organism. The instructions are encoded in the genes which consist of deoxyribonucleic acid (DNA) and exist in a linear sequence on chromosomes in the nucleus of cells. This human genome project would cost about $1 billion between now and 1995. The actual planned ordering and sequencing involves coordinated processing of some three billion bases from a reference human genome and the consequent more complete understanding of human genes should lead to great advancement in the diagnosis and treatment of human diseases.

Another biological advance lies in monoclonal antibodies based on proteins produced by white blood cells in response to foreign substances, such as a virus. Monoclonal antibodies are ideal for diagnosing infectious diseases. This technique, along with sophisticated computers, should make possible much more rapid diagnoses and, consequently, earlier selection of therapy.

Time restraints have obviously limited my description today to only a small sampling of the benefits of basic research to human welfare. Future investment in basic research, such as in the Superconducting Super Collider, will surely continue to contribute to the benefit of mankind.

In the course of my remarks I have described some of the benefits to mankind that have accrued from basic research conducted with accelerators and alluded to some results expected through use of the SSC. Others to follow me on the program will expand on this theme, especially that aspect that pertains to the more recent high energy accelerators that have opened and made it possible to develop the field of high energy or particle physics.

It is time to take the next step. First we had the 1930–1931 cyclotron of E. O. Lawrence which reached 0.08 MeV of energy built from scrap at a cost of about $1,000. Following many intermediate machines, a decade later we had here at Berkeley the 60-Inch Cyclotron (with which plutonium was discovered) which gave us 100 times more energy at 100 times more cost. This was followed 15 years later by the Bevatron (with which the antiproton, antineutron and many elementary particles were discovered, opening the field of high energy physics) which produced a 1,000-fold increase in energy at a 100-fold increase in cost. Then we went, nearly two decades later, to an accelerator with nearly another 100-fold increase in energy at a 20-fold increase in cost. (On a personal note I might add that I, in my role as Chairman of the Atomic Energy Commission, was intimately involved in the pursuit of authorization, funding, and site selection for this machine. Thus, I view with considerable interest the impending repetition of this process now with the SSC, two decades later.) During recent years the Europeans have overtaken and surpassed us in the field of high energy or elementary particle physics as a result of the accelerators they have built.

We in the United States have come this far with the exploration of the basic nature of matter. Now we have reached a limit set by existing accelerators. To go further we need the SSC. Dare we say that we can't take this next step — a factor of 20 in energy, where so much is to be learned — because of the question of cost? The vast majority of money expended in building such a project is cycled back into the economy and the support industries not only aid the economy but also motivate new developments in practical applications of science. Most importantly, the value of the expected discoveries in the basic nature of matter is so large it is impossible to quantify. We cannot predict exactly which practical applications for society will come out of the new understanding of the basic nature of matter that will come from the SSC, but the whole history of the last 60 years makes us confident to predict that there will be many important applications of the new discoveries and of the new technologies needed for the experimentation.

Many of our brightest young physical scientists have traditionally been drawn to this frontier of basic research. The excitement about pioneering such a new area inspires young people to enter science and the cream of the crop to stay in it. I recall that the cyclotron at Berkeley served as an attractive inducement for my election to do my graduate work at Berkeley. The absence of such frontline capability in high energy physics in the United States could cause a "brain drain" of our prospective bright young scientists from basic research or cause them to leave for Europe where the more advanced facilities exist.

In conclusion, I want to reiterate that I think our country should and must take the next step to enable us to continue to investigate one of the most basic frontiers of science: the fundamental constitution of matter.

The Crisis in Pre-College Science and Math Education

Given at the Glenn T. Seaborg Center for Teaching & Learning Science & Mathematics, Marquette, Michigan, April 21, 1989.

"Our Nation is at Risk. Our once unchallenged preeminence in commerce, industry, science and technological innovation is being overtaken by competitors throughout the world ... If an unfriendly power had attempted to impose on America the mediocre educational performance that exists today, we might well have viewed it as an act of war. As it stands, we have allowed this to happen to ourselves ... We have, in effect, been committing an act of unthinking, unilateral educational disarmament."

These are the dramatic opening lines of the report, "A Nation at Risk: The Imperative for Educational Reform" that the members of our National Commission on Excellence in Education handed to President Ronald Reagan in a ceremony at the White House in April 1983.

The Commission was created in August 1981 by then-Secretary of Education Terrel H. Bell, whom the Seaborg Center is very fortunate to have as Chairman of its National Advisory Council. The Commission was charged with reporting on the quality of education in our country and making positive recommendations for remedying our deficiencies. What we learned in the course of our 20-month study and at public hearings across the country was so appalling that we decided to make our report as dramatic as possible to draw attention to these serious problems and to reach maximum readership.

Reprinted with permission from *Skeptical Inquirer* **14**, 270 (1989).

We succeeded in drawing almost unprecedented attention from educators, parents, public and press. It is now apparent that the educational crisis and the urgent need for educational reform are broadly perceived as being a top priority. In the last presidential election George Bush made education an important campaign issue, thus putting the need for educational reform at the top of the national agenda. Recently, President Bush reaffirmed his commitment to improving education in this country when he visited the exhibits of the 40 Westinghouse Science Talent Search winners last month. This was the first time in the 48-year history of the Westinghouse Science Talent Search that the President of the United States has made a tour of the students' exhibits. In his remarks President Bush encouraged the efforts of programs such as the Science Talent Search and said that he believed that these efforts, aided by the federal government, would result in significant changes in our educational system. He said that "As a nation, we have no natural resource more precious than our intellectual resources" and that he wants to make science education one of the most important investments for the future of our nation.

So, how bad is the crisis in education in this country?

Since the report "A Nation at Risk," there have been dozens of reports by a wide spectrum of American organizations that have emphasized and deplored the state of pre-college education in science and math in the United States today. To mention some of those issued during the last year:

- *Everybody Counts: A Report to the Nation of the Future of Mathematics Education*, National Research Council (National Academy Press, Washington, D.C., 1989)
- *The Forgotten Half: Pathways to Success for America's Youth and Young Families*, The William T. Grant Foundation Commission on Work, Family and Citizenship (November 1988)
- *1988 Education Indicators*, Ed. Joyce D. Stern, Assoc. Ed. Marjorie O. Chandler, U.S. Department of Education (1988)
- "Moral Education and Character," Ed. Ivor Pritchard, U.S. Department of Education (September 1988)
- *American Education: Making It Work*, "A Report to the President and the American People," William J. Bennett (April 1988). This is a follow-up report to "A Nation at Risk" assessing the progress that has been made in education since 1983 when "A Nation at Risk" was published.

- *The Forgotten Half: Non-College Youth in America*, The William T. Grant Foundation Commission on Work, Family and Citizenship (January 1988)
- *Women and Minorities in Science and Engineering*, National Science Foundation (January 1988)
- *The Condition of Education: Postsecondary Education*, Ed. Joyce D. Stern, Assoc. Ed. Marjorie O. Chandler, U.S. Department of Education (1988)
- *Undergraduate Science, Mathematics and Engineering Education*, National Science Board, NSB Task Committee on Undergraduate Science and Engineering Education (March 1986).

These reports indicate that while some progress has been made, there is still much work to be done to resolve the crisis in education.

In a recent survey conducted in six nations, 13-year-old American schoolchildren placed last in mathematics behind South Korea, Spain, Great Britain, Ireland, and Canada. Moreover, the high-ranking Koreans surpassed the Americans by a wide margin. 78% of Korean 13-year-old students have the ability to use intermediate math skills to solve two-step problems but only half as many Americans could do so. Moreover, 40% of the Korean children could understand and apply mathematical concepts, but less then one in 10 American schoolchildren could do so.

While most American 13-year-old schoolchildren can add, subtract, multiply, and divide, they are seriously lacking in cognitive skills such as reasoning, investigating, and estimating. However, many entry-level jobs today are demanding workers with high-order, more sophisticated skills for which these students are not being adequately prepared.

In another report released in September 1988, results showed that, compared with students from 13 other countries, 12th-grade American students scored in the lower range on mathematics achievements tests. In geometry American students did only slightly better than Hungary and Thailand; in algebra only Thailand was worse, and the United States ranked last in calculus, below both Hungary and Thailand. Overall the American students obtained only half as many points on tests as the high-ranking students from Hong Kong.

In science the statistics are just as grim. In the same six-nation survey of 13-year-old schoolchildren, Americans again scored below average. Ironically, the United States came in behind Spain which does not have a reputation for being in the forefront in science and technology. Again, as in math, when the 13-year-olds were tested for conceptual understanding, interpretation, and application of what they had learned, the performance of the American schoolchildren dropped dismally. At the highest level, 33% of Koreans had the skills to apply scientific principles, while only 12% of Americans could do so.

In a 13-nation survey of 12-graders, American students scored 13th in biology, 11th in chemistry, and 9th in physics.

In report after report American students consistently score in the lower end of the scale. It is evident that too many students are leaving our schools without adequate skills to be full participants in our increasingly technologically-oriented society. By 1993, those 13-year-old students, who today are performing so poorly in math and science, will be voting and entering the work force — the scientific and industrial leadership of our nation will be in their hands.

Already it is evident that the industrial supremacy of our nation is being seriously threatened. Currently only about 6% of American adults are scientifically literate. This greatly diminishes their productivity in a more technically-demanding work place.

Scientific illiteracy also affects their ability to function effectively as citizens and their ability to play an informed role in social and political decisions with scientific or technological content, such as issues in nuclear power, acid rain, the ozone layer, genetic engineering, chemical warfare, and so forth. The vitality of a democracy assumes a certain "core of knowledge" shared by everyone which serves as a unifying force.

There can be no doubt that scientific literacy, a solid understanding of science and mathematics, is now more important then ever before — and there is irrefutable evidence that the skills of our youth are not only not progressing with the increasing demands — but actually are deteriorating at an alarming rate.

This country cannot afford another generation of students who are ill-prepared to respond to the world-wide rapid growth of scientific knowledge and technological power. The nation's future depends on them. Therefore, we must improve general science and math education for all our young.

The task of guiding the intellectual (and often social) development of our young is an all-important one. We must begin to recognize teachers' contributions not only by adequately compensating them for their service, but also by giving them due respect which would motivate them to refine their skills and expand their knowledge to meet future challenges. While some teachers are eminently qualified, a significant number of them have little background in science or mathematics or have not had any involvement with these subjects for many years and have simply lost touch with changes in their fields.

According to a 1985–1986 national survey supported by the National Science Foundation and released in 1988, many science and math teachers feel they lack adequate training and are not qualified to teach. Indeed, the study showed that only about one-third of elementary science schoolteachers has taken a college chemistry course and only one in five has taken a college physics course. While 82% of them felt they were very well qualified to teach reading, only 66% felt they could teach math. In the science disciplines less than one-third of the teachers felt very well qualified to teach the life sciences and only 15% felt very well qualified to teach the physical or earth and space sciences.

At the high school level, less than one in three teachers included earth and space sciences in their undergraduate curriculum. The report also states that "more than half of all secondary science teachers have never had a college computer science course and almost half have had no college calculus."

However, most science and math teachers feel that they would enjoy teaching these subjects if they had adequate preparation. Whatever their situation, these teachers need opportunities to upgrade their math and science teaching skills. Thus, there is not only a need to increase the available pool of qualified math and science teachers, but there is a need to enhance the capabilities of those now teaching math and science.

The Lawrence Hall of Science is an institution committed to improving the quality of mathematics and science instruction for pre-collegiate students. For over two decades the Hall has dedicated its superior resources as part of the University of California to the continuing battle against educational mediocrity.

The Lawrence Hall of Science, of which I serve as Chairman, was conceived in 1958 and built in 1968 as a memorial to Ernest O. Lawrence, the University of California's first Nobel Laureate and inventor of the cyclotron. As a dynamic research and educational institution, the Hall continues today, 21 years after its dedication in 1968, to focus its efforts on three main objectives:

1. To improve the quality of mathematics and science instruction for the benefit of pre-collegiate students through the development of innovative math and science courses and accompanying curriculum materials and teacher training services;
2. To augment the mathematics and science instruction provided by our schools by offering special mathematics and science courses at the Hall; and
3. To enhance the knowledge, appreciation, and enjoyment of mathematics and science for the general public by providing the community with a math and science center.

In its efforts to improve instruction of math and science, the Hall has developed programs for students and teachers in their own schools. Two particular programs that have been recognized for their excellence are CHEMStudy, a comprehensive high school chemistry curriculum and The Science Curriculum Improvement Study (SCIS), an activity-oriented

Seaborg teaching an elementary school class at the Lawrence Hall of Science, Berkeley, California, USA. Reproduced by courtesy of the Lawrence Berkeley National Laboratory Photographic Services Department.

science program for children K-6 which was developed in the late 1960's and is now used in over 20% of the nation's elementary schools. Through its in-school programs, the Hall reaches over 122,000 children every year in the San Francisco Bay Area.

The Hall sets as a priority the development of programs that deal with the issue of attracting and retaining underrepresented students such as young women and minorities in mathematics and computer education. Additionally, research and development efforts also focus on designing educational programs and materials for specific target audiences with special needs, such as disabled students. Three programs of national importance are Mathematics, Engineering, Science Achievement, (MESA) a program serving 7,000 minority youth in California and at four other sites nationally, Science Activities for the Visually Impaired (SAVI), and Science Enrichment for Learners with Physical Handicaps (SELPH).

The Great Explorations in Math and Science (GEMS) is a curriculum development program funded by the Carnegie Corporation and The Mellon Foundation to finalize and disseminate the Hall's well-tested science and mathematics instructional activities and lesson plans. GEMS also includes some of our exhibits and their accompanying learning activities so these may now be used by other science centers and museums.

Industry Initiatives for Science and Mathematics (IISME) is a collaborative effort of the LHS and Bay Area industries designed to prevent attrition of our present science and mathematics teachers. This program selects excellent teachers to work in well-paid industry positions during the summer in order to familiarize the teachers and industry with each others' needs, problems, and successes. These teachers do real work — not busy work — and have made significant contributions to their host companies. These jobs augment the teachers' annual salaries, thus helping to prevent permanent teacher job-shifting to more lucrative industry positions. IISME placed about 100 teachers with host companies in 1988.

The Chemical Education for Public Understanding Program (CEPUP) has been developed to increase public knowledge of toxic substances and hazardous materials and their safe use, storage, and disposal. This will be accomplished by developing modules of materials for junior-high-school-age students and the general public; setting up an on-site computer-based information center; and having available printed information and audiovisual materials for business, industry, and the general public.

For over two decades, the Hall has also provided innovative leadership in pre-collegiate math and science education through the publication of major curricula. These learning materials are utilized by millions of students in the United States and around the world. Curricula and exhibits developed by the Hall are currently used by schools and science centers in over 30 countries. Each year, over 700 educators from around the world visit the Hall to learn new techniques to improve science and mathematics instruction.

For the effective utilization of these programs and materials, the Hall provides comprehensive teacher training workshops and seminars. One particularly exemplary program, EQUALS, focusses on helping K-12 educators acquire methods and materials to attract and retain underrepresented students in mathematics and computer education. Since 1966, in the state of California, over 150,000 teachers have participated in the Hall's teacher training programs.

Trained personnel of the Hall also design and manage an ever-changing curriculum of exhibits, public events, special events, classes, workshops, and seminars that are offered on a full-time, year round basis to the general public. In addition to participants in the Hall's specific programs, approximately 265,000 individuals visit the Hall annually.

The Lawrence Hall of Science has achieved national and international prominence as a result of its innovative and effective programs for improving mathematics and science instruction. However, despite the Hall's numerous discoveries and noted accomplishments, much more work still needs to be done.

There is a need for more institutions like the Hall. Therefore, I am deeply honored to be here today to participate in the dedication ceremonies of another institution committed to improving the education of our young — The Glenn T. Seaborg Center for Teaching and Learning Science and Mathematics.

As part of Northern Michigan University (NMU), the Seaborg Center is in a unique position to provide, through its services, resources, and programs, quality education for students and teachers of the Upper Peninsula.

The on-campus location and educational focus of the Seaborg Center will permit interaction between K-12 teachers and NMU faculty through joint grade school/university conferences. The Center will also have displays of demonstrations, teaching concepts, and experiments, along with a Model Classroom facility for the testing and demonstration of these new materials and teaching methods. Many of these new teaching materials will be developed and produced on-site in the Center's own production facility.

Pre-kindergarten through pre-college students have specialized needs which the Center addresses. Programs like the Michigan Math Early Placement Test and the Seaborg Summer Science Academy have already shown rapid growth. There are plans to greatly expand services to students to include an interactive museum, learning centers, and year-round classroom offerings in science and math specialty areas.

The Seaborg Center, while still a young Center of Excellence, is rapidly gaining a national reputation which has funding agencies coming to the Center with specific requests. The Lawrence Hall of Science at the University of California at Berkeley, the Kellogg Foundation, and the Public Service Satellite Consortium have all sought out the Seaborg Center and Northern Michigan University to direct or participate in programs which will impact not only the Upper Peninsula, but teachers and students throughout the United States.

We at the Lawrence Hall of Science eagerly look forward to future collaborations between the Hall and the Seaborg Center and to joining forces to achieve the goal of educational reform so urgently needed in our country today.

Cooperation Between the U.S. and the U.S.S.R. in the Peaceful Uses of Atomic Energy

Presented at the International Conference "Fiftieth Anniversary of Nuclear Fission," Leningrad, U.S.S.R., October 16–20, 1989.

The decades of the 1960's saw a marked expansion of cooperation between the United States and the Soviet Union for the peaceful uses of atomic energy. In my opinion, this development constituted one of the most encouraging elements in the international scene.

Until 1955 contacts between American and Soviet nuclear scientists were virtually nonexistent, as indeed (after World War II) were U.S.–U.S.S.R. contacts in other fields, except as required in formal intergovernmental relations. Then, in July 1955, the discussions of the Heads of Government meeting in Geneva led to the declaration of a policy with the following aims:

> To lower the barriers which now impede the interchange of information and ideas between our peoples.
>
> To lower the barriers which now impede the opportunities of people to travel anywhere in the world for peaceful, friendly purposes, so that all will have a chance to know each other face to face.
>
> To create conditions which will encourage nations to increase the exchange of peaceful goods throughout the world.

Reprinted with permission from *Fiftieth Anniversary of Nuclear Fission: Proceedings of the International Conference* (Nova Science Publishers, Commack NY, 1993).

Lawrence Berkeley Laboratory Report No. LBL-27859, Berkeley, California (1989).

In connection with these objectives, it was noted that atomic science "possesses a tremendous potential for helping raise the standards of living and providing greater opportunity for all the world."

The task of planning ways to implement this policy was entrusted to the Foreign Ministers and their staffs in meetings which began in October 1955. The following year, the policy of the U.S. government to encourage mutually beneficial contacts with eastern European countries — including the Soviet Union — was declared in a statement by President Eisenhower on June 29, 1956. Soon thereafter, negotiations were initiated that led to the signing in January 1958, by U.S. Ambassador William S. B. Lacy and Soviet Ambassador G. N. Zaroubin, of the first "Agreement Between the United States of America and the Union of Soviet Socialist Republics on Exchanges in the Cultural, Technical, and Educational Fields."

Meanwhile, in August 1955, the Geneva Conference on Peaceful Uses of Atomic Energy had provided the setting for the first significant contacts and the first impetus toward collaboration between East and West in this field. In September 1956, both the United States and the Soviet Union were among the 81 signatories of the Statute establishing the International Atomic Energy Agency. During the next two years, some visit exchanges between the United States and Soviet Union by nuclear scientists took place on an informal basis. Initial steps toward formal cooperation in peaceful nuclear applications followed, pursuant to the 1958 Lacy–Zaroubin agreement. Among the various programs provided for under this agreement was an exchange of exhibits in the peaceful uses of atomic energy. In addition, there were many scientific and technical visits and exchange projects in which USAEC and/or USAEC contractor employees participated. The 1958 Geneva Conference offered additional opportunities for interesting contacts.

The Lacy–Zaroubin agreement was succeeded by a similar one signed in Moscow on November 21, 1959, which specified exchanges to be carried out in 1960–1961. Three days later USAEC Chairman John A. McCone and his Soviet counterpart Professor Vasily Semenovich Emelyanov, Director of the Main Administration on the Use of Atomic Energy, signed in Washington a "Memorandum on Cooperation between the United States and the Soviet Union in the Field of the Utilization of Atomic Energy for Peaceful Purposes." This Memorandum, known as the McCone–Emelyanov Memorandum, specified four principal areas in which visits and information were to be exchanged during 1960 and 1961 and the procedures to be followed in arranging them. It also declared the parties' agreement to consider the feasibility and the possibility of making new scientific instruments available on a reciprocal basis. In order to assure that the IAEA and its members would "benefit to the fullest from this effort for further development of the peaceful uses of the atom," it was also agreed that the Agency would be given all reports and the results of the exchanges carried out. It would also be asked, to the extent possible, to assist in the consideration of possible joint projects by sponsoring meetings, symposia, or studies considered necessary for such planning.

The provisions of this Memorandum had been developed gradually in the course of numerous personal discussions between principal U.S. and U.S.S.R. officials concerned with nuclear energy. In Vienna in May 1959, Dr. Isidor I. Rabi (U.S. representative to the IAEA Scientific Advisory Committee) and Dr. John A. Hall (USAEC assistant General Manager for International Affairs) had broached to Emelyanov the idea of an exchange on controlled fusion and nuclear power. Further discussions on the subject of an information exchange were held in the Soviet Union two months later by Professor Emelyanov and Admiral Hyman G. Rickover, who was accompanying then-Vice President Richard Nixon on his visit to the Soviet Union. When Premier Nikita S. Khrushchev visited the United States in September 1959, Emelyanov came with him in order to discuss peaceful uses cooperation with Chairman McCone. Further talks were possible during a trip Chairman McCone made to the Soviet Union in October with a group of U.S. nuclear scientists, and during a reciprocal visit to the United States by Professor Emelyanov and a party of eight at the end of the same month. It was during this visit that the McCone–Emelyanov Memorandum was signed.

I have mentioned these details because of the vital part the many meetings and talks played in reaching initial agreement on U.S.–U.S.S.R. cooperation in peaceful nuclear applications. While useful and constructive in all of our cooperative activities, such personal contacts and discussions are particularly important in relationships especially likely to be affected by international political developments. They have, therefore, been essential to the continuation and expansion of our cooperation with the Soviet Union.

At the time I assumed my USAEC responsibilities, something over a year after the signing of the McCone–Emelyanov Memorandum, some of the cooperation planned had been initiated. Despite the renewed international tension of that period, satisfactory visit exchanges had taken place in the fields of high energy physics and controlled thermonuclear fusion. Furthermore, a small group of U.S. and Soviet scientists had met at the American Institute of Physics in New York on September 16, 1960, to consider the scientific desirability and feasibility of large accelerators with energy in excess of 300 BeV. A number of individual scientific exchanges conducted outside the Memorandum on Cooperation had also been arranged or were being explored, including one in theoretical high energy physics. Efforts by the U.S. toward implementation of the other exchanges covered by the Memorandum, however, had been unsuccessful. One of my first tasks, therefore, was to try to carry out the agreed-on programs and develop realistic plans for cooperation in the years ahead. I was deeply convinced of the value of this cooperation. It had long been my view that scientific and technical exchanges could serve to increase understanding and friendship between the Soviet Union and the United States. I had some personal knowledge of such exchanges; before and during my tenure as Chancellor at Berkeley I had received many Soviet bloc nationals. I was eager to promote further U.S.–U.S.S.R. contacts, which I was sure would be mutually beneficial.

In February 1961, the Soviet organization concerned with peaceful nuclear applications (which in June 1960 had been renamed the "State Committee on the Utilization of Atomic Energy," or SCAE) was still directed by Professor Emelyanov. I had first known Emelyanov years before as a fellow chemist, having met him when he visited Berkeley in the early 1950's, and I had seen him in the intervening years at the First and Second Geneva Conferences on Peaceful Uses of Atomic Energy. With our previous cordial associations as a foundation, I looked forward to a constructive relationship with him.

As there was no chance of an immediate meeting, I wrote Emelyanov on March 3, 1961, reviewing the status of our cooperation and the various exchange possibilities. This followed a meeting with U.S. Ambassador to the Soviet Union, Llewellyn Thompson, on February 16th, at which he had agreed to discuss these matters with Emelyanov and Chairman Khrushchev. I shall not attempt to give the detailed background pertinent to these proposals; suffice it to say that they had been the subject of extensive correspondence and discussions, which had not yet led to the necessary agreement on details.

The IAEA Fifth General Conference in September 1961 afforded me the first opportunity to discuss our exchanges in person with Emelyanov who, in addition to serving as the Soviet member of the IAEA Scientific Advisory Committee and on the Board of Governors, also headed his country's delegation to the Conference. On several occasions, however, our conversation centered entirely on IAEA matters not closely related to my present subject of our cooperation with the Soviet Union. This was the case, for example, at a dinner given by Indian Ambassador Arthur S. Lall on September 29th, at which IAEA questions of particular interest to U.S., Soviet, and Indian delegations were discussed, especially the imminent election of a new Secretary-General (on which we strongly disagreed). Among the guests at this dinner, incidentally, was former Foreign Minister Vyacheslav M. Molotov. Since September 1960 he had served as Permanent Soviet Representative to the IAEA with the rank of Ambassador, though apparently without any real representative authority. The 1961 General Conference was the last that he attended.

After the talks in Vienna, discussion of topics of mutual U.S.–U.S.S.R. interest in the nuclear area was renewed in correspondence with the principal officials of the SCAE (State Committee on Atomic Energy of the Soviet Union) — at first with Professor Emelyanov, and then with Andronik M. Petrosyants, who became SCAE Chairman in February 1962. (Emelyanov remained with the organization for several years as a Vice Chairman.) Plans for the Third Geneva Conference on the Peaceful Uses of Atomic Energy were among the subjects on which we exchanged views; however, generally our letters concerned the exchange program. The exchanges in reactor development and waste disposal were scheduled definitely for March and April 1962, but the Soviet Union was unable to accommodate our delegations, so the visits did not take place. Aside from frustrations such as these, we were confronted by the problem of planning for the future. On March 8, 1962, the third "overall" U.S.–U.S.S.R. Agreement on Exchanges was signed in Washington, providing for exchanges in 1962 and 1963. It was time to look toward negotiation of a new Memorandum on Cooperation in Peaceful Uses of Nuclear Energy and agreement on exchanges that could be carried out.

Virtually no real progress on future planning had been made when I returned to Vienna in September 1962 for the Sixth IAEA General Conference. I was hopeful that conversations with the Soviet representative there would lead to some positive action regarding our exchanges. The atmosphere seemed more favorable than that of 1961. In general the U.S.S.R. delegates appeared to be more interested in cooperation, demonstrating fuller support of the IAEA than previously and apparently seeking to avoid any major differences or difficulties. On several occasions Emelyanov and I had cordial talks that ranged far from the subject of our exchanges but in the final instance did lead to a decision on action.

Not long after the meetings in Vienna, we did receive the expected proposals from the Soviet Union for future exchanges. We considered these generally satisfactory and an excellent basis for a mutually beneficial program. Most points of difference were soon resolved in my correspondence with Petrosyants and in a meeting of USAEC and SCAE representatives. As soon as I felt confident that agreement could be achieved, I invited Petrosyants and Emelyanov to meet with us in Washington to settle time details and sign the new Memorandum here. In reply, recalling that the previous Memorandum had been signed in Washington, Petrosyants cordially invited me and my colleagues to come to the Soviet Union for the event and for visits to Soviet research facilities. I accepted with great pleasure, and plans were made for our visit to take place in May 1963.

In addition, on the flight to Moscow, we had an unexpected last-minute companion: Vitalii I. Goldanskii, Chief of the Radiation and Nuclear Chemistry Group at the Chemical Physics Institute of Moscow (Fig. 1). Goldanskii, an outstanding scientist with whom I was well acquainted, had been visiting U.S. nuclear research facilities including the USAEC Brookhaven National Laboratory. On learning of our projected trip, he had asked me only a few days earlier whether he could "hitch a ride" home with me. I had agreed without hesitation. While officials of both the U.S. State Department and the U.S.S.R. Embassy in Washington may have been somewhat taken aback by this impromptu arrangement, neither side raised any objection.

Our trip was indeed a memorable one (Fig. 2). Everywhere, we were cordially received and treated with the warmest hospitality. Our hosts accepted unhesitantly the itinerary we had proposed, and even included some additional sites they thought would interest us. Throughout the trip, we were accompanied either by Chairman Petrosyants or by one of his vice chairmen. Our journey marked a considerable number of "firsts," starting with record flying time to Moscow with the fine pilots and crew

Fig. 1. Arrival of the U.S. scientific delegation to the U.S.S.R. at Moscow's Shermetyevo Airport. Left foreground are U.S. Ambassador to the Soviet Union Foy Kohler, Seaborg, and State Committee on Atomic Energy [SCAE] (Soviet Union) Chairman Andronik M. Petrosyants, 19 May 1963. Reproduced by courtesy of Arnold R. Fritsch.

Fig. 2. Seaborg and Petrosyants signing the U.S.–U.S.S.R. Agreement for Cooperation, Moscow, 21 May 1963. Reproduced by courtesy of Arnold R. Fritsch.

of the presidential plane, "Air Force One," and continuing through milestones that were more significant from our point of view. We were the first foreign group to visit the Soviet Reactor Testing Station at Ulyanovsk/New Melekess and the site of the 70-BeV High-Energy Accelerator at Serpukhov; the first western visitors since the war to visit the Radium Institute in Leningrad; and the first foreign group to see many industrial reactors and certain other scientific equipment.

On May 21, 1963, all of our delegation went with Morokhov to the U.S.S.R. Academy of Sciences, where I introduced our group, and President Mstislav V. Keldysh introduced his group. Later, Manson Benedict and I presented to N. N. Semenov a certificate of membership in the U.S. National Academy of Sciences on behalf of NAS President Frederick Seitz. Albert Ghiorso and I presented a mendelevium folio to Keldysh (Fig. 3). Albert Crewe and I presented CP-1 graphite pieces to Keldysh, Mikhail D. Millionshchikov (Vice President of U.S.S.R. Academy of Sciences), and Morokhov (SCAE). Academicians Tamm, Aleksandr

Fig. 3. Presentation of the mendelevium folio to President Mstislav V. Keldysh of the U.S.S.R. Academy of Sciences, Moscow, 21 May 1963. From L to R: Igor D. Morokhov (First Deputy Chairman, SCAE), Mikhail D. Millionschchikov (Vice President, U.S.S.R. Academy of Sciences), Seaborg, Keldysh, Albert Ghiorso (scientist, Lawrence Radiation Laboratory, Berkeley, CA, USA), and in foreground with back to camera, Dimitrii V. Skobeltsyn (Director, Lebedev Physics Institute). Source unknown.

P. Vinogradov, V. I. Spitsyn, Skobeltsyn, and Lev A. Artsimovich were also present.

On May 29, 1963, I had an appointment in the Kremlin for an hour and a quarter with Leonid Ilyich Brezhnev, Chairman of the Presidium of the U.S.S.R. Supreme Soviet. (Two days before, my Soviet hosts proudly told me they had made an appointment for me with an important official of the Soviet government, namely the president. When I indicated some hesitation or lack of knowledge of this individual, they hastened to assure me that this presented a very unusual opportunity for me, because the president, a man named Leonid Brezhnev, was destined in their opinion to play a very important role in the future of the Soviet government.)

Brezhnev seemed a personable man of pleasant appearance, about 55 or so, with more a Western manner than most Soviet officials. He first asked me if this was my first visit to the Soviet Union. When I replied that it was, he said he thought it was a good start and added: "Good relations require frequent visits." He agreed with me that this was particularly true in the field of science because of the international aspect of scientific research, which makes science an excellent vehicle for continuing contact and the development of good relations.

He asked me about Soviet atomic power plants and whether I thought that Soviet scientists were working in the proper direction. I replied that the Soviet approach was very similar to ours, and that similar problems were being encountered, such as corrosion. I said that cooperation in solving such problems would help both countries, and Brezhnev replied that he would welcome such cooperation. He agreed with my statement that there was a place for atomic power now in the economies of both countries, particularly in areas where other sources of power were scarce and expensive.

As a memento of my visit, I presented him with a square, transparent paperweight, containing (as I explained) a small piece of the original graphite taken from the reactor (CP-1) in which the world's first self-sustaining chain reaction was achieved on December 2, 1942, in Chicago, and Brezhnev responded to this gift with great warmth. After thanking me, Brezhnev said he wanted to leave the subject of science. He said that I would doubtless meet with President Kennedy upon my return and that he wanted me to tell the President that Khrushchev meant what he said about peaceful coexistence and peaceful cooperation in his speeches, addresses, and documents sent to the President.

Brezhnev's manner was warm and friendly; he seemed to display a good understanding of science and technology when he spoke of the details of the work in various institutes. All in all, he made a favorable impression of a man who wanted to get along with the United States. My talk with him was perhaps even more interesting in retrospect, since his replacement of Khrushchev as First Secretary of the Communist Party of the Soviet Union occurred less than a year and a half later. I think it is worth mentioning that although at the time of our meeting a number of people regarded him as a mere figurehead, the opinion was growing among certain experts (including Ambassador Kohler) that he was assuming a position of increasing importance and that — I was told prophetically — he might actually be the successor to Khrushchev. I had the impression, which I could not document, that he spoke as though he anticipated his future role in the government.

Sooner than I expected, I visited the Soviet Union again, in August of the same year, 1963. Unable at that time to achieve agreement on general and complete disarmament, the United Kingdom, the United States, and the Soviet Union had reached agreement on the Limited Test Ban Treaty, banning nuclear weapons tests in the atmosphere, in outer space, and under water. The treaty was to be signed in Moscow, and I was invited to be a member of the U.S. delegation attending this event.

From 11 a.m. to noon on Monday, August 5, 1963, the U.S. delegation visited Nikita Khrushchev in his Kremlin office, which is long and narrow. We sat at a table with a green felt top, like a pool table. There were windows on the west side, pictures of Lenin and Marx, an electric clock on Khrushchev's desk, bookcases, and two telephones at the conference table (but I saw only one telephone at Khrushchev's desk). Gromyko, Kuznetsov, Smirnovsky, and Dobrynin were present. After greetings by Khrushchev and Rusk, Khrushchev said the Test Ban Treaty was only a first step.

At 4.30 p.m. we attended the historic signing of the Limited Nuclear Test Ban Treaty, in Catherine's Hall, by Rusk, Gromyko and Home simultaneously, followed by speeches by Gromyko, Rusk, Home and U Thant (Fig. 4). I stood just behind Khrushchev and he and I tipped our champagne glasses together for toasts at least five times. About 50–60 press representatives and photographers were present.

At about 5.15 p.m. we attended a huge reception in Georgian Hall (magnificent!) where Khrushchev pulled a prepared speech out of his

Fig. 4. Signing of the Limited Test Ban Treaty (LTBT), Georgian Hall, the Kremlin, Moscow, August 5, 1963.

Seated L to R: Dean Rusk (U.S. Secretary of State), Andrei Gromyko (Soviet Foreign Minister), Lord Home (British Foreign Secretary).

Standing L to R (partial list): Senator George Aiken (partly out of picture), Senator William Fulbright (leaning forward), Alexander Akalovsky, Senator Hubert Humphrey, Senator Leverett Saltonstall (behind Humphrey), UN Ambassador Adlai E. Stevenson, UN Secretary-General U Thant, Soviet Premier Nikita Khrushchev, (unidentified individual), Soviet Deputy Foreign Minister Valerin Zorin, Soviet Ambassador to U.S. Anatoly Dobrynin (behind and slightly left of Zorin), William Bundy (behind and slightly left of Dobrynin), British Lord Privy Seal Edward Heath, (unidentified individual), and Seaborg.

UPI photo, reproduced by courtesy of the Bettmann Archive.

pocket and delivered it. I had a chance to talk to Brezhnev, Petrosyants, Gromyko, Kuznetsov, Dobrynin, Zorin, Tsarapkin, and Voroshilov (of the military).

During my May visit to the Soviet Union I had invited Chairman Petrosyants and his colleagues to visit nuclear facilities in the United States. Upon my return from Moscow I sent him a formal invitation, suggesting the coming fall. He accepted, and plans were developed accordingly. The visit took place during the period from November 16–December 3, 1963.

I was disappointed by the fact that the physicist Georgiy N. Flerov, who was scheduled to participate in the visit and for whom a visa had been issued, did not come. Members of the delegation indicated that his absence was due to ill health. Either I, another USAEC Commissioner, or the USAEC General Manager accompanied the Soviet delegation during almost the entire trip.

The tour proceeded according to plan until Friday, November 22nd. I was with the group, visiting facilities at the Lawrence Radiation Laboratory at Berkeley, when word came of President Kennedy's assassination. The effect of the terrible tragedy made it impossible to proceed with the events scheduled for the day. After considering various aspects, including particularly the physical safety of our guests, I decided to cancel the balance of the activities planned for Friday and Saturday and have the group go to Yosemite National Park until Sunday evening. The original schedule was resumed on Monday with a visit to the National Reactor Testing Station in Idaho. There, at the suggestion of the Soviet delegation, television sets were installed at the EBR-II building and at the Central Facilities cafeteria, to permit viewing of the funeral preliminaries and part of the service. I, of course, had returned to Washington and the members of Petrosyants groups were very touched to see me as one of the officials attending the service at Arlington Cemetery.

I cannot speak too warmly of the comportment of our visitors at this difficult time. They made the necessary schedule changes and spur-of-the-moment arrangements easier by their unhesitating cooperation. Even more important, perhaps, they were genuinely sympathetic to the United States and saddened, sincerely sharing our grief.

After completing their tour of major unclassified nuclear facilities in the United States, the Soviet delegates came to Washington. They met with us at USAEC Headquarters on November 30th for a review of their impressions, a discussion of ways to implement our exchange program, and a comprehensive briefing on the U.S. civilian power development program. Following our meeting, during the luncheon we gave in honor of our Soviet guests, Chairman Petrosyants and I continued our conversation regarding future exchanges.

As I had hoped, the successful accomplishment of my visit to the Soviet Union in May 1963 and Petrosyants' return visit to the United

States heralded the start of a continuing series of exchanges in accordan with the program established in our Memorandum on Cooperation. During the next three years, exchanges of delegations of scientists were completed in many areas.

Areas for exchange visits included solid state physics, controlled thermonuclear reactions, radioactive waste disposal, power reactor development, radioneurological research, and low-energy physics. In addition, a U.S. team of medical tracer specialists visited medical facilities in the Soviet Union. Several long-term reciprocal assignments were made, involving specialists in high energy physics and controlled thermonuclear reactions. A modest reciprocal USAEC–SCAE exchange of unclassified documents on a monthly basis proceeded satisfactorily. Aside from all these activities conducted pursuant to our Memorandum, scientists of each country continued to visit nuclear facilities in the other, in conjunction with attendance at conferences or through visits arranged by other organizations participating in the overall U.S.–U.S.S.R. exchange program.

In addition to the various activities summarized above, in June 1964, a cooperative program for the exchange of technical information on nuclear desalination was initiated. A meeting was held in Washington on this subject on July 14, 1964. This was followed by a Soviet tour of U.S. desalting and reactor facilities. A formal U.S.–U.S.S.R. agreement on cooperation in the general field of desalination, including the use of nuclear energy, was signed in Moscow on November 18, 1964, during a visit of U.S. desalting experts to the Soviet Union. In 1966, this agreement was extended to remain in effect for two more years.

There were several occasions during this period when I was able to discuss the progress of our exchange program with Petrosyants and/or Emelyanov. In the course of the Third Geneva Conference on the Peaceful Uses of Atomic Energy (at which Emelyanov served as President), both of them, as well as Nikolai S. Khlopin of the U.S.S.R. Institute of Atomic Energy and Igor V. Tikhonov of the SCAE, were among the representatives of 15 countries for whom I had the pleasure of hosting an overnight visit on the U.S. nuclear ship *Savannah*, then in Sweden (Fig. 5). There were, of course, numerous other opportunities for talk during the Geneva Conference; I recall a particularly pleasant and relaxing lunchtime when Mr. and Mrs. Emelyanov and my wife and I were among the guests of German Ambassador and Mrs. von Keller at their charming home, "Villa Primerose," in suburban Geneva-Vesenaz.

My trip to the Soviet Union in September 1969 was one of my most interesting and rewarding as Chairman of the USAEC. Occasioned by four special events — the Thirteenth General Conference of the IAEA in Vienna, a symposium convened in Stockholm by the Nobel Foundation on the role of science in society, the 100th Anniversary Commemoration in Leningrad of Mendeleev's formulation of the periodic system of the elements, and the opening in Bucharest of the USAEC's "Atoms-in-Action" exhibit — this trip afforded an opportunity to visit nuclear facilities in a number of countries including the Soviet Union and other East European countries.

As discoverer of the periodic table, Mendeleev is an authentic hero of Russian science, and the Leningrad symposium was a major event. I had received a special invitation to attend, undoubtedly because of my role in extending and establishing the validity of the table with respect to a

Fig. 5. Petrosyants, Helen Seaborg, and Seaborg visit to NS *Savannah* in the Baltic Sea, 3 September 1964. Reproduced by courtesy of the U.S. Department of Energy (formerly the U.S. Atomic Energy Commission).

number of new man-made elements. Another factor was probab Russian scientists' satisfaction over the fact that as long ago as 1955 (when U.S.–U.S.S.R. relations were somewhat cool) we in the United States, after discovering element 101, had named that element mendelevium. In any case, I accepted my invitation with pleasure and keen anticipation.

Before our tour of the metrology Institute we met in the office of Professor V. O. Arutunov, Director of the Institute. Present were B. N. Oleinik (Deputy Director), L. K. Kayak (Division Chief, Linear Measurements), M. F. Yudin (Division Chief, Ionizing Radiations), A. I. Kartashev (Division Chief, Optics), I. A. Yaritsyna (Chief of Laboratory, Ionizing Radiations), F. M. Karavayev, S. A. Smolitch (Scientific Secretary), and B. I. Ignatyev (translator). Arutunov explained the work of the Institute. The Institute is organized in some 60 divisions and was noted as having responsibilities similar to our National Bureau of Standards. Among the divisions are theoretical, magnetics, thermal, optics, and ionizing radiation. There are about 300 control laboratories throughout the Soviet Union that are a part of the Metrology Institute and have the responsibility of verifying reference standards and calibration of measuring equipment. It was stated their control laboratories have the "right and duty to stop production in a plant if a violation of the standard is detected."

We toured the Mendeleev rooms. We saw the appointment calendar (stack of sheets) in Mendeleev's study with the sheet for January 11, 1907, on top, with a note in Mendeleev's handwriting (his last in this appointment calendar) saying he had an appointment with the Minister of Industry for 1 p.m. — he became sick, didn't return to the office in the intervening days, and died on February 14, 1907.

One of my most treasured possessions is an autographed copy of one of Mendeleev's books. In August 1959, when I was serving as Chancellor of the University of California at Berkeley, I received this very precious gift. The book was sent to me by Emmanuel Tsipelzon, who described himself as "an old Moscow secondhand bookseller." The accompanying message read:

> During this visit to the U.S.S.R., Vice President, U.S.A., R. Nixon informed us that before his departure for the Soviet Union he was visited by his friend, professor of chemistry, Mr. Seaborg, who named the 101st element, discovered by him, of the D. I. Mendeleev Periodic Table after this great Russian chemist.

> In this friendly act of the American scientist each Soviet citizen discerns a great respect toward our people and its culture, as well as one of the steps toward the liquidation of the absurd, according to Nikita Sergeevich Khrushchev, tense state of "cold war" between two great nations.
>
> May I present to you, in commemoration of your remarkable discovery and your noble act, the book by Dimitri Ivanovich Mendeleev *Fundamentals of Chemistry* with his autograph.

The autograph reads, "To my deeply appreciated colleague, Dr. N. I. Bistrov, in commemoration of saving my son, D. Mendeleev, 1889." Mr. Tsipelzon wrote that Mendeleev presented this book to Bistrov when he arrived back in Moscow, having left the meeting of the Royal London Society where he was to read a paper because he heard that his son was deathly ill, and discovered that thanks to Bistrov, his son was already on the road to recovery.

Arutunov then led us on a tour of the Institute. This consisted of very brief 5–10 minute stops at several rooms where specific standard investigations were conducted. Among these were light frequency work for timing devices using krypton in the optics department under Kartashev, voltage and ampere calibration data in the electrical measurement department under P. N. Goryunov, and dosimeter work in the neutron calibration department under Yaritsyna. We also were shown the official U.S.S.R. standard for the meter and kilogram which were located in a special vault which required three separate keys, each in the possession of a different person, to open.

Once back at my desk after this extremely interesting trip, I was anxious for the earliest possible implementation of the U.S.–U.S.S.R. exchange plans discussed in Moscow. Prompt arrangements for the proposed joint high energy physics experiments seemed especially desirable. I wrote Petrosyants in this connection on November 12, 1969, pointing out the need to insure that the new Memorandum on Cooperation (necessary to cover exchanges in 1970 and 1971) contained appropriate provisions. This time there were no long delays. The fourth U.S.–U.S.S.R. Memorandum on Cooperation in the Peaceful Uses of Atomic Energy was signed in Washington (by Abraham Friedman for the USAEC and Ivan Smolin for the SCAE) on February 10, 1970, as an annex to the seventh overall U.S.–U.S.S.R. Agreement on exchanges, signed the same day. Later Chairman Petrosyants and I signed copies of the memorandum and exchanged them by mail.

In line with our conversations and various letter exchanges, this latest Memorandum contained a section — Article VI — declaring our mutual agreement to conduct joint projects in high energy physics. Subsequently, a Protocol setting forth general procedures for collaborative activity pursuant to Article VI was developed and was signed in Washington on November 30, 1970. The Protocol stipulated that details regarding the joint experiments to be conducted would be spelled out in annexes in this document.

Actually, thanks to the Memorandum on Cooperation's Article IV ("Visits by Invitation" — a provision that, as I mentioned earlier, was first included in the 1968 Memorandum), it was possible to start the first of our joint projects even before the Protocol was signed. Pursuant to exploratory discussions held at the Serpukhov facility and in Moscow in the spring and fall of 1969 and the spring of 1970, a five-man U.S. group headed by Darrell Drickey of the University of California, Los Angeles, spent six months at Serpukhov (September 1970 to April 1971) collaborating with Soviet physicists in a study of pion–electron scattering. Work proceeded so satisfactorily on this project that after a short stay at home Dr. Drickey and several colleagues returned to Serpukhov in the summer of 1971 to gather additional data. Several other visits also took place in the field of high energy physics. For example, while preparations were under way for the U.S. team's stay at Serpukhov, two Soviet physicists were attending a summer seminar at the National Accelerator Laboratory at Batavia, Illinois; their assignment there was regarded as a useful preliminary to the eventual selection of a U.S.–U.S.S.R. collaborative project as part of the experimental program of the 200-BeV accelerator under construction at Batavia.

In addition to the cooperation in high energy physics, exchanges in other nuclear fields continued to expand during the 1970–71 period. A visit a Soviet reactor team had made to the United States in November 1969 was reciprocated in June 1970, when a U.S. reactor group of ten (representing industry as well as the USAEC) spent two weeks in the Soviet Union. As Petrosyants had promised me, our group was able to visit all the facilities requested, including the Shevchenko fast reactor and desalting project. In the area of controlled thermonuclear reactions (CTR), three U.S. turbulent heating specialists visited laboratories in Moscow and Kharkov in late 1970, and arrangements were made for three Soviet superconductivity specialists to visit the United States. Long-term assignments in the same general area (CTR and plasma physics) were arranged. Also worth noting, though not part of the collaboration covered by our Memorandum on Cooperation, is the fact that while Drickey and his group were at the Serpukhov Institute of High Energy Physics, Dr. Robert D. Sard of the University of Illinois was also working there as a member of a CERN group engaged in an experiment utilizing the boson spectrometer.

During my talks in Moscow in September 1969, Petrosyants proposed an exchange of high-level delegations during 1970. I concurred in this idea, and tentative plans were made accordingly. Unfortunately, schedule demands on both sides necessitated postponement of our trips until 1971. Early that year, arrangements were agreed on for the SCAE Chairman and a group of Soviet scientists to visit nuclear facilities in the United States, April 15–28, and for a U.S. group to travel with me in the Soviet Union in August.

I welcomed the Soviet delegates on their arrival in New York on April 14, 1971, and accompanied them during part of their stay in the United States (Fig. 6). Their travel and visits took place as scheduled. At its conclusion, I and the other USAEC Commissioners met with our guests for a review of their visits to U.S. facilities and related USAEC programs, a resumé of Soviet programs, discussion of future exchanges, and plans for my forthcoming visit to the Soviet Union.

On entering the Soviet Union for the fourth time, I had the newly-acquired status of Foreign Member of the U.S.S.R. Academy of Sciences. The distinction of election to the Academy had been conferred on me during the organization's annual General Assembly meeting in March. Knowing how few foreign scientists were admitted, I felt greatly honored when I received the cable by which President Keldysh and Secretary General Peive of the Academy notified me of my selection. I deeply appreciated the heart-warming messages of congratulations that came in the succeeding weeks from SCAE Chairman Petrosyants and many other leading Soviet scientists. My formal admission to membership occurred on

Fig. 6. Petrosyants riding bicycle around one-sixth of the main ring during visit to the National Accelerator Laboratory, Chicago, Illinois, USA, 16 April 1971. Source uncertain.

July 13th at the U.S.S.R. Embassy, where Ambassador Dobrynin presented me with the certificate of my election (Fig. 7).

Three weeks later I was on my way to Moscow. The warmth of the welcome accorded us was too clearly genuine to be credited to policy or protocol. All with whom we came in contact — scientists and laymen, officials and ordinary citizens — displayed great respect for us as scientists and as Americans and seemed to appreciate our manifestions of respect for them and their achievements. I was interested to note the extensive news coverage our activities received in all the cities we visited. There were articles in the local newspapers, usually on the front page, describing our delegation and our visits to the local institutes. There was also radio coverage in nearly all instances, and apparently the progress of our delegation was covered by Moscow radio as we travelled around the Soviet Union.

Our trip was especially worthwhile in enabling my colleagues and me to become acquainted with a broader range of Soviet institutions than had been possible for me on my previous visits to the Soviet Union. I think that the Soviet method of supporting science and other areas of learning

Fig. 7. Persons attending presentation ceremony for Seaborg to the membership in the Soviet Academy of Sciences at the Soviet Embassy, Washington, D.C., USA, 13 July 1971. L to R: (front row): David Seaborg, Dianne Seaborg, Helen Seaborg, and Mrs. Anatoly Dobrynin. (second row): Roger Batzel, Michael May, Justin Bloom, Stephen Seaborg, and Eric Seaborg (hidden by Helen Seaborg). (third row): Clarence Larson, Jane Larson, Em Rubin, and Herman Pollack (behind Julie Rubin). Reproduced by courtesy of the U.S. Department of Energy (formerly the U.S. Atomic Energy Commission).

through institutes has much to recommend it, especially from the standpoint of continuity of funding, though the system has the weakness that there is not so much cooperation between institutes and the universities as there should be with respect to the teaching of students.

In the nuclear field, I observed substantial progress since my first tour of Soviet atomic energy installations in May 1963. In reactor construction, particularly, I received an impression of significant advances and improvement since that time. In many ways, the trip of August 1971 represented a high point in the development of U.S.–Soviet cooperation in peaceful nuclear applications.

This visit to the Soviet Union, August 4–19, 1971, included detailed discussions and inspection of nuclear power reactors, visits to technical institutes of the State Committee on Atomic Energy (SCAE) and of Academies of Sciences (all Union and various individual Republics), a meeting with Chairman A. M. Petrosyants and senior officials of the SCAE, a number of lectures, and meetings with officials of scientific societies. A chartered Aeroflot jet aircraft was provided by the SCAE, which permitted visits to nine cities in ten days, with a total distance travelled of 12,110

Fig. 8. At the airport in Moscow, the U.S. delegation, with members of the Soviet State Committee on Atomic Energy, the crew members and flight attendants, 15 August 1971. L to R: M. V. Naidionov (SCAE International Relations Department), crew member, Ilgius C. Timerbulatov (Chief of Protocol Division, SCAE International Relations Department), flight attendant, V. F. Menshikov (SCAE International Relations Department), crew member, Seaborg, Captain Kzavchenko, crew member, head flight attendant, Aleksandr G. Meshkov (deputy chairman for reactors, SCAE), Edwin E. Kintner (Associate Director for Reactor Engineering, Division of Reactor Development and Technology, AEC), John J. Taylor (General Manager, Breeder Reactors Divisions, Westinghouse Electric Corporation), S. V. Patrakeev (Division Deputy Chief, SCAE International Relations Department), Julie Rubin, Em Rubin, Robert L. Hirsch (Controlled Thermonuclear Research Program, Division of Research), Maggie Hollingsworth, Helen Seaborg, Lombard Squires, Robert Duffield (Director, Argonne National Laboratory, Chicago, Illinois, USA), Robert Hollingsworth (General Manager, AEC), Joseph Lewin (Quality Assurance Coordinator for Special Research Facilities, Oak Ridge National Laboratory, Oak Ridge, Tennessee, USA), R. Cool, and Klara Biriukova (SCAE International Relations Department). Source unknown.

kilometers, or approximately 7,500 miles (Fig. 8). The cities visited, besides Moscow and environs, were Minsk, Leningrad, Melekess, Novosibirsk, Tashkent, Samarkand, Erevan, Tbilisi, and Shevchenko.

The schedule was as demanding as any I had undertaken in my official travels for the Atomic Energy Commission but the trip was unusually informative and rewarding. During the concluding four days of the trip, while in Moscow, I presented papers at two international meetings (Fifth International Conference on Organometallic Chemistry and Thirteenth International Congress of the History of Science) and also lectured to the staff of the Institute of Physical Chemistry. I also delivered a letter from Dr. Philip Handler, President of the U.S. Academy of Sciences, to Dr. M. V. Keldysh, President of the U.S.S.R. Academy of Sciences, which encouraged cooperation in future planning and experiments for large and high-cost accelerators required for high energy physics research; we reviewed plans for discussions involving U.S., Soviet and European experts on these subjects. A separate meeting was held with Dr. I. I. Artobolevski, President of the Znanie Society, the Soviet counterpart of the American Association for the Advancement of Science, of which I was president-elect; we discussed cooperative exchange arrangements between our two societies.

One nontechnical item of interest about the trip is worth noting. The special plane provided for our travel was a twin engine jet (TU 134) capable of carrying about 75 people and containing three separate compartments with seats and tables for meetings or working. There was a five-man crew and four hostesses. We were informed that Premier Kosygin frequently used the same plane.

Acknowledgment

This work was supported by the U.S. Department of Energy under Contract DE-AC03-76SF00098.

Education for the 21st Century

Remarks made at the Initiatives for Science and Engineering Seminar Series, Georgia Institute of Technology, Atlanta, GA, April 21, 1991.

As a nation, we Americans are pretty pleased with ourselves right now. All America is celebrating how relatively small was the number of American lives lost in the recent conflict in the Persian Gulf. This was made abundantly clear by the deluge of standing ovations when President Bush gave his last address to a Joint Session of Congress about a month ago. In that speech the President said:

> We hear so often about our young people in turmoil, how our children fall short, how our schools fail us, how American products and American workers are second class. Well, don't you believe it. The America we saw in Desert Storm was first-class talent. And they did it using America's state-of-the-art technology.

We are all proud of the bravery of our troops in the Persian Gulf and, by no means, do I wish to belittle their contributions or the successes of American technology like the Patriot missiles. Yet, I am not here today so that we can congratulate ourselves and trade pats on the back. I think that, in the wake of this stunning military victory, we are in danger of becoming a bit complacent. This glittering image of an America where youth are not in turmoil, where children do not fall short, and where schools do not fail us — well, don't you believe it — it is a mirage which will disintegrate into a cloud of desert dust as we approach the horizon of reality.

There is ample evidence that the warning given in 1983 by the National Commission on Excellence in Education, on which I served, is as timely today as it was in 1983. For the sake of our national security and the future of our nation we must take heed. The opening lines of our report, "A Nation At Risk" were:

> Our Nation is at Risk. Our once unchallenged preeminence in commerce, industry, science and technological innovation is being overtaken by competitors throughout the world ... the educational foundations of our society are presently being eroded by a rising tide of mediocrity that threatens our very future as a Nation and as a people. What was unimaginable a generation ago has begun to occur — others are matching and surpassing our educational attainments.
>
> If an unfriendly foreign power had attempted to impose on America the mediocre educational performance that exists today, we might well have viewed it as an act of war. As it stands, we have allowed this to happen to ourselves ... We have, in effect, been committing an act of unthinking, unilateral educational disarmament.

We all recognize that we live in a rapidly changing, increasingly high technology world. I have characterized our present age as that of the Third Revolution. The Revolution of Independence gave birth to our nation and established the democratic principles on which our classical concept of "equality of opportunity" — largely through education — is based. The Industrial Revolution rewarded the American spirit of inventiveness and made us leaders in the world's economy, blessed with an extremely high standard of living. The Third Revolution, the Revolution of Science, has already transformed how we understand our world — through the remarkable expansion of knowledge in a few decades — and is radically altering almost every aspect of our lives. Our response to the challenges of the revolution in science will, quite simply, decide our future. Our most valuable resources are our intelligence and ingenuity. As a nation, we pride ourselves on our history of pioneering new technologies; in the future much will depend not only on that capacity for innovation but also on our general preparedness to participate in the practice and production of those technological advances. The strength of our technological and scientific enterprise will determine our economic well-being, our security and our health and safety.

Since the report "A Nation At Risk" there have been numerous reports by a wide spectrum of American organizations that have emphasized and deplored the state of pre-college education in science and math in the

United States today. These reports indicate that while some progress has been made, there is still much work to be done to resolve the crisis in education.

- Some 23 million American adults and 13% of all 17-year-olds are functionally illiterate. Nearly 80% of 17-year-olds cannot write a persuasive essay, and only one-third can solve a mathematics problem requiring several steps.
- In a survey conducted in six nations, 13-year-old American school children placed last in mathematics — behind South Korea, Spain, Great Britain, Ireland, and Canada. Moreover, the high-ranking Koreans surpassed the Americans by a wide margin. Seventy-eight percent of Korean 13-year-old students have the ability to use intermediate math skills to solve two-step problems, but only half as many Americans could do so. Moreover, 40% of the Korean children can understand and apply mathematical concepts, but less than one in ten American school children could do so.
- In science the statistics are just as grim. In the same six-nation survey of 13-year-old school children Americans again scored below average. Ironically, the United States came in behind Spain which does not have a reputation for being in the forefront in science and technology.
- Again, as in math, when the 13-year-olds were tested for conceptual understanding, interpretation, and application of what they had learned, the performance of the American school children dropped dismally. At the highest level 33% of Koreans had the skills to apply scientific principles, while only 12% of Americans could do so.
- In another report released in September, 1988, results showed that, compared with students from 13 other countries, 12th grade American students scored in the lower range on mathematics achievement tests. In geometry, American students did only slightly better than Hungary and Thailand; in algebra only Thailand was worse; and the United States ranked last in calculus, below both Hungary and Thailand. Overall the American students obtained only half as many points on tests as the high-ranking students from Hong Kong. Among the 13 nations surveyed, 12th-grade American students scored 9th in physics, 11th in chemistry, and 13th in biology.

In report after report American students consistently score in the lower end of the scale.

- In many other industrialized nations, courses in mathematics, biology, chemistry, physics, and geography start in sixth grade and are required of all students. Students in many countries spend three times the amount of time — based on class hours — on these subjects as do science-oriented students in the United States. In other countries — for example, in England — it is not unusual for academic high school students to spend eight hours a day at school, 220 days per year. By contrast, the typical U.S. school day is six hours and the school year is 180 days. In our country 35 states require only one year of mathematics, and 36 require only one year of science for a diploma.
- The average achievement of high school students on most standardized tests is now lower than it was 30 years ago. SAT scores have suffered a virtually unbroken decline from 1963 to 1987, with average verbal scores falling 50 points and average mathematics scores falling 40 points. There has been a steady decline in science achievement scores of 17-year-olds since 1969. The number and proportion of students demonstrating superior achievement on the SAT's (those with scores of 650 or more) have dramatically declined.
- In 1989 it was estimated that in the Atlanta MSA (Metropolitan Statistical Area) there were 660,000 adults age 20 and older who might be considered functionally illiterate. This translates to a horrifying conclusion: 39% of the total adult population lacks literacy skills either to be employable or perform adequately the full range of work their job requires. (Of this group, 57% are male, 43% are female; 70% are white and 30% are other races.)
- Thirty-seven percent of Georgia's high school students fail to graduate. Among the group of students who entered kindergarten in Atlanta's public schools in 1975, 44% didn't graduate.
- According to the Atlanta Institute of Banking, out of every 100 applicants wanting employment in banks, only 10 or 15 are able to fill out applications properly or pass minimum screening tests.

The situation is just as bad in other parts of the country. It is evident that too many students are leaving our schools without adequate skills to be full participants in our increasingly technologically-oriented society. By 1993 those 13-year-old students, who were performing so poorly in math and science in 1988, will be voting and entering the work force — the scientific and industrial leadership of our nation will be in their hands.

Science plays a central role in the world of today. Research in basic science leads to advances in applied science and then to widespread practical applications of this acquired knowledge in the derived technology. Incremental scientific advances, as well as major discoveries, result in new technologies of great commercial importance. They can give us entire new industries, as in the case of advances in molecular biology. They can give us whole new ranges of products, as in the case of polymer chemistry. They can revolutionize other technologies and industries, as has been the case for the transistor and the laser.

Basic research leads to the creation not only of new products but also new industrial processes and manufacturing systems. These can greatly increase industrial productivity, reduce costs, and improve the quality of products. For example, advances in microelectronics are aiding the production of automobiles, steel, and many other manufactured goods. Discoveries in biology are influencing the processing and production of pharmaceuticals, foods, and chemicals.

This country cannot afford another generation of students that is ill-prepared to respond to the worldwide rapid growth of scientific knowledge and technological power. The nation's future depends on them. The National Science Foundation has predicted a shortfall of about half a million scientists and engineers by the end of the century. It is particularly frightening that this shortfall is happening at a time when overall educational standards in our country are dismally low in comparison to those of our major industrial competitors.

A large percentage of our scientists and engineers are going to be retiring in the coming years — to make this vivid, let me cite an NSF study which said that in 1987 over 50% of doctoral scientists and engineers were 45 years old or older compared with 40% in 1977. More dramatic, however, is the rate at which the college-age population is shrinking — a result of the low birth rates of the 1960's and 1970's. Currently, a little more than 5% of 22-year-old Americans earn B. S. degrees. In California this figure is an even lower 4% — I don't know what the figure is for Georgia. Unless that fraction is increased, NSF says there will be a shortage of 400,000 scientists and 275,000 engineers by the year 2006.

However, it's not just the future scientists and engineers who need a better foundation in math and science. A White House press release of February 26, 1990, entitled "National Goals for Education," expressed it this way: "What our best students can achieve now, our average students must be able to achieve by the turn of the century."

We must improve general science education for all of our young (and not only for those who plan to continue their education and become professional scientists, mathematicians or engineers) because we need a large number of scientifically literate, nonprofessional workers with the understanding and skills to manufacture, operate, and repair increasingly complex technological equipment. Future employment opportunities, necessary to replace jobs lost in our declining "smokestack industries," will be in areas requiring technical sophistication and will depend on a work force endowed with a practice in learning and the flexibility of mind to adapt to a society constantly changing. The old concept of a replaceable worker standing in a production line and doing one thing over and over is obsolete. The workplace demands workers who understand the automated equipment that they use and who can adjust and repair it. They must understand and apply the statistics of quality control and make decisions which require knowledge and judgment. The definition of "basic skills" is changing to include such areas as critical thinking, problem-solving, decision-making, reasoning, teamwork, adaptability, and computer literacy.

We must actively recruit young people from what have been traditionally underrepresented populations — women and minorities. Workforce 2000 reported that "White Males, thought of only a generation ago as the mainstays of the economy, will comprise only 15% of the net additions to the labor force between 1985 and 2000." Nationwide, in 1985 one in five 18-year-olds was black or Hispanic; in 2010 the ratio will be one in three. As of 1986, minority enrollment in K-12 schools was 46% in California and 40% in Georgia. Georgia has the fifth largest number of blacks and 64.6% of Atlanta's population is black. As Harold Hodgkinson, director of the Center for Demographic Policy of the Institute for Educational Leadership said, "Over 1.1 million Georgians are black. If the blacks in Georgia fail, Georgia will fail." If the nation is to meet the challenge of reversing the shortfall of engineers and scientists, it must begin to do so with members of these groups who are currently underrepresented in college preparatory mathematics and science courses.

In addition to the need for trained scientists, mathematicians and engineers, and nonprofessional workers with an understanding of complex technological equipment, we need widespread understanding of science among the general population. Support of the critical need for scientific literacy is best expressed by Thomas Jefferson's famous dictum, often-quoted and profoundly true:

> I know of no safe depository of the ultimate powers of society but the people themselves; and if we think them not enlightened enough to exercise their control with a wholesome discretion, the remedy is not to take it from them but to inform their discretion.

The vitality of a democracy assumes a certain "core of knowledge" shared by everyone which serves as a unifying force. I think that the most horrifying example I have heard is that in one survey of adult Americans, only 45% — less than half! — knew whether the Earth travels around the sun or the sun around the Earth! How basic a lack of understanding of the Universe in which we live is this! And these same people will make judgments on the issue of global warming and the commitment our nation will make to solving this potentially disastrous problem? Thirty-six percent of the adults in the same survey believed that boiling radioactive milk would make it safe! Citizens who don't know what radiation is are being asked to judge nuclear power, and citizens who know no chemistry are making decisions about toxic waste. It is fundamental to the effectiveness of our democratic system that our citizens be able to make informed judgments on the more and more complex issues of scientific and technological public policy. Decisions must be made which are of critical importance to our health and safety.

There can be no doubt that scientific literacy, a solid understanding of science and mathematics, is now more important than ever before — and there is irrefutable evidence that the skills of our youth are not only not progressing with the increasing demands — but actually are deteriorating at an alarming rate. While our nation's needs for both an educated citizenry and a technologically trained workforce have grown by leaps and bounds, our ability to satisfy those needs has diminished. We must act now to reverse the self-destructive trend.

We all have an important stake in the success of our education system, and every part of our society must be involved in meeting the challenge. Education is an investment, not an expense. The Committee for Economic Development reports that each year's class of dropouts costs the nation about $240 billion in crime, welfare, health care, and services. For every $1.00 spend on education, it costs $9.00 to provide services to dropouts. For example, about 80% of all prison inmates are school dropouts, and each inmate costs the nation about $28,000 per year.

There is good news, however. Trends in SAT scores are not universally discouraging. One positive trend is the narrowing of the gap between minority and non-minority students. Between 1978 and 1988, scores of black students rose 21 points on the verbal and 30 points on the math portions of the SAT. Native Americans, Asians and Hispanics also showed gains.

Nationwide, many proposals and programs have been developed by people committed to improving the quality of education in this country and the future prospects of our youth and of our economy.

In September 1989 governors from almost every state in the Union attended the summit conference convened by President Bush in Charlottesville, Virginia. Goal Number Four of the five National Education Goals established at that time relates directly to math and science education: "By the year 2000 U.S. students will be first in the world in science and mathematics education."

Over the past year the Committee on Education and Human Resources of the Federal Coordinating Council for Science, Engineering and Technology (FCCSET) has been working on defining how the federal government can contribute to meeting this National Education Goal. I would commend to you their recently released report, entitled "By the Year 2000: First in the World." The Committee, chaired by Admiral James D. Watkins (Secretary of Energy), had the benefit of participation from top leaders of federal agencies. As Admiral Watkins wrote in his letter conveying the report to Assistant to the President for Science and Technology Allan D. Bromley,

> This is a groundbreaking document. Never before has an Administration presented an interagency budget for mathematics, science, engineering and technology education. Along with this coordinated budget and priorities for future growth, this report includes another valuable resource for policy makers, educators and the public: a comprehensive inventory of mathematics and science education programs and activities across the entire Federal government — a snapshot of where the Federal government stands at the beginning of the decade.

I am extremely pleased by the appointment last month of Lamar Alexander as Secretary of Education and of David Kearns as his Deputy. Immediately following the publication of "A Nation At Risk," Mr. Alexander, then Governor of Tennessee, acted to push through far-reaching and effective educational reforms. He made it clear that education was a top priority for the State of Tennessee, and I am hopeful that he will be as dynamic a leader for change in the federal government. Mr. Kearns used his platform as chairman of the Xerox Corporation to bring out in the open some of the deficiencies of our emerging workforce, to advocate decisive action to improve every American's opportunities, and to urge the participation — using both financial and human resources — of the private sector. I am looking forward to seeing this new team in action.

The biggest obstacle Alexander and Kearns may face is the determination of the federal government to leave the bulk of the burden for financing the schools to the state. I keep insisting — and I hope someone is listening! — that no number of studies of existing problems, and no number of marvelous suggestions for solutions can make the difference. All of these reforms take DOLLARS.

On February 20, 1991, D. Allan Bromley, Director of the Office of Science and Technology Policy, testified before a House Committee regarding the President's FY 1992 research and development budget, as follows:

> The federal investments in science and mathematics education are larger than previously recognized: in fiscal year 1991, for example, the federal government spent $1.7 billion expressly for science, mathematics, engineering, and technology education at all levels.
>
> The FY 1992 budget proposes that the federal government increase this amount by over 13 percent, to about $1.94 billion, as part of an integrated interagency effort in science and mathematics education. Precollege education would receive the largest increase — $146 million, or 28 percent.

Education That Works: An Action Plan for the Education of Minorities, published in 1990, interprets the numbers a little differently:

> According to the U.S. Department of Education, funds appropriated in the federal budget for education programs increased from $34.3 billion

in 1980 to $43.3 in 1988 in actual dollars. But adjusting the figures for the 43.6 percent inflation in that period shows that in constant 1988 dollars, federal spending on education actually dropped 12.2 percent, from $49.3 to $43.3 billion. In other words, adjusting for inflation, federal spending on education in 1988 was $6 billion less than necessary to maintain programs at their 1980 levels.

Certainly, the federal role in education spending has always been relatively small. In 1990, of the $353 billion to be spent by the public and private sectors on education, federal expenditures will account for only about 6 percent of the total; private spending will account for 27 percent, and the balance will come from state and local governments.

Another thing about the FCCSET's committee report that encourages me is the recognition that a top priority should be to increase the supply of well-trained science and mathematics teachers. Half of the newly employed teachers of mathematics, science, and English are not qualified to teach these subjects, and fewer than one-third of U.S. schools have qualified physics teachers. In part as a result of this, 30% of our high schools offer no courses in physics, 17% offer none in chemistry, and 70% offer none in earth or space science.

According to the FCCSET Committee report,

> In the United States today there are 2.3 million public school teachers in grades K-12. The Department of Education estimates that over the next decade, we must hire 1.6 million new teachers, or an average of 160,000 teachers a year. Yet our primary source of new teachers, college students majoring in education, has fallen 55% since 1972. Today we are graduating only about half the teachers we will need to bridge the gap in the future. If it is becoming difficult to recruit teachers, it is even harder to retain them. Twenty percent of new teachers leave during their first year, and more than half leave before the sixth year. We are currently losing thirteen mathematics and science teachers for each one entering the profession.

The task of guiding the intellectual (and often social) development of our young is an all-important one. We must begin to recognize teachers' contributions not only by adequately compensating them for their service, but also by giving them due respect which would motivate them to refine their skills and expand their knowledge to meet future challenges. There are a number of vital new programs and proposals which address this need.

I would like to mention in particular the initiation of the Georgia Industrial Fellowships for Teachers (GIFT) program. Working with the nationwide Triangle Coalition, for which I serve as Chairman for Science and Engineering, and IISME, a very successful program in California headquartered in the Lawrence Hall of Science, for which I serve as Chairman — hear a little advertising here? — GIFT will start this summer with a pilot program for 10–12 Atlanta high school and middle school science and mathematics teachers, who will spend eight weeks working with a mentor in industry. Teachers will work in well-paid industry positions during the summer in order to familiarize the teachers and industry with each others' needs, problems, and successes. These teachers do real work — not busy work — and have made significant contributions to their host companies. The jobs augment the teachers' annual salaries, thus helping to prevent permanent teacher job-shifting to more lucrative industry positions. Georgia Tech will serve as educational coordinator for the program. Among the organizations committed to providing the fellowships are: the Coca-Cola Company, Equifax, MCI Telecommunications, Amdahl Corporation, Georgia Power, The Robinson–Humphrey Company, and Georgia Tech.

Another outstanding teacher support program based at Georgia Tech is SECME (Southeastern Consortium for Minorities in Engineering), which I understand is similar to our MESA program in California. Based at Georgia Tech since its inception in 1975, it consists of 28 universities, over 50 corporations and 280 schools in an eight-state southeastern region. Over 19,000 students in grades 6–12 participate in their schools' SECME program led by teams of teachers who have attended a SECME Summer Institute. The Summer Institute is a two-week graduate level course for mathematics, science and language arts teachers that promotes understanding of science and engineering, team work, and hands-on classroom activities. SECME results have been dramatic: 85% of the high school seniors since 1979 have continued their education, with 45% majoring in engineering or the sciences. SAT scores have consistently averaged 200 points above the national average for black students.

There are many interesting new curriculum development projects. Among the most fascinating to me is an initiative by the American

Association for the Advancement of Science (AAAS), Project 2061 (so named to make clear its goal of revolutionizing the teaching of science by the time of the arrival of the next Halley's Comet). Its report *Science for All Americans* (1989) makes fascinating reading: its represents Phase I (Goals) of the project by attempting to define what basic core of science knowledge should be included in the education of all young Americans. Phase II (Formulation) makes recommendations on new science curricula, instructional materials, testing methods, teacher training, school organization, and educational research and development programs. Phase III (Implementation) will probably take a good deal longer to accomplish!

Another idea which has been picked up and further developed by the National Science Teachers Association (NSTA) is called "Scope and Sequence." Students in grades 7 through 12 would study biology, chemistry, physics and earth sciences each year, concentrating on phenomenological studies in 7th and 8th grades, empirical studies in 9th and 10th grades, and theoretical studies in 11th and 12th grades.

I would like to conclude by emphasizing the important role that science centers can play in changing the face of education in our country. Needless to say, I will use the University of California's Lawrence Hall of Science and a description of its far-reaching programs as an example.

The Lawrence Hall of Science is an institution committed to improving the quality of mathematics and science instruction for pre-collegiate students. For over two decades the Hall has dedicated its superior resources as part of the University of California to the continuing battle against educational mediocrity.

The Lawrence Hall of Science was conceived in 1958 and built in 1968 as a memorial to Ernest O. Lawrence, the University of California's first Nobel Laureate and inventor of the cyclotron. As a dynamic research and educational institution, the Hall continues today, 23 years after its dedication in 1968 to focus its efforts on three main objectives:

1. To improve the quality of mathematics and science instruction for the benefit of pre-collegiate students through the development of innovative math and science courses and accompanying curriculum materials and teacher training services;
2. To augment the mathematics and science instruction provided by our schools by offering special mathematics and science courses at the Hall; and

3. To enhance the knowledge, appreciation, and enjoyment of mathematics and science for the general public by providing the community with a math and science center.

For over two decades, the Hall has also provided innovative leadership in pre-collegiate math and science education through the publication of major curricula. Over 20 million students and 100,000 teachers in the U.S. use LHS-produced materials. Curricula and exhibits developed by the Hall are currently used by schools and science centers in over 30 countries. Each year, over 700 educators from around the world visit the Hall to learn new techniques to improve science and mathematics instruction. For the effective utilization of these programs and materials, the Hall provides comprehensive teacher training workshops and seminars. It instructs 20,000 teachers each year to improve their science and mathematics knowledge as well as their instruction techniques.

The Hall has about 300,000 visitors each year. It provides classes for over 50,000 students a year at the Hall and for another 125,000 children in schools within a 100-mile radius.

I would like to briefly describe a few of our many programs.

Curriculum Development:

- Chemical Education Material Study (CHEM Study) is an integrated program of written materials and films developed to improve the teaching of chemistry at high school level. It emphasizes the discovery method and the experimental approach to learning. Materials from this program have been translated into 17 foreign languages, and are used throughout the world.
- Great Explorations in Math and Science (GEMS) is a curriculum development program funded by the Carnegie Corporation and the Mellon Foundation to finalize and disseminate the Hall's well-tested science and mathematics instructional activities and lesson plans. GEMS also includes some of our exhibits and their accompanying learning activities so these may now be used by other science centers and museums. To date 29 GEMS guides have been published, as well as two teacher handbooks, one leader's handbook, two exhibit guides, two assembly guides, and one parent's guide.

- Science Curriculum Improvement Study (SCIS) is a course content improvement project that has developed a sequential, articulated elementary school science program. This program is based upon the structure of science as seen by contemporary scientists, is consistent with a current view of the intellectual development of children, and reflects the experience of elementary teachers working with preliminary SCIS units. Developed in the late 1960's, it is now used in over 20% of the nation's elementary schools.
- Health Activities Project (HAP K-8) is a science-based health program for students in schools. The primary goal is to help children understand that they are in control of their own well-being. The program is experience-based with extensive participatory activities including experiments, data collection, and analysis.
- Chemical Education for Public Understanding Program (CEPUP) has been developed to increase public knowledge of toxic substances and hazardous materials and their safe use, storage, and disposal. This is accomplished by developing modules of materials for junior-high-school-age students and the general public; setting up an on-site computer-based information center; and having available printed information and audio-visual materials for business, industry, and the general public.

Teacher Training:

- Science for Science Teachers (S4ST) provides upgrading opportunities for junior high school science teachers by increasing their science knowledge. UC Berkeley faculty and master teachers use a practical experiment-based approach to help these teachers return to their science classroom better prepared to develop positive attitudes among their junior high school students.
- Institute for Chemical Education (ICE) teaches the fundamentals of chemistry to high school chemistry teachers during six-week summer institutes. Part of a national ICE program, the Hall works with teachers to teach the fundamentals and to increase teachers' familiarity with the use of instrumentation in chemistry. Lectures, discussions, and problem-solving sessions are also used to improve those teachers' abilities to teach chemistry.
- EQUALS encompasses programs for teachers, counselors, administrators, and parents. It promotes participation of students and adults, particularly

women and girls, in mathematics courses and encourages their interest and involvement in math-based fields of study and work. LHS provides teacher training through programs such as EQUALS to over 10,000 teachers each year.

- Family Math teaches parents how to help their children with math at home, informs parents of the role mathematics plays in their children's studies and career choices, and creates a family enjoyment of mathematics. The project provides training and curriculum materials to help parents and teachers establish Family Math classes in their schools.

Student programs:

- ACCESS/CCPP is an effort of UC Berkeley and the Oakland and San Francisco School Districts to strengthen the capacity of the Districts' secondary schools to prepare minority students for college. University staff work daily at school sites providing a wide range of technical assistance, training, and student support to improve mathematics, English, and science curriculum and instruction; school management policy and organization; counseling; and parent involvement.
- Center for Multisensory Learning (CML) specializes in science and computer education for disabled students. CML staff conduct workshops for disabled students at the Hall, train teachers to use the SAVI/SELPH science program in their classrooms, and organize special events that enhance opportunities for disabled students to enjoy science and technology experiences.
- Mathematics, Engineering, Science Achievement (MESA) program is designed to increase the number of underrepresented minorities who graduate from high school with the mathematics, science, and English necessary to pursue math-based curricula in college. MESA's Pre-College Program operates in over 200 high schools and serves about 8,000 students. It provides study assistance; academic, university, and career advising; field trips; summer programs; and scholarship incentive awards. MESA's university-level program, the Minority Engineering Program (MEP), is designed to increase the number of underrepresented minorities who receive B. S. degrees in engineering and related fields. MEP operates on 15 UC and CSU campuses. It provides matriculation and university admissions assistance, tutoring, freshman transition programs, academic

and personal counseling, career development and summer jobs, and financial aid and scholarships. The retention rate of MEP students is dramatically better than that of students who do not receive such support.

We must work together, employing our considerable resources, to ensure a prosperous future. Whatever the expense of improving science and mathematics education, this is an investment in the future we must make. Excellence is costly. But in the long run, mediocrity costs far more.